Introduction to Experimental Infrared Spectroscopy

Introduction to Experimental Infrared Spectroscopy

Fundamentals and Practical Methods

Edited by

MITSUO TASUMI

Professor Emeritus, The University of Tokyo, Japan

and

AKIRA SAKAMOTO

Associate Editor, Professor, Aoyama Gakuin University, Japan

WILEY

Registered office
John Wiley & Sons Ltd, The Atrium, Southern Gate, Chichester, West Sussex, PO19 8SQ, United Kingdom

For details of our global editorial offices, for customer services and for information about how to apply for permission to reuse the copyright material in this book please see our website at www.wiley.com.

Library of Congress Cataloging-in-Publication Data

Introduction to experimental infrared spectroscopy / edited by Mitsuo
Tasumi, Professor Emeritus, The University of Tokyo, Japan and Akira Sakamoto, Associate Editor, Professor, Aoyama Gakuin University, Japan.
   pages cm
 Includes bibliographical references and index.
 ISBN 978-0-470-66567-1 (pbk.)
 1. Infrared spectroscopy. I. Tasumi, Mitsuo, 1937– editor. II. Sakamoto, Akira, 1965– editor. III. Title.
 QD96.I5S4513 2015
 535.8′42–dc23

             2014012251

A catalogue record for this book is available from the British Library.

ISBN: 9780470665671

Set in 10/12pt TimesLTStd by Laserwords Private Limited, Chennai, India
Printed and bound in Malaysia by Vivar Printing Sdn Bhd

1   2015

Contents

List of Contributors

Yoshiaki Hamada Studies of Nature and Environment, The Open University of Japan, Japan

Takeshi Hasegawa Institute for Chemical Research, Kyoto University, Japan

Koji Masutani Micro Science, Inc., Japan

Shigeaki Morita Department of Engineering Science, Osaka Electro-Communication University, Japan

Seizi Nishizawa Research Center for Development of Far-Infrared Region, University of Fukui, Japan

Isao Noda Department of Materials Science and Engineering, University of Delaware, USA

Shukichi Ochiai S. T. Japan, Inc., Japan

Yukihiro Ozaki Department of Chemistry, School of Science and Technology, Kwansei-Gakuin University, Japan

Akira Sakamoto Department of Chemistry and Biological Science, College of Science and Engineering, Aoyama Gakuin University, Japan

Hirofumi Seki Toray Research Center, Inc., Japan

Shigeru Shimada Bruker Optics K. K., Japan

Hideyuki Shinzawa National Institute of Advanced Industrial Science and Technology (AIST), Japan

Masao Takayanagi United Graduate School of Agricultural Science, Tokyo University of Agriculture and Technology, Japan

Mitsuo Tasumi Professor Emeritus, The University of Tokyo, Japan

Preface

Measurements of infrared spectra have been providing useful information for over half a century now for a variety of scientific and industrial research studies, as well as for more practical purposes. This situation will continue in the foreseeable future.

This book is intended to be a handy guide to those who have limited or no experience in infrared spectroscopic measurements but are interested in using at least one of the methods for their research, or are required to use it for some practical purpose. The book consists of 22 chapters; these cover almost all the information available on presently used measurement methods of infrared spectroscopy, together with six appendices, which will help readers to understand the contents of the chapters without necessarily having to consult other books.

Infrared spectroscopy broadly means the science of spectra relating to infrared radiation, namely, electromagnetic waves in the wavelength region occurring intermediate between visible light and microwaves. In practice, it has three branches, namely, near-infrared, mid-infrared, and far-infrared spectroscopy. When one simply refers to infrared spectroscopy, it is often meant to indicate mid-infrared spectroscopy because structural information on the target matter is mostly readily available from its mid-infrared spectrum. For this reason, most chapters of this book are primarily concerned with mid-infrared spectroscopy with some references to the near- and far-infrared regions when necessary. Only Chapters 18 and 19 specifically treat near- and far-infrared spectroscopy, respectively.

Most chapters are described from the standpoint of users of an infrared spectrometer and its accessories. Although Chapters 4, 5, and 19 were written by authors who once worked with manufacturers of infrared spectrometers, these authors were strongly requested by the Editor to make their descriptions easy to understand for those who wish to learn about infrared spectrometers for the first time.

All chapters and appendices are written by Japanese researchers. This is simply because this book is based on a book written in Japanese and published in Japan in 2012. However, this book is not a translation of the Japanese book. In preparing it, four chapters in the original Japanese book were deleted and many paragraphs and sentences replaced, changed or deleted, and some new descriptions and explanations added, in order to make it more easily understandable, up to date, and practically useful. Appendix F, which did not exist in the Japanese book, was written by Isao Noda, originator of the two-dimensional correlation spectroscopy technique, and added to the present book.

Infrared spectroscopy has a long history, which is briefly described in Chapter 1. Its measurement methods have experienced great changes over the past several decades, and improvements of the methods are still occurring. Lasers, not only as sources of infrared radiation but also as elements for nonlinear spectroscopy such as sum-frequency generation

(SFG), have already introduced many new aspects in infrared spectroscopic measurements. This trend will continue in future. However, the Editor believes that many descriptions in this book will not lose their value or relevance, even if infrared lasers more useful than the ones currently existing become available. Also, the book's content is relevant for other more specialized experimental set-ups such as infrared measurements based on utilizing infrared emission radiation from synchrotron orbital resonance (SOR); such facilities are already available at several large-scale, often government, laboratories.

As for the technical terms, their symbols, and units used in this book, the Editor asked the authors to follow the recommendations of the Physical and Biophysical Chemistry Division of the International Union of Pure and Applied Chemistry (IUPAC), which were published as "Quantities, Units and Symbols in Physical Chemistry" in 2007 (known as the *Green Book*).

All the chapters and appendices, except for Chapter 21 and Appendix F (which were either arranged or written by Isao Noda), were arranged by the Editor using original manuscripts written in either Japanese or English as their base. English expressions of all the chapters and appendices were carefully examined and amended by John Chalmers, based on the consultant contract with the Publisher. The Editor is most grateful for John's serious efforts to make this book useful for readers all over the world. Akira Sakamoto, Associate Editor, paid much attention to keep the forms of the figures within a certain format, in order to make different chapters easy to follow and compare and give an impression of consistency.

The Editor wishes to express his gratitude to the authors for their cooperation, and dealing with many detailed requests by the Editor when writing their original manuscripts. Finally, the Editor thanks the team of John Wiley & Sons, Chichester, UK, in charge of publishing this book, for their valuable efforts in bringing this book to its present form.

30 May, 2014
Mitsuo Tasumi
Study Sapiarc
Tokyo, Japan

Part I

Fundamentals of Infrared Spectroscopic Measurements

1
Introduction to Infrared Spectroscopy

Mitsuo Tasumi
Professor Emeritus, The University of Tokyo, Japan

1.1 Introduction

Infrared spectroscopy is a useful tool for molecular structural studies, identification, and quantitative analyses of materials. The advantage of this technique lies in its wide applicability to various problems in both the condensed phase and gaseous state. As described in the later chapters of this book, infrared spectroscopy is used in chemical, environmental, life, materials, pharmaceutical, and surface sciences, as well as in many technological applications. The purpose of this book is to provide readers with a practical guide to the experimental aspects of this versatile method.

In this chapter, introductory explanations are given on an infrared absorption spectrum and related basic subjects, which readers should understand before reading the later chapters, on the assumption that the readers have no preliminary knowledge of infrared spectroscopy.

As is well known, visible light is absorbed by various materials and the absorption of visible light is associated with the colors of materials. Blue materials absorb radiation with a red color, and red materials absorb radiation with a blue color. The wavelengths of radiation with a red color are longer than those with a blue color. A diagram showing quantitatively the absorption of visible light at different wavelengths from violet to red is called a *visible absorption spectrum*. The visible absorption spectrum closely reflects the color of the material from which the spectrum is measured.

The wavelengths of infrared radiation are longer than those of radiation with red color. Radiation with red color has the longest wavelengths among visible light, the wavelength of which increases from violet to red. Infrared radiation, though not detectable by human eyes, is absorbed by almost all materials. An infrared spectrum is a plot quantitatively showing the absorption of infrared radiation against the wavelength of infrared radiation.

Introduction to Experimental Infrared Spectroscopy: Fundamentals and Practical Methods,
First Edition. Edited by Mitsuo Tasumi and Akira Sakamoto.
© 2015 John Wiley & Sons, Ltd. Published 2015 by John Wiley & Sons, Ltd.

It is usually possible to observe an infrared absorption spectrum from any material except metals, regardless of whether the sample is in the gaseous, liquid, or solid state. This advantage makes infrared spectroscopy a most useful tool, utilized for many purposes in various fields.

Measurements of infrared spectra are mostly done for liquid and solid samples. In the visible absorption spectra of liquids and solids, only one or two broad bands are typically observed but infrared absorption spectra show at least several, often many relatively sharp absorption bands. Most organic compounds have a significant number of infrared absorption bands. This difference between the visible and infrared absorption spectra is due to the different origins for the two kinds of spectra. Visible absorption is associated with the states of electrons in a molecule. By contrast, infrared absorptions arise from the vibrational states of atoms in a molecule. In other words, the visible absorption spectrum is an electronic spectrum and the infrared spectrum is a vibrational spectrum. Vibrational motions of atoms in a molecule are called *molecular vibrations*.

At present, measurements of infrared spectra are widely performed in materials science, life science, and surface science. In these fields, the states of targets of research are usually liquids or solids. This book primarily aims at describing the fundamentals of infrared spectroscopy and practical methods of measuring infrared spectra from various samples in the liquid and solid states.

1.2   Fundamentals of Infrared Spectroscopy

A basic knowledge of infrared spectroscopy that readers should have before performing infrared measurements is briefly described in this section.

1.2.1   The Ordinate and Abscissa Axes of an Infrared Spectrum

It has been known for a long time that vapors, liquids, crystals, powder, glass, and many other substances absorb infrared radiation. The wavelength region of infrared radiation is not strictly defined but the wavelength regions generally accepted for near-infrared, mid-infrared, and far-infrared radiation are as follows: 700 nm to 2.5 μm for near-infrared, 2.5–25 μm for mid-infrared, and 25 μm to 1 mm for far-infrared.

The absorption intensity is taken as the ordinate axis of an infrared spectrum. The wavelength can be used as the abscissa axis of an infrared spectrum. At present, however, it is customary, in the mid-infrared region in particular, to use the wavenumber as the abscissa axis instead of the wavelength. The wavenumber is the number of light waves per unit length (usually 1 cm) and corresponds to the reciprocal of the wavelength. The wavenumber used as the abscissa axis of an infrared spectrum is always expressed in units of cm^{-1}. In this book, the abscissa axis of an infrared spectrum is always designated as "Wavenumber/cm^{-1}." It should be mentioned, however, that the wavelength is often used as the abscissa axis in the near-infrared region, if a near-infrared spectrum is measured as an extension of a visible absorption spectrum.

There are publications in which the higher wavenumber (corresponding to the shorter wavelength) is placed on the left side of a spectrum, whereas it is placed on the right side in other cases. This inconsistency has occurred because infrared spectra published before the 1950s used the wavelength as the abscissa axis and placed the longer wavelength (corresponding to the lower wavenumber) on the right side. Following this tradition, many infrared spectra published since the 1960s also have placed the higher wavenumber on the left side and the lower wavenumber on the right side. In recent years, however, infrared spectra in publications which feature the direction of the abscissa axis oppositely have been increasing in number.

The wavenumber, which is the number of light waves per centimeter as mentioned above, corresponds to the frequency divided by the speed of light. Therefore, the wavenumber is proportional to the energy E of a photon as expressed in the following equation:

$$E = hc\tilde{\nu} \tag{1.1}$$

where h is the Planck constant, c the speed of light, and $\tilde{\nu}$ the wavenumber of infrared radiation. This proportionality between E and $\tilde{\nu}$ is the reason why the wavenumber is now used as the abscissa axis of an infrared spectrum. In Appendix A relations closely associated with Equation (1.1) are explained in detail.

The above-mentioned wavelength regions of infrared radiation correspond to the wavenumber regions of about $14\,000-4000\,\mathrm{cm}^{-1}$ for near-infrared, $4000-400\,\mathrm{cm}^{-1}$ for mid-infrared, and $400-10\,\mathrm{cm}^{-1}$ for far-infrared.

The wavenumber region of $400-10\,\mathrm{cm}^{-1}$ for far-infrared corresponds to the frequency region of $(12-0.3) \times 10^{12}\,\mathrm{Hz}$ or $12-0.3\,\mathrm{THz}$. This means that the far-infrared region approximately coincides with the terahertz frequency region. For this reason, the term *terahertz spectroscopy* is recently being increasingly used in place of far-infrared spectroscopy. However, the term *far-infrared spectroscopy* is considered a better designation because of its consistency with other optical spectroscopies.

1.2.2 The Intensity of Infrared Radiation

As infrared radiation is an electromagnetic wave, electromagnetic theory is applicable to it. In this theory, the intensity of an electromagnetic wave irradiating an area is defined as the average energy of radiation per unit area per unit time. In this book, according to the tradition of spectroscopy, the term *intensity* is used for this quantity. It is worth pointing out, however, that the term *irradiance* is increasingly used in other fields instead of "intensity." This quantity is given in units of $\mathrm{W\ m^{-2}}$ ($= \mathrm{J\ s^{-1}\ m^{-2}}$), although its absolute value is rarely discussed in infrared spectroscopy except when lasers are involved. The intensity I is proportional to the time average of the square of the amplitude of the electric field E. In vacuum, I is expressed as

$$I = \varepsilon_0 c_0 \langle E^2 \rangle_t \tag{1.2}$$

where ε_0 and c_0 denote, respectively, the electric constant and the speed of light in vacuum, and the symbol $\langle\ \rangle_t$ means time average. This relationship will be mentioned later in Section 1.2.4.

1.2.3 Lambert–Beer's Law

Let us consider the absorption of infrared radiation which occurs when an infrared beam passes through a sample layer. As shown in Figure 1.1, a collimated infrared beam with

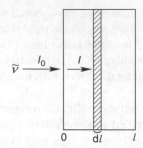

Figure 1.1   *Absorption of infrared radiation by a sample layer with thickness l.*

intensity I_0 at wavenumber $\tilde{\nu}$ irradiates a sample with thickness l at right angles to its surface. If the sample is transparent to the infrared beam, the infrared beam passes through the
sample without losing its intensity. Here, reflection of the infrared beam at the surface of
the sample is not considered. If the sample absorbs the infrared radiation of wavenumber $\tilde{\nu}$,
the infrared intensity decreases as the beam passes through the sample. If the amount of the
absorption by the thin layer dl in Figure 1.1 is expressed by $-dI$ (the minus sign reflects
the fact that dI is a negative quantity corresponding to an intensity decrease), the following
equation holds.

$$-dI = \alpha I dl \qquad (1.3)$$

where α is a proportionality constant representing the magnitude of absorption (called the
absorption coefficient) and I is the intensity of the beam entering the thin layer dl. Integration of Equation (1.2) gives the following equation.

$$\ln I = -\alpha l + a \qquad (1.4)$$

where the integration constant a must be equal to $\ln I_0$ (I_0 is the intensity at $l = 0$). Then,
the following two equations are derived.

$$\ln\left(\frac{I_0}{I}\right) = \alpha l \qquad (1.5)$$

$$I = I_0 \exp(-\alpha l) \qquad (1.6)$$

It should be remembered that each of I_0, I, and α is a function of $\tilde{\nu}$ and, strictly speaking, should be expressed, respectively, as $I_0(\tilde{\nu})$, $I(\tilde{\nu})$, and $\alpha(\tilde{\nu})$. Equation (1.5) means that
the absorption intensity expressed as $\ln(I_0/I)$ is proportional to the thickness of the sam-
ple. This relation is known as *Lambert's law*. The quantity $\ln(I_0/I)$ is called *absorbance*
(denoted by A) or *optical density* (*OD*). The quantity I/I_0 expressed in a percentage scale
is called *transmittance* ("transmission factor" is also recommended by IUPAC [1]). It is
also used to express the absorption intensity. The absorption intensity expressed in either
absorbance or transmittance is taken as the ordinate axis of an infrared spectrum.

   If the sample is a solution, the absorption coefficient α in Equations (1.3)–(1.6) should
be replaced by εc_s, where ε, a function of $\tilde{\nu}$, is a proportionality constant and c_s is the
concentration of a solute in the solution. This means that the absorbance is proportional to
both the thickness (sometimes called the *cell pathlength*) of the solution and the concentration of the solute. This relation, known as *Lambert–Beer's law*, holds for most dilute

solutions, but deviations from this law may occur in various cases, particularly in concentrated solutions (see Section 3.6.5). In practice, the natural (or Napierian) logarithm in Equation (1.5) is often replaced by the common (or decadic) logarithm, as will be described in Section 3.4.1. The relation that the absorbance is proportional to the concentration of the solute is called *Beer's law*.

1.2.4 Complex Refractive Index

The relation in Equation (1.6) can also be derived in the following way. The electric field E of an electromagnetic wave (or an infrared beam) traveling in the x direction in vacuum may be expressed as

$$E = A \exp 2\pi i \left(\frac{x}{\lambda} - vt \right) \tag{1.7}$$

where A is a vector representing the amplitude of the light wave, v the frequency of the light wave, t time, and λ the wavelength of the light wave. In Appendix B the meaning of the right-hand side of Equation (1.7) (and a closely related form of $A \exp 2\pi i(vt - x/\lambda)$) is discussed in more detail.

If the light wave travels in a medium other than vacuum, it is necessary to introduce the complex refractive index $\hat{n}$ defined as

$$\hat{n} = n + ik \tag{1.8}$$

where n is the (real) refractive index, and k, which is called the *absorption index* or the *imaginary refractive index*, is a constant to express the absorption of light by the medium. The wavelength λ in the medium is related to the wavelength λ_0 in vacuum by the equation $\lambda = \lambda_0/\hat{n}$. By substituting $\lambda_0/\hat{n}$ for λ in Equation (1.7), the following equation is derived.

$$E = A \exp \left(-\frac{2\pi kx}{\lambda_0} \right) \exp 2\pi i \left(\frac{nx}{\lambda_0} - vt \right) \tag{1.9}$$

As is clear from Equation (1.9), the amplitude of the light wave decreases in proportion to the factor $\exp(-2\pi kx/\lambda_0)$. Since the intensity is proportional to the square of the amplitude of the electric field, the ratio of the intensity I at $x = l$ to I_0 at $x = 0$ is given by the following equation.

$$\frac{I}{I_0} = \frac{\left[|A| \exp(-2\pi kl/\lambda_0) \right]^2}{|A|^2}$$

$$= \exp \left(\frac{-4\pi kl}{\lambda_0} \right) \tag{1.10}$$

If $4\pi k/\lambda_0$ in Equation (1.10) is equated with α, Equation (1.10) is identical with Equation (1.6). In some literature, the complex refractive index is defined as $\hat{n} = n - ik$. In this case, the right side of Equation (1.7) should be given as $E = A \exp 2\pi i(vt - x/\lambda)$.

The complex refractive index is an important quantity. It shows that refraction and absorption are closely related with each other and these should always be treated as a pair. It may be said that the purpose of experimental spectroscopy is to determine quantitatively the real and imaginary parts of the complex refractive index for any material. However, this purpose is not easy to fulfill, because it is experimentally difficult to separate absorption from reflection (and refraction) completely. For example, a cell is usually used to measure a transmission infrared spectrum from a liquid or solution. In this case, reflection occurs at the cell window surfaces, making the determination of I_0 difficult. If the sample is a dilute solution, this problem may be solved by using a cell containing solvent only and measuring the intensity of infrared radiation transmitted from this cell. The intensity measured in this way may be used as I_0. There is no convenient, reliable method applicable to concentrated solutions, liquids, solid films, and so on. In most practical applications, however, measurements of absorption in a qualitative manner give useful information, so that the quantitative determination of the complex refractive index is usually set aside.

1.2.5   Signal-to-Noise Ratio

Here, mention is made of the measure of rating the quality of an observed spectrum. The purpose of spectral measurements is to determine the quantity of the radiation absorbed, reflected, or emitted by the sample. This quantity is called the *signal*. In practice, the signal is always accompanied by noise arising from various sources. The magnitudes of the signal and noise are usually denoted by S and N, respectively. The signal-to-noise ratio of an observed band, which is often abbreviated as the S/N ratio, SN ratio, or SNR, is used to rate the quality of an observed spectrum; that is, a spectrum with a higher signal-to-noise ratio has a higher quality.

1.3   Origin of Infrared Absorption

According to quantum mechanics, a molecule has discreet energy levels for electronic motions, molecular vibrations, and molecular rotations. If the molecule is irradiated by electromagnetic waves ranging from ultraviolet light to microwaves, the molecule absorbs the energies of the electromagnetic waves and undergoes a transition from a state of a lower energy to another state of a higher energy. Such a transition by absorbing the energy of an electromagnetic wave (or a photon) must satisfy certain conditions. The energy of a photon given in Equation (1.1) should coincide with the energy difference between an energy level of the molecule (usually the ground-state level) and a higher energy level (an excited-state level). In addition to this requirement for the coincidence of energy, whether the transition in the molecule occurs by interaction with the electromagnetic wave is governed by specific rules based on quantum mechanics and group theory, which are called *selection rules* [2–4]. Due to the selection rules, transitions in a molecule are grouped into allowed and forbidden cases, and only allowed transitions give rise to absorptions of the electromagnetic waves including infrared radiation.

Almost all infrared absorptions arise from the transitions between the energy levels of molecular vibrations. In this sense, infrared absorption spectra, together with Raman spectra and neutron inelastic scattering spectra, constitute vibrational spectra. Infrared absorption spectra observed from gases are accompanied by fine structures due to the transitions associated with rotational energy levels. As a result, vibration–rotation spectra are observed from gaseous samples. In liquids and solids, free rotation of molecules does not occur, so that infrared spectra observed from liquids and solids are "pure" vibrational spectra.

Infrared absorption occurs by electric interaction of molecules with infrared radiation. If the dipole moment of a molecule changes with a molecular vibration, infrared radiation with the frequency equal to the frequency of the molecular vibration is absorbed by the molecule. In quantum mechanical terms, a molecular vibration has energy levels specified by the vibrational quantum number v, and usually the transition from the ground vibrational state ($v = 0$) to the first excited vibrational state ($v = 1$) occurs with an infrared absorption. In other words, the molecule is vibrationally excited by absorbing the energy of infrared radiation. The absorption arising from the $v = 0$ to $v = 1$ transition is called the *fundamental tone*. Overtones correspond to the transitions from $v = 0$ to $v = 2, 3, 4, \ldots$, but the probabilities for these transitions are small. Generally, the overtone absorptions are much weaker in intensity than the fundamental absorption. It should be mentioned, however, that near-infrared absorptions arise from either overtones or combination tones, the latter being transitions from the ground state to excited vibrational states involving two or more vibrations of a polyatomic molecule. Absorptions corresponding to transitions between excited vibrational states are called *hot bands*. Unless the temperature of the sample is very high, hot bands are not observed.

1.4 Normal Vibrations and Their Symmetry

If an atom in a molecule is displaced from its equilibrium position, a force is exerted on the atom to help it restore to its equilibrium position. This restoring force is closely approximated by the Hooke's-law force in most cases; the restoring force is proportional to the displacement in magnitude and works in the direction of decreasing the displacement. A vibration in such a force field (harmonic force field) is called a *harmonic vibration*. A molecule consisting of N atoms has $(3N - 6)$ degrees of vibrational freedom [$(3N - 5)$ for a linear molecule]. Molecular mechanics shows that there are $(3N - 6)$ or $(3N - 5)$ molecular vibrations which are independent of each other and have no contribution from either the translation or rotation of the molecule as a whole. These are called *normal vibrations*. The frequencies and patterns of normal vibrations can be calculated if the molecular structure and force constants are known [3–5]. At present, such calculations are performed by using an appropriate program package of quantum chemical computations [6].

If a molecule has any symmetry, its normal vibrations are classified by the symmetry. To discuss in detail the relationship between molecular symmetry and normal vibrations, group theoretical analysis is required. Each molecule belongs to one of about 20 point groups. Each point group has symmetry species specified by symmetry operations, and normal vibrations are classified into symmetry species. Whether or not the normal vibrations of a molecule belonging to a symmetry species of a point group have possibilities of absorbing infrared radiation is determined by group theory. If a normal vibration has a possibility of absorbing infrared radiation, it is called an *infrared-active vibration* (or *mode*). In other words, it is an allowed case and gives rise to an infrared absorption. If a normal vibration has no such possibility, it is an infrared-inactive vibration; it is forbidden in infrared absorption.

In most applications of infrared spectroscopy, materials to be studied seldom contain molecules of high symmetry. It may therefore seem that the group theoretical analysis is of little value in practice. However, many molecules have substructures which are nearly symmetric. It may be worth pointing out that the group theoretical analysis helps understand the spectral feature due to the vibrations occurring mainly in such a substructure.

Normal vibrations form a solid basis for understanding molecular vibrations. It should be remembered, however, that they are conceptual entities in that they are derived from the harmonic approximation which assumes a harmonic force field for molecular vibrations. Deviations from this approximation (i.e., deviations from Hooke's law) exist in real molecules, and the energy levels of a molecular vibration are determined by not only the harmonic term but also higher-order terms (anharmonicities) in the force field function. Although the effects of anharmonicities on molecular vibrational frequencies are relatively small in most molecules, normal (vibrational) frequencies derived in the harmonic approximation do not completely agree with observed frequencies of fundamental tones (fundamental frequencies). However, a fundamental frequency is frequently treated as a normal frequency on the assumption that the difference between them must be negligibly small.

1.5   Group Vibrations and Characteristic Absorption Bands

In theory, a normal vibration involves vibrational motions of all atoms in a molecule. In many cases, however, the contribution of a vibration localized within a particular group of atoms in a molecule is dominant in a normal vibration. Such a vibration is called a *group vibration* or a *characteristic vibration* (of the group). The frequency of a group vibration, being essentially free from the effect of neighboring groups, occurs within a relatively narrow range. If an absorption due to such a group vibration has a high intensity, it is called a *characteristic absorption band*, which indicates the presence of the group. Some representative group vibrations and the approximate wavenumbers of the infrared absorption bands arising from them are listed in Table 1.1. The characteristic absorption bands provide clues for identifying compounds or their substructures by infrared spectroscopy [7–9].

1.6   Brief History of Infrared Spectroscopy

In this section, how infrared spectrometry (the method of measuring infrared spectra) has been developed since the beginning of the twentieth century is briefly described. Further information may be found elsewhere [10].

1.6.1   Before the Mid-1960s

Development of a reliable method of measuring infrared spectra was an important subject of research in physics in the beginning of the twentieth century. William W. Coblentz (1873–1962) made a major contribution to the instrumentation of early infrared spectrometers and the compilation of the infrared spectra of many organic compounds in the period before 1930. At that time, alkali halide crystals were used as the prism for dispersing infrared radiation.

Table 1.1   *Group vibrations and their characteristic absorption wavenumber.*

Group vibration	Classification	Wavenumber/cm^{-1}
OH stretch	Free	~3600
	H-bonded	3000–2500
CH stretch	Olefin	~3080
	Methyl	~2960, ~2870
	Methylene	~2925, ~2850
SH stretch	–	2600–2550
C=O stretch	Acyl chloride	~1810
	Ester	~1735
	Aliphatic aldehyde	~1730
	Aliphatic ketone	~1715
	Aromatic aldehyde	~1705
	Carboxylic acid	~1700
	Amide (peptide)	1700–1630
	Aromatic ketone	~1690
	Quinone	~1670
Benzene ring stretch	–	1610–1590
CCl stretch	RCH$_2$Cl	760–700 or 690–650
	R$_1$R$_2$CHCl	700–670 or 640–600
	R$_1$R$_2$R$_3$CCl	640–610 or 580–550
CH$_3$ bend	Asymmetric or degenerate	~1460
	Symmetric	~1380
CH$_2$ bend	Scissor	~1450
CH out-of-plane bend	RHC=CH$_2$	~990, ~910
	RHC=CHR (*trans-*)	~960
	R$_1$R$_2$C=CH$_2$	~890
	R$_1$R$_2$C=CHR$_3$	~820
	Monosubstituted benzene	~740
	Disubstituted benzene (*o-*)	~750
	Disubstituted benzene (*p-*)	~800

During World War II, great progress was made in infrared-related technology, particularly in electronics and detectors. This resulted in the commercial production of recording infrared spectrophotometers. In the postwar period, the recording infrared spectrophotometer soon became an instrument indispensable for chemical research in not only universities but also research laboratories of industrial companies. In recognition of Coblentz's early contributions to infrared spectroscopy as well as to foster the understanding and application of infrared spectroscopy, the Coblentz Society was formed in 1954 in USA. This Society is still active and now has members from all over the world. In the mid-1950s, the use of gratings as the dispersive element began, and gratings replaced prisms by the mid-1960s.

1.6.2   After the Mid-1960s

Spectrometry using prisms and/or gratings as the means for dispersing radiation of continuously varying wavelengths is called *dispersive spectrometry*. At present, this type of spectrometry is not used in the infrared region (particularly in the mid-infrared region), except in special cases (see Section 1.6.3). Instead, interferometric (or interferometer-based) spectrometry is common in the infrared, and it is usually referred to as *Fourier transform infrared* (*FT-IR*) spectrometry, because the Fourier transform of an interferogram measured by the interferometer is an infrared spectrum.

The principle of FT-IR spectrometry was established by the 1950s, but its practical use had to wait for the advent of high-speed computers, which enabled the fast Fourier transform. In the mid-1960s, the first commercial FT-IR spectrometer became available, and it was after the mid-1970s that FT-IR spectrometers became widely used. As will be discussed in Chapters 4–6 the advantages of FT-IR spectrometry over dispersive spectrometry are now well established from the viewpoint of instrumentation theory. From a practical viewpoint, the greatest advantage is the high reproducibility of observed interferograms, which ensures high S/N ratios of an observed spectrum through accumulation of the observed interferograms. As a result of this advantage, minute spectral differences of 10^{-4} or less in the absorbance scale between two spectra can easily be detected.

In this book, most chapters deal with FT-IR spectrometry and its applications to various methods of infrared spectroscopic measurements. Only terahertz spectrometry in Chapter 19 and a large part of time-resolved infrared spectrometry in Chapter 20 are laser-based measurements. This shows how widely FT spectrometry is used at present in the measurements of vibrational spectra.

1.6.3 Future

Although it is not easy to foresee future developments in infrared spectrometry, there is a possibility of revival of dispersive spectrometry in which a grating is combined with a planar array detector. Such a spectrometer is expected to be handy and small in size, as it needs no mechanism for moving an optical component. In fact, spectrometers in this category are already being produced and may become a convenient tool for practical analysis in various fields. Finally, a completely new type of spectrometry may happen if a laser widely tunable in the infrared and easy to operate should come into existence.

References

1. International Union of Pure and Applied Chemistry, Physical and Biophysical Chemistry Division (2007) *Quantities, Units and Symbols in Physical Chemistry*, 3rd edn, RSC Publishing, London.
2. Herzberg, G. (1945) *Infrared and Raman Spectra of Polyatomic Molecules*, D. Van Nostrand, Princeton, NJ. A reprint version is available from Krieger Publishing, Malabar.
3. Wilson, E.B. Jr., Decius, J.C. and Cross, P.C. (1955) *Molecular Vibrations: The Theory of Infrared and Raman Vibrational Spectra*, McGraw-Hill, New York, A paperback version is available as one of Dover Books on Chemistry, Dover Publications, New York.
4. Woodward, L.A. (1972) *Introduction to the Theory of Molecular Vibrations and Vibrational Spectroscopy*, Oxford University Press, Oxford.

5. Groner, P. (2002) Normal coordinate analysis, In: *Handbook of Vibrational Spectroscopy*, Vol. **3** (eds J.M. Chalmers and P.R. Griffiths), John Wiley & Sons, Ltd, Chichester, pp. 1992–2011.

6. Matsuura, H. and Yoshida, H. (2002) Calculation of vibrational frequencies by Hartree–Fock-based and density functional theory, In: *Handbook of Vibrational Spectroscopy*, Vol. **3** (eds J.M. Chalmers and P.R. Griffiths), John Wiley & Sons, Ltd, Chichester, pp. 2012–2028.

7. Lin-Vlen, D., Colthup, N.B., Fateley, W.G. and Grasselli, J.G. (1991) *The Handbook of Infrared and Raman Characteristic Frequencies of Organic Molecules*, Academic Press, Boston.

8. Vidrine, D.W. (1997) *IR Correlations*, http://www.vidrine.com/vcorr.htm (accessed December 2013).

9. Shurvell, H.F. (2002) Spectra–structure correlations in the mid- and far-infrared, In: *Handbook of Vibrational Spectroscopy*, Vol. **3** (eds J.M. Chalmers and P.R. Griffiths), John Wiley & Sons, Ltd, Chichester, pp. 1783–1837.

10. Sheppard, N. (2002) The historical development of experimental techniques in vibrational spectroscopy, In: *Handbook of Vibrational Spectroscopy*, Vol. **1** (eds J.M. Chalmers and P.R. Griffiths), John Wiley & Sons, Ltd, Chichester, pp. 1–32.

2

Sample Handling and Related Matters in Infrared Spectroscopic Measurements

Akira Sakamoto

Department of Chemistry and Biological Science, College of Science and Engineering, Aoyama Gakuin University, Japan

2.1 Introduction

In this chapter, sample handling and related matters in performing infrared spectroscopic measurements are broadly described. More specific methods of spectral measurements are treated in Part II of this book. Most infrared spectroscopic measurements do not require difficult sample handling techniques. This is one of the reasons why infrared spectroscopy is widely used for a diverse range of analytical purposes.

2.2 Points to Note in the Laboratory

The relative humidity in the room for infrared spectroscopic measurements should be lower than 50%, because, as described in Section 2.3, some infrared transparent materials used for spectral measurements are hygroscopic.

When infrared spectroscopic measurements are performed, attention should always be paid to the presence of water vapor and carbon dioxide in the air and water existing in samples; these tend to adversely affect measured spectra, because they have strong absorptions which may overlap the absorption bands of samples to be examined. Water vapor shows intense and sharp vibration–rotation bands over wide infrared regions above $3300 \, cm^{-1}$,

Introduction to Experimental Infrared Spectroscopy: Fundamentals and Practical Methods,
First Edition. Edited by Mitsuo Tasumi and Akira Sakamoto.
© 2015 John Wiley & Sons, Ltd. Published 2015 by John Wiley & Sons, Ltd.

2000–1300 cm^{-1}, and below 400 cm^{-1}. Carbon dioxide has strong absorptions around 2350 and 670 cm^{-1}. If no precaution is taken to remove the water vapor and carbon dioxide from within the optical compartment of a spectrometer, the intensities of the infrared beam in these wavenumber regions decrease in the optical path, possibly causing inaccuracy in the measured absorption intensities of samples. It is therefore advisable to purge the optical compartment of a spectrometer with either dried air or nitrogen or to evacuate air from the same part at least while the spectral measurements are performed. For a condensed state sample, it is important to completely remove any moisture or unbound water existing in the sample. Even a small amount of moisture present in a sample exhibits very broad bands over almost the entire mid-infrared region, particularly around 3400 and 1650 cm^{-1}, giving rise to an undesirable distortion in the observed spectrum of the sample.

2.3   Mid-Infrared-Transparent Materials

To measure an infrared absorption spectrum from a liquid or solution sample, it is necessary to use windows (polished plates of certain crystals) or a cell for containing the sample. The material used for the windows must be transparent to the infrared radiation and appropriate for the purpose of the measurement to be performed. Some characteristics of representative infrared transparent materials are given in Table 2.1. More information is available elsewhere [1–3]. The windows are commercially available usually in the form of polished plates varying in size and thickness.

Table 2.1   Infrared transparent materials.

Material (chemical formula)	Transparent wavenumber region/cm^{-1} (thickness 2 mm)	Refractive index[a]	Solubility in water and other properties
Quartz (SiO$_2$)	40 000–2500	1.53 (1 μm)	Insoluble
Lithium fluoride (LiF)	83 000–1100	1.39 (5 μm)	Almost insoluble
Calcium fluoride (CaF$_2$)	77 000–830	1.40 (5 μm)	Insoluble
Barium fluoride (BaF$_2$)	40 000–670	1.45 (5 μm)	Almost insoluble
Sodium chloride (NaCl)	40 000–380	1.49 (10 μm)	Soluble
Potassium chloride (KCl)	47 000–330	1.45 (10 μm)	Soluble
Silver chloride (AgCl)	24 000–370	1.90 (10 μm)	Insoluble, UV sensitive, plastic
Potassium bromide (KBr)	40 000–250	1.52 (10 μm)	Soluble
Cesium iodide (CsI)	40 000–130	1.73 (10 μm)	Soluble, slightly plastic
Zinc sulfide (ZnS)	17 000–830	2.25 (10 μm)	Insoluble
Zinc selenide (ZnSe)	20 000–500	2.40 (10 μm)	Insoluble, orange
KRS-5 (TlBr$_{0.42}$I$_{0.58}$)	16 600–220	2.37 (10 μm)	Almost insoluble, orange, toxic
Silicon (Si)	8300–100	3.41 (10 μm)	Insoluble, dull gray or black
Germanium (Ge)	5000–590	4.00 (10 μm)	Insoluble, dull gray or black
Diamond (type IIa) (C)	40 000–15	2.41 (0.6 μm)	Insoluble, hard

[a]Wavelength for measuring refractive index in parentheses.

Polished plates of KBr and NaCl are the most commonly used as infrared transparent materials. It is necessary to avoid using them in humid conditions, because they are hygroscopic to some extent. While not in use, they should be kept in a dry box with an auto-recycling desiccant or in a glass desiccator (or something similar) containing dried silica gel. For handling the polished KBr plates, it is better to wear rubber or plasticizer-free protective plastic gloves. Their polished flat surfaces easily become opaque due to their hygroscopicity and are damaged due to their softness. To make these plates flat and transparent, they should be polished over a flat surface of glass tightly covered with a sheet of fine cloth. It helps to make the cloth moist by dropping a small amount of alcohol or water on it; apparatus for such polishing is commercially available.

For examining samples as aqueous solutions, water-insoluble CaF_2, BaF_2, or ZnSe plates should be employed as windows.

The powder of KBr (spectroscopic grade) is employed to form KBr disks (or pellets) of solid samples, as described in detail in Section 2.4.3.

2.4 Measurements of Infrared Spectra from Samples in the Gaseous, Liquid, and Solid States

2.4.1 Gaseous Samples

Generally speaking, molecules in the gaseous state are free from strong intermolecular interactions. Therefore, the observation of the infrared spectrum of a sample in the gaseous state can give useful information to the studies of structure and other properties of the free molecule of the sample.

If a sample is gaseous at room temperature, its infrared spectrum is easily measurable by using a cell for a gaseous sample (simply called a *gas cell*). The simplest gas cell has a short pathlength, for example, 10 cm. If a liquid or solid sample has a vapor pressure high enough at room temperature, it is not difficult to measure its infrared spectrum in the same way as that employed for a sample that is gaseous at room temperature.

Although a variety of gas cells are commercially available, it is not difficult to make a simple one in the laboratory, particularly if facilities for glasswork and metalwork are near at hand. A simple short-path gas cell consists of a tube (about 4 cm in diameter, usually made of Pyrex® glass) and windows (of the crystal plates of KBr, for example) on both ends of the tube. A gas cell can be connected, through an inlet valve fixed to the cell tube, to an evacuation system that is also used for introducing a gaseous sample and controlling its pressure. The windows are attached to the cell tube simply with an appropriate adhesive that is resistant to the gaseous sample, or by using metal flanges that hold a circular window with a gasket or an O-ring placed between the inner flange and the window (Figure 2.1). The vapor pressure of the sample in the cell should be controlled to obtain an appropriate intensity for an absorption.

For a sample of a gaseous mixture, Lambert–Beer's law (see Section 1.2.3) may be rewritten in the following form:

$$A(\tilde{v}) = \alpha_{gX}(\tilde{v})lp_X \tag{2.1}$$

where $A(\tilde{v})$ is the absorbance at wavenumber $\tilde{v}$, $\alpha_{gX}(\tilde{v})$ is the absorption coefficient of component X in a gaseous sample at $\tilde{v}$, l is the pathlength, and p_X is the partial pressure

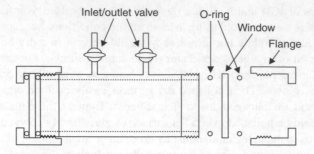

Figure 2.1   *Schematic of a short pathlength gas cell.*

of component X. The absorbance of a component in a gas mixture depends not only on its partial pressure, but is also a function of the total pressure. Therefore, although 70 hPa or less is a partial pressure at which many gases in a 10 cm pathlength cell yield useful spectra, a common practice is to keep the total pressure constant by adding a nonabsorbing gas such as nitrogen up to a standard pressure, such as about 10^3 hPa (1 atm).

When the vapor pressure of the sample is low or the absorption coefficient of a band is small, it is necessary to employ a long-path cell having a multiple-reflection optical system [4]. Various types of long-path cells are commercially available, but such a cell should be handled with great care to keep clean the surfaces of its multiple-reflection mirrors and to maintain their positioning properly.

The method of matrix isolation is frequently used to obtain infrared spectra from molecules isolated in the matrices of inert gases at a low temperature [5]. To measure an infrared spectrum from a matrix-isolated sample, the gaseous sample diluted with a large amount of an inert gas (Ar, Kr, Ne, etc.) is deposited onto a plate (usually CsI) kept at a low temperature (about 15 K for Ar matrices and about 8 K for Ne matrices, see Section 2.5.1). Since interactions between a matrix-isolated molecule and surrounding inert-gas atoms are negligibly small, an infrared spectrum obtained by this method represents a "pure" vibrational spectrum of isolated molecules without rotational fine structures or their envelopes. Matrix isolation is a very useful technique for measuring infrared spectra from unstable species, such as free radicals, reaction intermediates, metastable species, and van der Waals clusters, trapped in inert-gas matrices. It is also applied to studies of intermolecular hydrogen bonding.

2.4.2   Liquid and Solution Samples

If a sample is a nonvolatile liquid, the simplest way of sample handling is to make a thin liquid film by squashing a few drops of the liquid between two parallel KBr or any pair of infrared transparent windows. By this method, it is often possible to obtain infrared spectra useful for qualitative purposes. However, the thickness of the liquid film cannot be easily controlled by this method.

For more quantitative purposes, a cell for a liquid sample (simply called a *liquid cell*) should be used. A liquid cell basically consists of two parallel windows, a spacer between the two windows, gaskets, and a cell holder made of metal. The spacer is used to determine the thickness of the liquid layer within an aperture in the spacer. The spacer is usually made

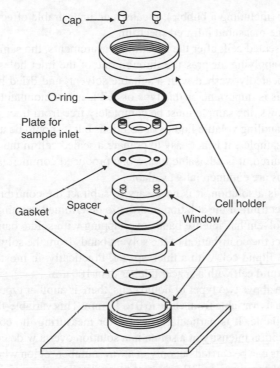

Cap

O-ring

Plate for
sample inlet

Gasket   Spacer

Window

Cell holder

Figure 2.2   *Schematic of a demountable liquid cell.*

of a sheet of Teflon® (polytetrafluoroethylene) or polyethylene. Spacers of various kinds are commercially available.

There are two types of liquid cells, namely, a demountable cell and a sealed cell. A demountable cell is schematically shown in Figure 2.2. Demountable cells are commercially available. Since it can be assembled before use and may be disassembled after use for cleaning, and so on, it is conveniently used for treating viscous liquids and colored samples that may stain the cell. The pathlength may be varied by choosing an appropriate spacer. By contrast, the pathlength is always maintained to a fixed value in a sealed cell, in which a spacer is held between two windows with some adhesive.

The cell thickness l can be determined by measuring interference fringes for an empty cell. Interference fringes are observed if the inner surfaces of the two windows of a cell are parallel and clean. The cell thickness l is given by the following equation:

$$l = \frac{m}{2(\tilde{\nu}_i - \tilde{\nu}_j)} \tag{2.2}$$

where m corresponds to the number of fringe peaks in the wavenumber region between $\tilde{\nu}_i$ and $\tilde{\nu}_j$ ($\tilde{\nu}_i > \tilde{\nu}_j$). For further information, see Section 3.5.

A liquid or solution sample is injected into the cell with a syringe through one of the holes for sample inlet drilled through the cell frame and the window plate. The aperture in the spacer should be fully filled with the liquid sample. If even a small part of the aperture is occupied by air (including a bubble), it causes an undesirable effect on the percentage transmittance of the measured infrared spectrum.

In the case of a sealed cell, after the spectral measurements, the sample should be carefully removed by applying air pressure through one of the inlet holes. The inside of the cell should be repeatedly washed with a suitable solvent, and dried by flowing nitrogen gas through it. This is important, as this type of cell is not demountable. If KBr plates are used for the windows, the sample must be completely free from traces of water. A sealed cell is useful for handling volatile liquids or solutions but it should be used with care when handling viscous samples. It is not easy to prepare a sealed cell in the laboratory. If a cell of this type is required, it is advisable to use one produced commercially; sealed cells of various pathlengths are commercially available.

When a sample is a solution, it is necessary to subtract the contribution of the solvent bands from the spectrum of the solution in order to determine the unhindered spectrum of the solute. If the solvent has intense bands overlapping with solute bands, it is not easy to completely subtract the contribution of the solvent bands from the solution spectrum. This limits the use of a liquid cell with a thick spacer. Practically, in most instances, it is not advised to use a liquid cell with a spacer thicker than 100 μm.

In addition to the above two types of liquid cells, there is another type, in which the pathlength is continuously variable from close to 0 to 10 mm. This variable-thickness cell is also commercially available. It is particularly useful for measuring the concentration dependence of the absorption intensity of a solute in a solution over a wide concentration range. Such measurements can be carried out only in a wavenumber region where the solvent does not have a strong absorption band. Because of this limitation, the variable-thickness cell is more useful in the near-infrared than in the mid-infrared region.

For examining aqueous solutions, it is necessary to use a liquid cell with water-insoluble window materials such as CaF_2, BaF_2, and ZnSe. Since water (H_2O) has strong and broad absorptions, as mentioned above, in the regions around 3400 and 1650 cm^{-1}, only thin spacers (thinner than 20 μm) can be used to obtain a solute spectrum of reasonable quality by subtracting the effect of water absorptions from an observed spectrum. This means that a high solute concentration is required; in other words, it is practically impossible to obtain a complete mid-infrared spectrum of a solute in a dilute aqueous solution. D_2O is often used instead of H_2O, particularly for measuring spectra in the region of 1800–1500 cm^{-1}. In this case, a thicker spacer (up to 50 μm) may be used. When using D_2O, it must always be remembered that the H–D exchange reaction takes place very rapidly. Therefore, it is better to perform sample preparations in a dry box and to purge the sample compartment of the spectrometer with dried air or nitrogen.

For the purpose of measuring the infrared spectra of unstable species such as free radicals in solution, it is advisable to place a small Fourier transform infrared (FT-IR) spectrometer into an inert-gas glovebox system, in which the concentrations of oxygen and water vapor can be kept at <0.1 ppm. In such an experimental setup, it is possible to generate unstable species in appropriate solvents and measure their infrared spectra in solution [6]. Preferably, a small ultraviolet/visible spectrometer should also be placed within the glovebox system. Then, it will be possible to determine the concentration of the unstable species under study by additionally measuring its ultraviolet/visible absorption spectrum, provided the molar absorption coefficient of the ultraviolet/visible absorption band of the unstable species is known.

2.4.3   Solid Samples

Solid samples for infrared spectroscopic measurements are frequently in the form of a powder. It is impossible to measure directly an infrared transmission spectrum from a powder sample, because powder strongly scatters the infrared light. To measure an infrared transmission spectrum from a powder sample, it is a usual practice to disperse the powder sample in a medium (KBr, liquid paraffin, etc.) to reduce the scattering.

Since KBr has some plasticity and can coalesce under pressure, it is possible to form a transparent disk (or pellet) from a mixture of finely powdered, dried KBr and a powder sample by applying appropriate pressure to the mixture in a vacuum. A small amount (1–3 mg) of a sample is ground to fine powder in a mortar (for example, an agate mortar). About 300 mg of fine powder or small cubic crystals of KBr (commercially available) is added to the mortar and also ground to a fine powder. The mixed powder of the sample and KBr should be completely homogeneous. A disk-molding apparatus is used to make a disk (about 10 mm in diameter and about 1 mm thick) from the mixture by applying appropriate pressure in an evacuated condition. Disk-molding apparatus (a press and a die consisting of a barrel and rams made of stainless steel) are commercially available. If only a small amount of a sample is available, it is advisable to make a disk of 2 mm in diameter by using a microdisk molder. For measuring an infrared spectrum of good quality from a microdisk, a beam condenser may be used to focus the infrared beam on the microdisk. The transparency of the disk formed depends on various factors: the degree of mixing of the sample and KBr, the particle size of the mixed powder, moisture adsorbed, and so on. Hygroscopic samples should be treated in a dry box where air is replaced with flowing nitrogen gas. Trial and error may be needed to obtain a transparent disk. However, it is often possible to measure an infrared transmission spectrum of tolerable quality from a disk that is not completely transparent to the eye.

The Nujol paste (or mull) method is also useful for measuring infrared spectra from powder samples. A small amount of liquid paraffin (usually called *Nujol*) is dropped onto a sample of about 10 mg of finely ground powder in an agate mortar, and the mixture is well ground until it becomes paste-like. The paste-like mixture is then spread evenly and held between two plates of KBr or any infrared transparent material. Nujol has infrared bands in the regions of 2960–2850, 1460, 1380, and 720 cm^{-1}, but they are not so strong as they hinder the observation of the bands of the sample. They can be subtracted by taking a difference between the absorbance infrared spectrum of the sample and that of Nujol itself multiplied by an appropriate factor. Any other nonvolatile liquid such as Fluorolube® (fluorinated hydrocarbon oil) having no bands in the above regions of Nujol bands may be used to observe the bands of the sample in those regions.

A solid film may be made of a solid sample soluble in a volatile solvent by drying a solution on a flat plate of an infrared transparent material or by dropping the solution onto a flat plate in a spin coater. Good films can be made by these methods, particularly for some polymers. A polymer sample may also be made into a thin film by applying pressure on it when it becomes soft by heating. Film-forming apparatus for such a purpose are commercially available. A thin film of a solid sample may be obtained on a flat plate by vapor deposition in a vacuum chamber. A thin film of an organic compound with a low melting point may be formed by once melting the sample between two infrared transparent plates or on a single plate and then gradually cooling the melted sample. This method would fail, however, if the sample easily crystallizes to become a strong scatterer.

In addition to the methods described above, the diffuse reflection method (Chapter 12), the attenuated total reflection (ATR) method (Chapter 13), and the infrared microscopy methods (Chapter 16) can be used for measuring infrared spectra from solid samples. The ATR method is particularly suitable for rubbery samples, and it is also useful for powder samples as well as films and liquid- and paste-like samples on substrate plates. The diffuse reflection method may be used to obtain infrared spectra from powder samples appropriately mixed with KCl or KBr powder. The infrared microscope method is useful for examining a sample such as a single fiber or a small crystal.

Further information on the subjects treated in this section may be found in Ref. [7].

2.5   Measurements of Infrared Spectra at Various Conditions

2.5.1   Infrared Spectroscopic Measurements at Low Temperatures

Infrared spectroscopic measurements at low temperatures down to the liquid-nitrogen temperature (77 K) may be performed by using cryostats such as those shown schematically in Figures 2.3a, b. Each cryostat is a Dewar vessel having a bath for liquid nitrogen. The cryostat in Figure 2.3a (single window type) is used for solid samples and that in Figure 2.3b (double window type) is usually used for liquid samples. In the cryostat in Figure 2.3a, a solid sample is firmly fixed to a copper frame under the bottom of the coolant bath with a packing material of high thermal conductivity (for example, a thin indium plate). The chamber between the coolant bath and the outer vessel is evacuated to hinder heat conduction, and then liquid nitrogen is introduced into the bath. The cryostat in Figure 2.3b is used for measuring the infrared spectra of liquid and solution samples. In this case, the sample is placed in the sample chamber which is filled with nitrogen gas, so that no evaporation of the sample occurs. The temperature of the sample is monitored by a thermocouple and controlled by a heater placed under the coolant bath.

For infrared spectroscopic measurements at temperatures lower than 77 K, liquid helium is needed as the coolant instead of liquid nitrogen. The lowest temperature attainable is about 4 K. For matrix-isolation spectroscopy, a cryostat combined with a helium compressor for circulating cooled helium is commonly used. The lowest temperature attainable with such a system is about 15 K, which enables the use of argon, krypton, and xenon, as well as nitrogen for matrix materials.

2.5.2   Infrared Spectroscopic Measurements at High Temperatures

Infrared spectroscopic measurements at temperatures up to about 200 °C for a gas sample in a 10-cm cell described in Section 2.4.1 may be performed by controlling the electric current in a flexible heater wound around the cell; however, rapid heating and cooling may develop cracks in the window plates.

High-temperature measurements for liquid and solution samples and KBr disks may be made simply by attaching a heater to the cell frame. Temperature control is possible for commercially available demountable liquid cells. Some cryostats of the type as described

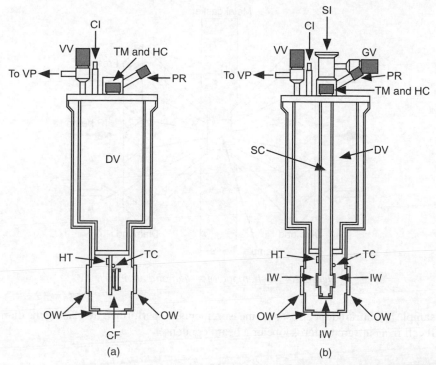

Figure 2.3 *Schematics of cryostats. (a) Cryostat for solid samples. CI, coolant inlet; TM and HC, wires (in socket) for temperature monitoring and heater control, respectively; PR, pressure release valve; VV, vacuum valve; VP, vacuum pump; DV, Dewar vessel; TC, thermocouple; HT, heater; OW, outer window; CF, copper frame. (b) Cryostat for liquid samples. SI, sample inlet; CI, coolant inlet; GV, gas inlet valve; PR, pressure release valve; TM and HC, wires (in socket) for temperature monitoring and heater control, respectively; VV, vacuum valve; VP, vacuum pump; DV, Dewar vessel; SC, sample chamber; TC, thermocouple; HT, heater; IW, inner window; OW, outer window.*

in Section 2.5.1 can be used for not only low-temperature measurements but also high-temperature measurements.

2.5.3 Infrared Spectroscopic Measurements under High Pressure

The simplest way to measure infrared spectra of samples under high pressure is to use a diamond anvil cell [8], which is commercially available (see Figure 2.4). A sample, if necessary together with a medium for leveling out pressure, is tightly put into a hole bored into a metal plate of about a few hundred micrometers in thickness. The hole for the sample should have a diameter of about half the surface diameter of the diamond anvil. The sample is placed between a pair of diamond anvils, and pressure is applied to it by turning a screw on the cell. The highest pressure attainable is over 100 GPa. The pressure is monitored by observing the shift of the wavelength of fluorescence from a small ruby crystal immersed in

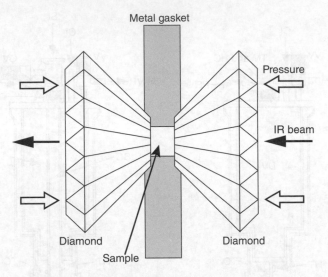

Figure 2.4   *Schematic of a diamond anvil cell.*

the sample [8]. Infrared spectroscopic measurements are performed by placing the diamond anvil cell in an infrared microscope or a beam condenser.

2.5.4   Polarized Infrared Spectroscopic Measurements

As described in Sections 1.2.2 and 1.2.3, a normal vibration of a molecule has a possibility of absorbing infrared radiation if the dipole moment of the molecule changes with the normal vibration under consideration. The dipole moment is a vector quantity having a direction fixed in the molecule. It is possible to express analytically the ith normal vibration of a molecule by a corresponding normal coordinate Q_i. If the dipole moment $\boldsymbol{\mu}$ of the molecule changes with the ith normal vibration, $d\boldsymbol{\mu}/dQ_i$ has a nonzero value.

Infrared absorption occurs by interaction of a normal vibration with the electric field E of the infrared radiation, and the interaction of the ith normal vibration with the electric field is expressed by a scalar product $(d\boldsymbol{\mu}/dQ_i)E$. This means that the infrared absorption by the ith normal vibration depends on the angle θ between the two vector quantities, $d\boldsymbol{\mu}/dQ_i$ and E. The scalar product may then be expressed as

$$\left(\frac{d\boldsymbol{\mu}}{dQ_i}\right)E = |\boldsymbol{\mu}'||E|\cos\theta \qquad (2.3)$$

where $\boldsymbol{\mu}'$ denotes $d\boldsymbol{\mu}/dQ_i$.

If natural (or unpolarized) infrared radiation is used for spectroscopic measurements, Equation (2.3) has no significance because the direction of E cannot be uniquely defined; that is, if x, y, and z represent a set of space-fixed coordinates, and the natural infrared radiation travels in the z direction, its E lies in any direction in the xy plane. However, it is possible to obtain a linearly polarized (or plane-polarized) radiation for which E takes either the x or y direction (see Section B.2 of Appendix B), by passing the natural radiation through a polarizer. Infrared polarizers are made of fine conducting grids placed on infrared transparent substrates such as Ge, ZnSe, or KRS-5. If the spacing between the grids is much smaller than the wavelength of the infrared radiation, the component of the natural radiation with E parallel to the grid is reflected, and only the perpendicular component is transmitted. The polarized infrared radiation produced in this way has a high degree of polarization (99% in the case of a Ge substrate).

For a gaseous, liquid (solution), and powder sample, Equation (2.3) is meaningless, because molecules in the gaseous and liquid states are moving in all directions and at the same time rotating independently, and molecules in a powder sample are randomly oriented. Only when molecules are spatially fixed in a unique orientation, can the angle θ between μ' of a molecular vibration and E of the polarized infrared radiation be defined. Polarized infrared absorption does not occur in the case of $\theta = 90°$, even if μ' is large in magnitude.

Practically, polarized infrared measurements are most useful for studying molecular orientation in polymer films. It is often possible to obtain a uniaxially oriented thin polymer film by stretching the film or by any other suitable method. A polarized infrared measurement is performed for such a uniaxially oriented film twice: first for the infrared beam polarized parallel to the axis of orientation and then for the perpendicularly polarized beam. The direction of E can be changed by rotating the polarizer. In practical measurements, it is advisable to place the orientation direction of the sample at an angle of 45° to the vertical line perpendicular to the spectrometer base, and to measure spectra with E at ±45° to this line. This is a precaution against a possible difference in the throughput of the spectrometer for infrared radiation polarized parallel and perpendicularly to the spectrometer base. In this process, the same area of the sample must be irradiated by the beams polarized parallel and perpendicularly to the orientation direction. For each arrangement of the parallel and perpendicular polarization, the reference (background) spectrum of the emission from the infrared source (see Section 3.3) should be measured, and the transmittance spectrum of the sample for either polarization should be calculated by using the corresponding reference spectrum in order to avoid any undesirable effects which may arise from any difference between the polarization reference spectra. It is usually simpler to measure the polarized spectra of an oriented sample by fixing the orientation direction of the polarizer at either +45° or −45° to the vertical direction and by rotating the sample by 90°. In this case, the measurement of the reference spectrum is made only once.

As an example of an infrared polarization measurement, the pair of polarized infrared spectra measured for a uniaxially oriented film of isotactic polypropylene are shown in Figure 2.5. The sample was a strand taken from a commercially available thin polypropylene tape, such as that routinely used for binding purposes. The two spectra (solid and dotted curves) in Figure 2.5 are very different from each other, indicating that the polymer molecules are highly oriented. The molecule of isotactic polypropylene has a threefold helical structure, so that its normal vibrations are classified into the totally symmetric A species and the degenerate E species according to an analysis based on group theory [9]. Normal vibrations in the A species are expected to show absorption bands polarized parallel to the molecular axis coinciding with the stretched direction of the strand, while those in the E species should show bands polarized perpendicularly to the molecular axis. The bands belonging to the A and E species can be clearly identified in Figure 2.5, and detailed assignments of individual bands were performed on the basis of such data [9].

If the peak absorbances for the infrared radiation polarized parallel and perpendicularly to the axis of orientation (denoted, respectively, as $A_{max}^{//}$ and $A_{max}^{\perp}$) are different from each other, the sample film has infrared dichroism, and the quantity $A_{max}^{//}/A_{max}^{\perp}$ is called the

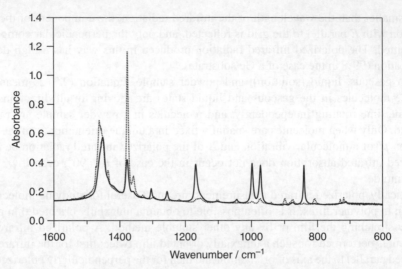

Figure 2.5 *Polarized infrared spectra of a stretched strand (tape) of isotactic polypropylene. Solid and dotted curves correspond, respectively, to the spectra measured with the infrared radiation polarized parallel and perpendicularly to the axis of molecular orientation of isotactic polypropylene (stretched direction of the strand).*

dichroic ratio. The dichroic ratio is utilized not only for assigning observed absorption bands but also for estimating qualitatively or even quantitatively the degree of orientation of polymer molecules in the film. Studies of molecular orientation in synthetic polymers by polarized infrared spectroscopy are reviewed in detail in Refs [10–12].

References

1. Griffiths, P.R. and de Haseth, J.A. (2007) *Fourier Transform Infrared Spectrometry*, 2nd edn, John Wiley & Sons, Inc., Hoboken, NJ.
2. Vidrine, D.W. (2002) Optical materials for infrared spectroscopy, In: *Handbook of Vibrational Spectroscopy*, Vol. **1** (eds J.M. Chalmers and P.R. Griffiths), John Wiley & Sons, Ltd, Chichester, pp. 368–382.
3. Vidrine, D.W. (2002) *Optical Materials*, http://www.vidrine.com/iropmat.htm (accessed December 2013).
4. Hanst, P.L. (2002) Long path gas cells, In: *Handbook of Vibrational Spectroscopy*, Vol. **2** (eds J.M. Chalmers and P.R. Griffiths), John Wiley & Sons, Ltd, Chichester, pp. 960–968.
5. Willson, S.P. and Andrews, L. (2002) Matrix isolation infrared spectroscopy, In: *Handbook of Vibrational Spectroscopy*, Vol. **2** (eds J.M. Chalmers and P.R. Griffiths), John Wiley & Sons, Ltd, Chichester, pp. 1342–1351.
6. Sakamoto, A., Harada, T. and Tonegawa, N. (2008) A new approach to the spectral study of unstable radicals and ions in solution by the use of an inert gas glovebox system: observation and analysis of the infrared spectra of the radical anion and dianion of *p*-terphenyl. *J. Phys. Chem. A*, **112**, 1180–1187.

7. Hannah, R.W. (2002) Standard sampling techniques for infrared spectroscopy, In: *Handbook of Vibrational Spectroscopy*, Vol. 2 (eds J.M. Chalmers and P.R. Griffiths), John Wiley & Sons, Ltd, Chichester, pp. 933–952.

8. Polsky, C.H. and Van Valkenburg, E. (2002) The diamond anvil cell, In: *Handbook of Vibrational Spectroscopy*, Vol. 2 (eds J.M. Chalmers and P.R. Griffiths), John Wiley & Sons, Ltd, Chichester, pp. 1352–1360.

9. Miyazawa, T., Ideguchi, Y. and Fukushima, K. (1963) Molecular vibration and structure of high polymers. IV. A general method of treating degenerate normal vibrations of helical polymers and infrared-active vibrations of isotactic polypropylene. *J. Chem. Phys.*, **38**, 2709–2720.

10. Chalmers, J.M. and Everall, N.J. (2002) Qualitative and quantitative analysis of polymers and rubbers by vibrational spectroscopy, In: *Handbook of Vibrational Spectroscopy*, Vol. 4 (eds J.M. Chalmers and P.R. Griffiths), John Wiley & Sons, Ltd, Chichester, pp. 2389–2418.

11. Bokobza, L. (2002) Molecular orientation studies of polymers by infrared spectroscopy, In: *Handbook of Vibrational Spectroscopy*, Vol. 4 (eds J.M. Chalmers and P.R. Griffiths), John Wiley & Sons, Ltd, Chichester, pp. 2496–2513.

12. Marcott, C. and Noda, I. (2002) Dynamic infrared linear dichroism spectroscopy, In: *Handbook of Vibrational Spectroscopy*, Vol. 4 (eds J.M. Chalmers and P.R. Griffiths), John Wiley & Sons, Ltd, Chichester, pp. 2576–2591.

3

Quantitative Infrared Spectroscopic Analysis

Shukichi Ochiai
S. T. Japan, Inc., Japan

3.1 Introduction

Analysis for the purpose of accurately determining the quantity of a chemical species existing in a sample is called *quantitative analysis*. Quantitative infrared spectroscopic analysis mainly deals with the intensity of an infrared absorption band. In this chapter, basic aspects of quantitative spectroscopic infrared analysis for a target substance (the analyte) in solution samples are described. The subjects to be described include the characteristics of a Fourier transform infrared (FT-IR) spectrometer, the relation between percentage transmittance and absorbance, Lambert–Beer's law on the relationship between the intensity of an infrared band and the concentration of a sample, the use of a working curve in quantitative analysis, and the origins of deviations from Lambert–Beer's law.

3.2 Characteristics of an FT-IR Spectrometer

In quantitative analysis using infrared spectra obtained by an FT-IR spectrometer, the reliability of analysis depends on the capability of the FT-IR spectrometer. Some basic characteristics of an FT-IR spectrometer are summarized here.

3.2.1 Accurate Determination of Wavenumbers

Wavenumbers measured by an FT-IR spectrometer are determined by the translation of the movable mirror, the movement of which is determined by using the interferogram measured

Introduction to Experimental Infrared Spectroscopy: Fundamentals and Practical Methods,
First Edition. Edited by Mitsuo Tasumi and Akira Sakamoto.
© 2015 John Wiley & Sons, Ltd. Published 2015 by John Wiley & Sons, Ltd.

for the He−Ne reference laser (see Chapter 5). The wavenumbers determined in this way are accurate enough as long as spectral measurements are performed at a resolution of 0.1 cm^{-1} or a resolution lower than this, which is used in most spectral measurements for samples in the condensed state. Each of the multiple measurements of infrared spectra from the same stable sample under the same condition gives exactly the same result. The wavenumber of the peak maximum position of an absorption band is always reproduced without fail. This accurate determination of wavenumbers ensures reliability in quantitative infrared analysis. A useful absorbance difference spectrum can be obtained between two spectra, as the wavenumbers in both of the two spectra are accurately determined (see Chapter 6).

3.2.2 High Wavenumber Resolution over a Wide Spectral Range

In infrared spectra, two or more bands are often located in close proximity. To resolve these bands, the spectral measurement should be performed at an appropriately high enough wavenumber resolution. In this book, the term *wavenumber resolution* is used to express resolution in wavenumber in an observed spectrum, in order to differentiate it from "spatial resolution" used in microscope and imaging measurements (Chapters 16 and 17) and "time resolution" used in the time-resolved measurements (Chapter 20).

In any spectral region, the wavenumber resolution obtained by an FT-IR spectrometer is given as the reciprocal of the maximum optical path difference between the fixed and movable mirrors in a Michelson interferometer. This wavenumber resolution applies to all the spectral range observable by an FT-IR spectrometer in use. The user of an FT-IR spectrometer can determine the wavenumber resolution by choosing the maximum distance of travel of the movable mirror. The wavenumber resolution ($\delta\tilde{v}$ in cm^{-1}) is related to the maximum optical path difference (D in cm) as

$$\delta\tilde{v} = \frac{1}{D} \tag{3.1}$$

The width of an infrared band is expressed by a half-band width, or more specifically, by a full width at half maximum (FWHM) (half the absorption maximum) which is shown in Figure 3.1. To measure a satisfactorily resolved infrared spectrum, the wavenumber resolution of an FT-IR spectrometer must be set at a value (in cm^{-1}) smaller than the FWHM of the bands to be observed. An infrared absorption band of a sample in the condensed

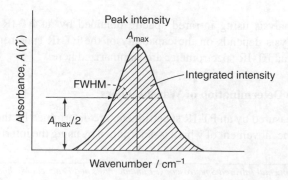

Figure 3.1 *Intensity and bandwidth of an infrared absorption band.*

phase usually has an FWHM of $10\,\text{cm}^{-1}$ or more. Thus, it is both practical and appropriate to employ a wavenumber resolution of 4, 2, or $1\,\text{cm}^{-1}$ for the spectral measurements for samples in the condensed phase. Note that a smaller number in wavenumber means a higher spectral resolution.

3.2.3   Accurate Measurement of Intensities

By using an FT-IR spectrometer, it is usually easy to measure an infrared spectrum with a high signal-to-noise (S/N) ratio. As explained in Chapter 5, accumulation of interferograms enhances the S/N ratio. The absorption intensity of a band obtained by ordinary spectrum-measuring procedures of the FT-IR spectrometer in use may be considered to be accurate enough for most analytical purposes. There are two ways of expressing the intensity of an absorption band as shown in Figure 3.1; one is the intensity of the absorption peak (peak intensity) and the other is the area of an entire band (integrated intensity).

3.3   Transmittance Spectrum and Absorbance Spectrum

An infrared spectrum may be defined as a sample-dependent change induced on the intensity distribution of infrared radiation emitted by a source over the entire infrared region. The intensity distribution (spectrum) of the mid-infrared radiation emitted by a source is shown in Figure 3.2a. This spectrum does not have absorptions by any sample (except for absorptions of atmospheric water vapor and carbon dioxide), and it is denoted as the reference spectrum (sometimes called the *single-beam background spectrum*) $B_e^R(\tilde{v})$. $B_e^R(\tilde{v})$ is determined by the characteristics of the emission from the source, the degree of transmission of the infrared beam in the optical system including the efficiency of the interferometer, and the capabilities of the detection system. The spectrum of the infrared radiation changes after passing through a sample placed in the beam within the sample compartment of the FT-IR spectrometer, and the resultant spectrum is denoted as $B_e^S(\tilde{v})$. As shown in Figure 3.2b, $B_e^S(\tilde{v})$ has a number of absorption bands due to the sample.

To obtain the infrared absorption spectrum of the sample, it is necessary to calculate $T(\tilde{v})$ given as

$$T(\tilde{v}) = \frac{B_e^S(\tilde{v})}{B_e^R(\tilde{v})} \tag{3.2}$$

$T(\tilde{v})$, expressed in percentage as shown in Figure 3.2c, is called the *transmittance spectrum*. The common logarithm of $1/T(\tilde{v})$ given in the following equation is called the *absorbance spectrum $A(\tilde{v})$*; this is shown in Figure 3.2d.

$$A(\tilde{v}) = -\log_{10}T(\tilde{v}) = \log_{10}\frac{B_e^R(\tilde{v})}{B_e^S(\tilde{v})} \tag{3.3}$$

The intensity in absorbance of a band of the analyte is used for quantitative analysis because it is proportional to the concentration of the analyte in samples (usually solutions) as explained in the next section.

When a solution is analyzed, the spectrum obtained at first is $B_e^{Sln}(\tilde{v})$, where the superscript stands for "solution." To obtain the absorbance spectrum of the solute, $A(\tilde{v})^{Slt}$, it is

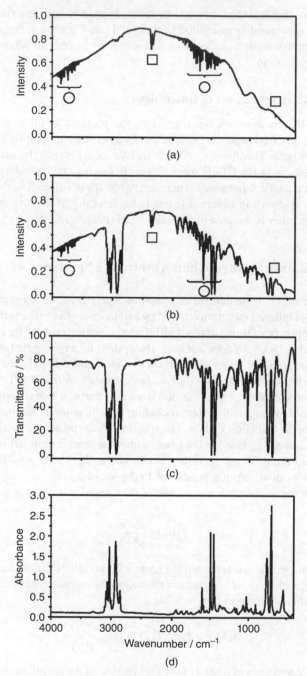

Figure 3.2   *Transmittance spectrum and absorbance spectrum: (a) spectrum of infrared radiation emitted from a source $B_e^R(\tilde{v})$; (b) spectrum of infrared radiation emitted from the source after passing through a sample $B_e^S(\tilde{v})$; (c) transmittance spectrum $T(\tilde{v})$; and (d) absorbance spectrum. ○, bands due to H_2O; □, bands due to CO_2.*

practical to calculate a difference between the absorbance spectrum of the solution $A(\tilde{\nu})^{\text{S ln}}$ and that of the solvent $A(\tilde{\nu})^{\text{Slv}}$ in the following way.

$$A(\tilde{\nu})^{\text{Sln}} = \log_{10} \frac{B_e^R(\tilde{\nu})}{B_e^{\text{Sln}}(\tilde{\nu})} \tag{3.4a}$$

$$A(\tilde{\nu})^{\text{Slv}} = \log_{10} \frac{B_e^R(\tilde{\nu})}{B_e^{\text{Slv}}(\tilde{\nu})} \tag{3.4b}$$

$$A(\tilde{\nu})^{\text{Slt}} = A(\tilde{\nu})^{\text{Sln}} - k_d \times A(\tilde{\nu})^{\text{Slv}} \tag{3.4c}$$

In Equation (3.4c), the coefficient k_d should be adjusted so that no solvent bands appear in $A(\tilde{\nu})^{\text{Slt}}$.

3.4 Fundamentals of Quantitative Infrared Spectroscopic Analysis

3.4.1 Lambert–Beer's Law

The transmittance $T(\tilde{\nu})$ and the absorbance $A(\tilde{\nu})$ of an analyte in a solution sample are related to its molar absorption coefficient $\varepsilon(\tilde{\nu})$ and to the concentration c of the analyte and the pathlength of a liquid cell l by the following two equations, both of which express the well-known Lambert–Beer's law (see Section 1.2.3).

$$T(\tilde{\nu}) = 10^{-\varepsilon(\tilde{\nu})cl} \tag{3.5a}$$

$$A(\tilde{\nu}) = \varepsilon(\tilde{\nu})cl \tag{3.5b}$$

If the concentration c is given in mol m^{-3} (or mol L^{-1}) and the pathlength l is in meter, the molar absorption coefficient $\varepsilon(\tilde{\nu})$ has a unit of m^2 mol^{-1}. Equation (3.5b) provides the basis of quantitative analysis for solution samples, as the absorbance $A(\tilde{\nu})$ is proportional to the concentration c. The transmittance $T(\tilde{\nu})$, which is not proportional to the concentration c, cannot be used for quantitative analysis.

As described in Section 3.2.3, there are two ways of expressing the intensity of a band, namely, the peak intensity A_{max} and the integrated intensity S, which are defined as

$$A_{\text{max}} = \varepsilon_{\text{max}}cl \tag{3.6}$$

$$S = \int_{\text{band}} A(\tilde{\nu})d\tilde{\nu} = \int_{\text{band}} \varepsilon(\tilde{\nu})cl\,d\tilde{\nu} \tag{3.7}$$

In Equation (3.6), the peak intensity A_{max} refers to the absorbance at the peak maximum position (see Figure 3.1), and ε_{max} is the molar absorption coefficient at this position. As indicated in Equation (3.7) and Figure 3.1, the intensity S is obtained by integrating the absorbance over an entire band. From a theoretical viewpoint, the integrated intensity corresponds to the true intensity of a band but it is often difficult to obtain a correct value of the integrated intensity because of overlapping of bands. Furthermore, it is not easy to define the range of integration. For practical quantitative analysis, the peak intensity A_{max} is commonly used, because it is easily measurable and less affected by band overlap.

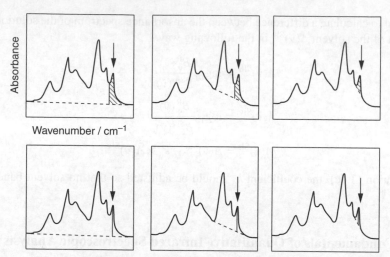

Absorbance

Wavenumber / cm⁻¹

Figure 3.3   *Various ways of drawing baselines. The integrated intensities of a band marked with an arrow are taken in the upper three spectra and the peak intensities in the lower three spectra.*

3.4.2   Intensity Measurements for Overlapping Bands

It often happens that A_{max} of the key band cannot be readily measured as a result of the band overlapping with a band or bands of other substances. In such a case, it is necessary to test the following methods to find a good solution. (i) Overlapping bands may be decomposed by applying a curve-fitting program (usually available as data processing software within an FT-IR spectrometer) with appropriately assumed band shapes. If the result of curve fitting is satisfactory, A_{max} of the separated key band is expected to be reliable. By this process, the integrated intensity S can also be obtained. (ii) It may be rewarding to draw various baselines as shown in Figure 3.3 and measure the apparent values of S and A_{max} (indicated, respectively, by slanted line filled areas and vertical lines in Figure 3.3), and test if any of these S or A_{max} values has a linear relationship with the concentration of the target substance in accordance with Lambert–Beer's law.

3.5   Working Curve for Quantitative Analysis

The purpose of quantitative infrared analysis for a solution sample is to determine the concentration c of the analyte in the solution. Equation (3.6) shows that the concentration c can be obtained by measuring the peak absorbance A_{max}, if the molar absorption coefficient ε_{max} and the pathlength l are known. The pathlength l of a liquid cell can be experimentally determined as described later, but ε_{max} is not known in most cases. It should be noted that ε_{max} depends on the solvent, temperature, pH, and so on. It is therefore necessary to make a working (calibration) curve for quantitative analysis as follows: (i) model solutions having various known concentrations of the analyte are so prepared that the concentrations extend over the range of concentrations of the analyte in the solutions to be analyzed; (ii) the

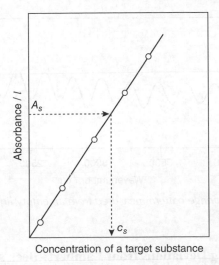

Figure 3.4 *Working curve for quantitative analysis.*

infrared spectra of the model solutions are measured by using the same liquid cell; (iii) the absorbance spectra of the analyte are obtained by the procedure described in Section 3.3 (Equations (3.4a)–(3.4c)); (iv) a key band of the analyte most useful for quantitative analysis is chosen; (v) the values of A_{max} of the key band in the absorbance spectra of the analyte are measured; and (vi) the values of A_{max} obtained are plotted against the corresponding c values. If cells of different pathlengths are used in step (ii), each A_{max} value must be divided by the pathlength l of the cell used for the corresponding spectral measurement. Of course, the reproducibility of the A_{max} values should be checked in an appropriate manner. If the absorbance spectra of the analyte follow Lambert–Beer's law, the working curve obtained by the above procedure becomes a straight line as shown in Figure 3.4.

In using the linear relationship between the A_{max} value and the concentration of the analyte for quantitative analysis, it is advisable to limit its use in the range of $A_{max} \leq 1$. This is particularly important when an MCT (mercury cadmium telluride) detector ($Hg_{1-x}Cd_xTe$, a mixture of HgTe and CdTe) is used, as the response of the detection system may become nonlinear at high radiation levels (see Section 3.6.2). Molecular association, which may cause the nonlinear relationship, tends to occur in concentrated solutions (see Section 3.6.5).

The pathlength of a liquid cell can be determined from the sinusoidal fringes, which appear in the infrared spectrum of an empty cell as a result of interference of the incident radiation with the radiation reflected inside the cell twice between the two parallel windows. An observed fringe pattern is shown in Figure 3.5. If m interference maxima are observed between two wavenumbers $\tilde{v}_i$ and $\tilde{v}_j$ ($\tilde{v}_i > \tilde{v}_j$), the pathlength l is given by the following equation, where n is the refractive index of the substance in the cell. In the case of an empty cell, n is equal to 1.0.

$$l = \frac{m}{2n(\tilde{v}_i - \tilde{v}_j)} \tag{3.8}$$

If $\tilde{v}_i$ and $\tilde{v}_j$ are given in cm^{-1}, l is obtained in units of cm.

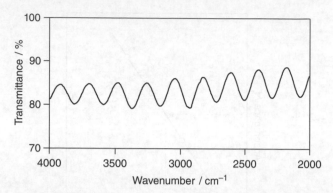

Figure 3.5   *Interference fringe pattern measured from an empty liquid cell with a pathlength of 23 μm.*

3.6   Factors Causing Deviations from Lambert–Beer's Law

Deviations from Lambert–Beer's law are caused by both instrumental problems and samples from which infrared spectra are measured. The factors causing the deviations are discussed in the following subsections.

3.6.1   Wavenumber Resolution in an FT-IR Spectrum

As mentioned in Section 3.2.2, it is advisable to measure an FT-IR spectrum at a wavenumber resolution smaller than the FWHMs (in cm^{-1}) of bands in the spectrum. For solution samples, spectral measurements are usually performed at a wavenumber resolution of 4 or 2 cm^{-1}. Even in such conditions, however, it is possible that the wavenumber of the peak of a band slightly differs from its true value. The reason for this small deviation is illustrated in Figure 3.6. The recorded digitized values of absorbance obtained from an FT-IR spectrometer are given discretely along the wavenumber axis; they are usually given at intervals of 1 cm^{-1} for a spectrum measured at 2 cm^{-1} resolution and at intervals of 2 cm^{-1} for that measured at 4 cm^{-1} resolution. In the spectrum obtained in this way, the correct value of

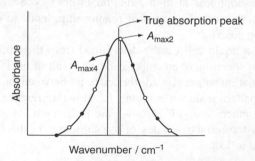

Figure 3.6   *Peak absorbances measured at two different wavenumber resolutions. Points measured at 2 cm^{-1} resolution are indicated with both filled circles (●) and open circles (○), and those at 4 cm^{-1} resolution are with filled circles (●) only.*

A_{max} may not be recorded; in the spectrum measured at $4\,cm^{-1}$ resolution in Figure 3.6, the peak absorbance A_{max4} deviates from the correct A_{max}, although A_{max2} in the spectrum measured at $2\,cm^{-1}$ resolution almost agrees with A_{max}.

If a wavenumber resolution higher than $1\,cm^{-1}$ is employed in the case mentioned above, the correct value of A_{max} should be obtained. However, for the same number of scans, a spectral measurement at a higher wavenumber resolution results in a lower S/N ratio, and a longer measurement time is required to make the S/N ratio higher by accumulating more interferograms (see Chapter 5). In other words, it is time consuming to employ too high a wavenumber resolution for a spectral measurement. Thus, it is a good practice to compare the values of A_{max} measured at different wavenumber resolutions and then decide an optimum wavenumber resolution at which a correct A_{max} value can be obtained without wasting time for a spectral measurement. In addition to employing an optimum wavenumber resolution, it is advisable to use the method of zero filling (see Chapter 5) to derive a smooth spectral curve from discretely located data points in an absorbance spectrum. By following such a procedure, the measured A_{max} value becomes close to the true one. Needless to add, the same wavenumber resolution should be employed for making a working curve for quantitative analysis.

3.6.2 Nonlinearity in Detector Response

An infrared detector converts the incident infrared radiation to electric signals. If the electric signal from a detection system (a detector and associated electronics) is not proportional to the intensity of the incident infrared radiation, in other words, if the detector response is nonlinear, this nonlinearity causes distortion in the measured interferogram. As a result, the infrared spectrum calculated from the distorted interferogram has inaccurate intensities, which may lead to deviations from Lambert–Beer's law. Such nonlinearity does not occur with a detection system with a pyroelectric TGS (triglycine sulfate) detector, but it may arise in the detection system with a photoconductive MCT detector.

Whether an MCT detector is functioning normally or not can be decided by examining the spectrum of the infrared radiation from the source measured by the MCT detector. MCT detectors cannot detect the radiation in the far-infrared region of $400-0\,cm^{-1}$. Accordingly, the signal in this region of the spectrum obtained by a normally functioning MCT detector should be at zero level as shown in Figure 3.7a. By contrast, if the MCT detector used is not functioning normally, a spurious signal exists in the wavenumber region below $400\,cm^{-1}$ as shown in Figure 3.7b. Quantitative analysis should not be carried out when the MCT detector in use is in such a condition.

The nonlinearity in detector response is particularly apparent when the intensity of the infrared radiation detected by the MCT detector is high. A practically useful method to avoid the effect of the nonlinearity is to reduce the intensity of the infrared radiation to a desirable degree by inserting a metallic mesh (available from the manufacturer of an FT-IR spectrometer) in the infrared beam from the source.

3.6.3 Absorptions of Water Vapor and Carbon Dioxide

As shown in Figure 3.2a,b, absorption bands due to water vapor and carbon dioxide exist in the spectra $B_e^R(\tilde{v})$ and $B_e^S(\tilde{v})$. In principle, these bands should cancel themselves

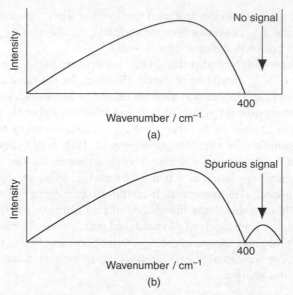

Figure 3.7   *Schematic spectrum of infrared radiation emitted from a source and detected by an MCT detector: (a) normal detection and (b) defective detection.*

mathematically in the transmittance spectrum $B_e^S(\tilde{v})/B_e^R(\tilde{v})$. In practice, however, these bands often remain in the transmittance spectrum, because of a small difference in the quantities of water vapor and carbon dioxide arising from slightly changed measuring conditions between $B_e^R(\tilde{v})$ and $B_e^S(\tilde{v})$, particularly when spectral measurements are carried out at a high resolution. If these bands overlap with bands of the analyte in the sample, the value A_{max} of the band of the analyte should be measured with caution. The absorption bands of water vapor in the $1700–1500\,cm^{-1}$ region often overlap with those of organic compounds. As mentioned previously, it is important to purge the optical compartment of the FT-IR spectrometer with dried air or nitrogen gas to remove the water vapor in the air.

3.6.4   Stray Light

Infrared radiation emitted from bodies other than the source of an FT-IR spectrometer may affect the measured spectrum. Such radiation is called the *stray light*. It may cause an error in quantitative analysis. The effect of the stray light can be eliminated by the following procedures. (i) An interferogram is measured as usual with a sample placed in the sample compartment. This interferogram is called $F^S(x)$. (ii) A mirror or a metallic plate is placed in front of the source to block the infrared beam completely, and an interferogram is measured in the same way as in (i). This interferogram is called $F^{S^*}(x)$. (iii) The spectrum $B_e^S(\tilde{v})$ is calculated from the difference interferogram, $F^S(x) - F^{S^*}(x)$ (see Chapter 6). (iv) The same procedures as (i) and (ii) are performed without the sample, and $F^R(x)$ and $F^{R^*}(x)$ are obtained. (v) The reference spectrum $B_e^R(\tilde{v})$ is calculated from the difference interferogram, $F^R(x) - F^{R^*}(x)$. (vi) The absorbance spectrum A_{max} is calculated from $B_e^S(\tilde{v})$ and $B_e^R(\tilde{v})$ obtained above by using Equation (3.3).

It may not be easy to follow the procedures described above but the absorbance spectrum obtained in this way should be useful, particularly when the key band for quantitative analysis is located in the wavenumber region below $1000\,cm^{-1}$.

3.6.5 Molecular Association

If various kinds of molecular association such as intermolecular hydrogen bonding, solvation, micelle formation, dimerization, and the formation of a charge-transfer complex, exist in a sample, the absorbance A_{max} may not be directly proportional to the concentration of the analyte c; in other words, Equation (3.6) may not hold for such a sample. Molecular association can occur not only between the molecules of an analyte (solute) but also between the molecules of the analyte and those of a solvent. The concentration of the analyte and temperature play a crucial role in such a sample. In a dilute solution, intermolecular hydrogen bonding between the molecules of the analyte is practically nonexistent, but that between the molecule of the analyte and surrounding solvent molecules may still exist.

Even if the results obtained for model (calibration) samples deviate from Lambert–Beer's law, the observed relationship between the A_{max} value and the concentration of the analyte may still be suitable for a fit-for-purpose practical quantitative analysis. In such a case, it is useful to express the observed results with a nonlinear relationship between the A_{max} value and the concentration of the target substance. This nonlinear relationship can be used for quantitative analysis, regardless of the origin of the deviation from Lambert–Beer's law. In a limited range of the concentration of the analyte, the nonlinear relationship may sometimes be approximated by a linear relationship, which is often easier to use practically.

3.7 Other Remarks

Quantitative infrared analysis is needed not only for solutions but also for other sampling techniques, such as KBr disks and polymer films. For samples other than solutions, the descriptions given in the preceding sections basically apply. However, unlike the situation for which a liquid cell is used, pathlength may not be readily determinable for many samples. In such a case, the intensity of an absorption band of the analyte relative to that of another substance may be needed to determine the quantity of the analyte in a sample. For these samples, deviations from Lambert–Beer's law occur also from surface roughness, variations in particle size, and film thickness, degrees of dispersion in KBr disks, and so on. Measures for reducing the unwanted effects arising from such factors should be worked out case by case.

4

Principles of FT-IR Spectrometry

Koji Masutani
Micro Science, Inc., Japan

4.1 Introduction

Fourier transform infrared (FT-IR) spectrometry has been widely used since the 1970s, and almost all mid-infrared spectrometers in use at present are of the type consisting of a Michelson interferometer or a modified version of it for generating interferograms, which is coupled with a computer for performing the Fourier transforms of the interferograms. This chapter is intended as a guide to the understanding of the principles of FT-IR spectrometry for those who wish to apply this method to scientific research or various other purposes.

Since FT-IR spectrometry is based on the interference of waves of light (or radiation), first an account of this phenomenon is briefly given, before explaining the Fourier transform method by which an infrared spectrum is obtained from a measured interferogram. Some characteristics of FT-IR spectrometry, namely, wavenumber resolution, measurable wavenumber region, and accurate determination of wavenumbers are discussed. To facilitate the understanding of the description, which inevitably requires some mathematical formulations, many illustrations are provided.

4.2 Interference of Light

Light, including infrared radiation, is an electromagnetic wave. Due to the wave nature of light, interference occurs between two light waves of an identical frequency when they overlap each other. Young's experiment, which is known as the first demonstration of the

---

Introduction to Experimental Infrared Spectroscopy: Fundamentals and Practical Methods,
First Edition. Edited by Mitsuo Tasumi and Akira Sakamoto.

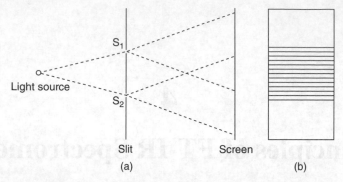

Figure 4.1 *Young's experiment. (a) Optical arrangement and (b) interference fringes on Screen.*

interference of two light waves, is schematically shown in Figure 4.1. In Figure 4.1a, a monochromatic light wave from the point source passes through Slits 1 and 2 (S_1 and S_2) and reaches the screen (Screen), where, as shown in Figure 4.1b, alternating bright and dark lines (or bands), namely, fringes are observed as a result of interference of the two light waves from S_1 and S_2.

The interference of two light waves in Young's experiment may be understood by depicting the light waves in the following way. The light waves propagating from the two slits are schematically shown in Figure 4.2a, where solid circles sections correspond to the crest positions of the light waves and dashed circles sections to the trough positions. The distance between two adjacent crests as well as that between two adjacent troughs along the propagation direction corresponds to a wavelength of the light. At the points where crests of the waves from the two slits overlap each other, the resultant crests become higher. At the points where troughs overlap each other, the resultant troughs become deeper. In other

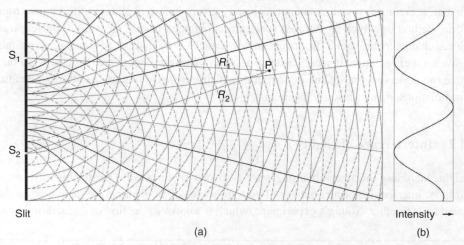

Figure 4.2 *Interference of light waves in Young's experiment. (a) Propagation of light waves from two slits, S_1 and S_2 and (b) light intensities on Screen.*

words, at these two kinds of points, the light waves additively strengthen each other (constructive interference). In contrast, at the points where crests of the wave from one slit overlap troughs of the wave from the other slit, the crests and the troughs cancel each other (destructive interference), resulting in disappearance of the light intensity.

The thicker solid lines in Figure 4.2a are formed by connecting jointly the points where the crests overlap each other and the points where the troughs overlap each other. Since the light waves from S_1 and S_2 strengthen each other at any point on these lines, the light waves traveling along the directions of these lines produce bright bands on the Screen. The thinner lines in Figure 4.2a are formed by connecting the points where the crests overlap the troughs. Contrary to the case of the thicker lines, the light waves from S_1 and S_2 cancel each other at any point on the thinner lines. Thus, the light waves traveling along the directions of these thinner lines produce dark bands on the Screen. As a result, the light intensities on the Screen are expected to change as shown in Figure 4.2b.

The qualitative explanation of Young's experiment given above may be formulated in the following way. The distances from S_1 and S_2 to a point (designated as P) located between the Slits and the Screen in Figure 4.1a are denoted, respectively, as R_1 and R_2. If P exists on the thicker lines in Figure 4.2a, the following relation holds:

$$R_1 - R_2 = n\lambda \tag{4.1}$$

where n is an integer (either positive or negative) and λ is the wavelength of the light. This simple equation corresponds to the fact that the two light waves overlap in phase at any point on the thicker lines in Figure 4.2a, where R_1 and R_2 are different by an integral multiple of the wavelength. Similarly, the following equation holds at any point on the thinner lines in Figure 4.2a, where the light waves overlap in antiphase (180° out of phase):

$$R_1 - R_2 = \left(n + \frac{1}{2}\right)\lambda \tag{4.2}$$

Interference of the two light waves can be understood more rigorously by taking the sum of the electric fields of the light waves. The electric field of a light wave with frequency v and wavelength λ propagating along the x direction at time t was given by Equation (1.7) in Section 1.2.4. If the electric fields at P of the light waves from S_1 and S_2 are denoted, respectively, by E_1 and E_2, they may be expressed as follows by taking the real part of the right-hand side of Equation (1.7). It is assumed that the light waves emitted from S_1 and S_2 are of equal strength and in phase as shown in Figure 4.2a.

$$E_1 = A \cos 2\pi \left(vt - \frac{R_1}{\lambda}\right) \tag{4.3a}$$

$$E_2 = A \cos 2\pi \left(vt - \frac{R_2}{\lambda}\right) \tag{4.3b}$$

The electric field E at P is given as the sum of E_1 and E_2.

$$E = A \cos 2\pi \left(vt - \frac{R_1}{\lambda}\right) + A \cos 2\pi \left(vt - \frac{R_2}{\lambda}\right) \tag{4.4}$$

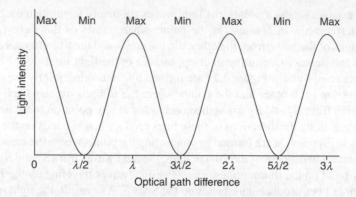

Figure 4.3   *Light intensities on Screen plotted against the optical path difference (OPD).*

The light intensity I at P is derived by taking the time average of E^2. Since the time averages of $\cos^2(2\pi\nu t)$, $\sin^2(2\pi\nu t)$, and $\cos(2\pi\nu t)\sin(2\pi\nu t)$ are equal to 1/2, 1/2, and 0, respectively,

$$I = A^2 \left[1 + \cos 2\pi \frac{(R_1 - R_2)}{\lambda} \right] \tag{4.5}$$

where $A = |A|$. As mentioned in Section 1.2.2, the right-hand side of this equation should be multiplied by a proportionality constant depending on the medium in which the light waves are passing (see Equation (1.2) in the case of vacuum). However, that constant is not included in this equation, because it is not important for understanding the issues treated in this chapter.

Equation (4.5) clearly indicates that the interference fringes depend only on the difference of the distances $R_1 - R_2$, which is called the *optical path difference* (*OPD*). The light intensity in Equation (4.5) is plotted against OPD in Figure 4.3, where the "Max" and "Min" positions correspond, respectively, to the conditions given by Equations (4.1) and (4.2).

4.3   Principles of FT-IR Spectrometry

In this section, it is shown that the interference of light can be used to derive the spectrum of the light. FT-IR spectrometry is a method for measuring the intensity of infrared radiation over a wide spectral region by the use of an interferometer. The Michelson interferometer, a representative two-beam interferometer, or its modification, is usually employed for this purpose.

4.3.1   The Michelson Interferometer

A setup of the Michelson interferometer is schematically shown in Figure 4.4. It has three main optical components, namely, beamsplitter BS, movable mirror M_1, and fixed mirror M_2. In the following discussion, mention is made of the two optical axes in the interferometer; one is the center line of the arm with M_1 and the other is the center line of the arm with M_2. In Figure 4.4, these optical axes are perpendicular to each other.

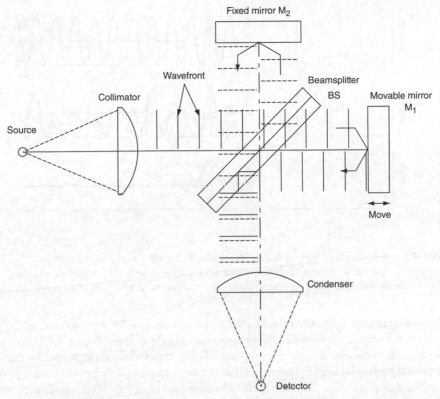

Figure 4.4　*Optical arrangements of the Michelson interferometer (schematically illustrated). Parabolic mirrors are commonly used instead of the lenses for Collimator and Condenser. Wavefronts of the infrared beam traveling toward M_1 are indicated by solid lines, and those traveling toward M_2 by dashed lines. Only half lengths of the wavefronts are shown to differentiate the forward and backward beams.*

Infrared radiation emitted from the point source becomes a collimated (i.e., parallel) beam after passing through the collimator, and travels to the BS. In Figure 4.4, the collimated beam is indicated by lines perpendicular to the direction of beam propagation that coincides with the optical axis normal to M_1. These lines indicate the wavefronts having equal phases in the waves.

The BS has the function of transmitting half of the incident infrared beam and reflecting the remaining half. The beams divided by BS are directed into the two arms of the interferometer, reflected back by M_1 and M_2, and join again at BS. Then both of them travel toward the detector (Detector).

M_1 scans back and forth along the optical axis. This mechanism, which is characteristic of the Michelson interferometer, has the role of changing the optical path length of the beam traveling from BS to M_1 to BS (the beam whose wavefronts are indicated by vertical solid lines in Figure 4.4) compared with the optical path length of the beam traveling from BS to M_2 to BS (the beam whose wavefronts are indicated by horizontal dashed lines in Figure 4.4); that is, the translation of M_1 makes it possible to generate an OPD between the two beams and thereby to cause interference of these beams.

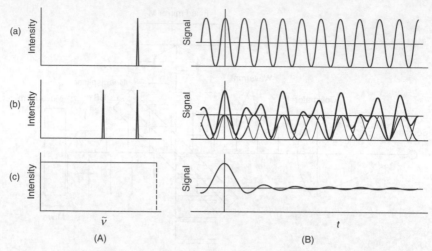

Figure 4.5 *Relation between emission spectra and corresponding interferograms. (A) Emission spectra and (B) interferograms corresponding to the emission spectra. (a) A monochromatic emission, (b) two different monochromatic emissions, and (c) a continuum emission.*

To understand the function of the Michelson interferometer, let us first consider the case where monochromatic infrared radiation enters the interferometer (Figure 4.5A(a)). In this case, as each of the two beams traveling from BS toward Detector (indicated by the solid and dashed horizontal lines in Figure 4.4) has the same phase everywhere in a plane perpendicular to the optical axis, the phase relation between the two beams is the same everywhere in the space between BS and Detector. Accordingly, the interference occurs in the same way everywhere in this space. This is a situation different from the case of Young's experiment in which the phase relation between the light waves from S_1 and S_2 varies on Screen in Figure 4.2. To derive an interference pattern like the one in Figure 4.3 with the Michelson interferometer, it is necessary to change the OPD between the two beams (BS–M_1–BS and BS–M_2–BS) by translating the movable mirror M_1.

It is easy to derive an equation corresponding to Equation (4.5) to express the interference in the Michelson interferometer. If the distances from a point on BS to M_1 and M_2 along the optical axes are denoted, respectively, by L_1 and L_2, the OPD is given as $2(L_1 - L_2)$. By substituting $2(L_1 - L_2)$ for $(R_1 - R_2)$ in Equation (4.5), the intensity I reaching Detector is given as

$$I = A^2 \left[1 + \cos 2\pi \frac{2(L_1 - L_2)}{\lambda} \right] \tag{4.6}$$

By replacing $2(L_1 - L_2)$ and $1/\lambda$ with x and $\tilde{v}$, respectively, Equation (4.6) is simplified to

$$I = A^2(1 + \cos 2\pi \tilde{v} x) \tag{4.7}$$

Equation (4.7) indicates that it is necessary to change x in order to generate an interference pattern. In practice, x is changed by moving M_1 at a constant speed v for a time period t, so that

$$x = 2vt \tag{4.8}$$

then,

$$I = A^2(1 + \cos 4\pi v \tilde{v} t) \tag{4.9}$$

This equation means that the intensity of the infrared radiation with wavenumber $\tilde{v}$ is modulated to a wave of frequency $2v\tilde{v}$. In this case, a corresponding signal is obtained from Detector, as shown in Figure 4.5B(a).

Next, a source which emits monochromatic infrared radiation at two wavenumbers $\tilde{v}_1$ and $\tilde{v}_2$ is considered (Figure 4.5A(b)). In this case, the intensities of infrared radiation having modulated frequencies $2v\tilde{v}_1$ and $2v\tilde{v}_2$ corresponding to their respective wavenumbers $\tilde{v}_1$ and $\tilde{v}_2$ are summed, and the signal from Detector becomes a wave synthesized from them as shown by the bold curve in Figure 4.5B(b).

Finally, if the source emits infrared radiation at all wavenumbers over a broad infrared region as shown in Figure 4.5A(c), the signal from Detector becomes the sum of waves having modulated frequencies in a wide range as shown in Figure 4.5B(c). This means that the interferogram obtained from the Michelson interferometer contains complete information on the input infrared radiation, and furthermore it should be possible to derive the spectrum of the input infrared radiation by properly analyzing the intensities of frequency components contained within the interferogram.

4.3.2   Interferogram and Its Fourier Transform

The emission from an ordinary thermal source contains radiation over a wide wavenumber range. However, the spectrum of the emission from the source is not as flat as that in Figure 4.5A(c). The intensity of the emission from the source is denoted by $B(\tilde{v})$ to indicate that it is a function of wavenumber $\tilde{v}$ (i.e., a spectrum). When such radiation enters the Michelson interferometer with an OPD (denoted by x), the intensity of the radiation reaching Detector $I(x)$ is given as

$$I(x) = \int_0^\infty B(\tilde{v})(1 + \cos 2\pi\tilde{v}x)\mathrm{d}\tilde{v} \tag{4.10}$$

The second term of Equation (4.10) is called the *interferogram*, which depends on x (or the corresponding time t) in a way similar to the curve shown in Figure 4.5B(c). The interferogram has the highest intensity at $x = 0$ because the two beams (BS–M_1–BS and BS–M_2–BS) of any wavenumber interfere with each other in phase at $x = 0$. The peak at $x = 0$ is called the *centerburst*; this position is sometimes referred to as *zero path difference* (*ZPD*). With increasing x, the intensity of the interferogram undulates because the two beams in the interferometer interfere with each other at various phases, and it gradually decays toward zero.

The interferogram is symmetric about $x = 0$ as is clear from Equation (4.10). In other words, the interferogram is an even function. Then, the interferogram, denoted by $F(x)$, may be expressed in the following form by extending the lower limit of the integral to $-\infty$.

$$F(x) = \int_{-\infty}^\infty B_e(\tilde{v}) \cos 2\pi\tilde{v}x\mathrm{d}\tilde{v} \tag{4.11}$$

where $B_e(\tilde{v})(\tilde{v} \geq 0) = B(\tilde{v})/2$ and $B_e(\tilde{v})(\tilde{v} < 0) = B(-\tilde{v})/2$. $B_e(\tilde{v})$ is a virtual function introduced for the purpose of mathematical treatment, but it is directly related to the

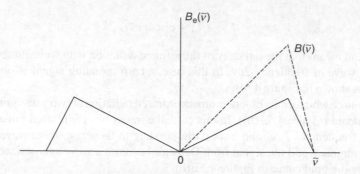

Figure 4.6   *Relation between $B_e(\tilde{v})$ and $B(\tilde{v})$.*

spectrum of the radiation from the source $B(\tilde{v})$. The relation between $B_e(\tilde{v})$ and $B(\tilde{v})$ is schematically illustrated in Figure 4.6. $B_e(\tilde{v})$ exists in both the positive and negative wavenumber regions, the $B_e(\tilde{v})$ in the negative wavenumber region being the mirror image of that in the positive wavenumber region. Like $F(x)$, $B_e(\tilde{v})$ is an even function.

Equation (4.11) indicates that the interferogram $F(x)$ is the Fourier cosine transform of $B_e(\tilde{v})$. Then, according to the theory of Fourier transforms, $B_e(\tilde{v})$ can be obtained as the (inverse) Fourier cosine transform of $F(x)$ as

$$B_e(\tilde{v}) = \int_{-\infty}^{\infty} F(x) \cos 2\pi\tilde{v}x \mathrm{d}\tilde{v} \qquad (4.12)$$

$F(x)$ and $B_e(\tilde{v})$ are a Fourier transform pair.

It is important to note in Equation (4.12) that the spectrum of the radiation from the source is obtained as the (inverse) Fourier transform of the observed interferogram, and an artifact spectrum also appears in the negative wavenumber region. This accounts for the term *Fourier transform* in FT-IR spectrometry or spectroscopy. As described later in this chapter and in subsequent chapters also, Fourier transforms are often used in various aspects of FT-IR spectroscopy.

It may be mentioned here that, in FT-IR spectrometry, radiation over a wide infrared region is simultaneously measured with a single detector. This is the first reason why FT-IR spectrometry is faster in spectral measurements in comparison with dispersive spectrometry with a grating and a scanning mechanism. This feature of FT-IR spectrometry, which is called the *multiplex advantage* or *Fellgett's advantage*, was emphasized during the early period of development of FT-IR spectrometry.

4.4   Characteristics of FT-IR Spectrometry

In this section, characteristics of FT-IR spectrometry, that is, wavenumber resolution, measurable spectral region, and accurate determination of wavenumbers, are discussed. As mentioned in Section 3.2.2, the term *wavenumber resolution* is used in this book instead of the commonly used terminology of just "resolution."

4.4.1   Wavenumber Resolution

Although a brief introduction to the wavenumber resolution in FT-IR spectrometry was given in Section 3.2.2, this subject is discussed here in greater detail.

4.4.1.1   Finiteness of the OPD and the Instrumental Function

Consider two collimated beams of monochromatic radiation with wavenumbers $\tilde{v}_1$ and $\tilde{v}_2$ which are in close proximity, and let these beams separately enter the Michelson interferometer. For each of the two beams, the dependence of the output signal from the detector on the OPD is measured. The output signal from the detector is proportional to the intensity I in Equation (4.7). When the OPD spans only a short distance from the position of $x = 0$ (i.e., when the mirror M_1 is made to move only a little), the output signal for $\tilde{v}_1$ does not significantly differ from that for $\tilde{v}_2$. With increasing OPD, however, a difference in the output signals for the two beams must be detectable due to the small phase difference $2\pi(\tilde{v}_1 - \tilde{v}_2)x$ between the two beams in Equation (4.7). This suggests that a higher wavenumber resolution is attainable with a longer OPD range. It seems reasonable to think that a phase difference of 2π between two lines would suffice to differentiate $\tilde{v}_1$ and $\tilde{v}_2$. This leads to the condition that $(\tilde{v}_1 - \tilde{v}_2)x = 1$. This simple equation becomes identical to the relation $\delta\tilde{v} = 1/D$ in Equation (3.1), if $(\tilde{v}_1 - \tilde{v}_2)$ and x are substituted with $\delta\tilde{v}$ and D, respectively.

Let us discuss in mathematical terms the relation between the wavenumber resolution and the maximum OPD in Equation (3.1). Before beginning the discussion, an explanation is given for the convolution theorem relating to Fourier transforms. This theorem is expressed by the following two equations. The bar above the function indicates the Fourier transform, and the symbols "·" and "*" indicate, respectively, ordinary multiplication and convolution.

$$\overline{A(x) \cdot B(x)} = \overline{A(x)} * \overline{B(x)} \tag{4.13a}$$

$$\overline{A(x) * B(x)} = \overline{A(x)} \cdot \overline{B(x)} \tag{4.13b}$$

These equations are useful for calculating the Fourier transforms of a product of functions and convolved functions in that the answer is obtained from the Fourier transforms of individual functions.

The operation of convolution is expressed by the following equation.

$$A(x) * B(x) = \int_{-\infty}^{\infty} A(t) \cdot B(x - t)\mathrm{d}t \tag{4.14}$$

Convolution is a mathematical operation used for expressing blurs in the optical image, the response of electrical signal, and so on. In Figure 4.7, the operation of convolution is pictorially shown. The picture of a spinning top in Figure 4.7a is out of focus, while that in Figure 4.7b is in focus. The circular area in Figure 4.7c is a spread shape obtained from a point object under the same condition as used for taking the out-of-focus picture of the top in Figure 4.7a. The blurred picture in Figure 4.7a is obtained by the convolution of the focused picture in Figure 4.7b with the spread shape of the point object in Figure 4.7c. Generally speaking, the work of a measuring apparatus is expressed by an instrumental function which expresses the response of the apparatus to an input signal given as an impulse. A result actually measured by an apparatus is the convolution of a sharp response signal having no blur with the instrumental function of the apparatus.

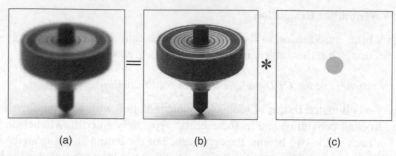

(a)   (b)   (c)

Figure 4.7 *Effect of convolution. The picture in (a) is the result of convolution of the pictures in (b) and (c). (a) Out-of-focus picture of a top, (b) in-focus picture of a top, and (c) spread picture of a point object.*

The convolution theorem described above is useful in deriving the relation between the wavenumber resolution and the maximum OPD in Equation (3.1). In Equation (4.12), the integration is performed for the entire OPD range from $-\infty$ to ∞, but the maximum OPD is a finite value D in a real spectral measurement. Then, $B_e(\tilde{\nu})$ in Equation (4.12) should be replaced by $B'_e(\tilde{\nu})$, which is given as

$$B'_e(\tilde{\nu}) = \int_{-D}^{D} F(x)\cos 2\pi\tilde{\nu}x\,\mathrm{d}x \qquad (4.15)$$

To derive a mathematical relation between $B'_e(\tilde{\nu})$ and $B_e(\tilde{\nu})$, it is necessary to use the rectangular function $\Pi_{2D}(x)$ and its Fourier transform $\Phi(\tilde{\nu})$. The shape of $\Pi_{2D}(x)$ is shown in Figure 4.8.

$$B'_e(\tilde{\nu}) = \int_{-\infty}^{\infty} F(x)\Pi_{2D}(x)\cos 2\pi\tilde{\nu}x\,\mathrm{d}x$$

$$= \overline{F(x)\cdot\Pi_{2D}(x)} = \overline{F(x)} * \overline{\Pi_{2D}(x)}$$

$$= B_e(\tilde{\nu}) * \Phi_{2D}(\tilde{\nu}) \qquad (4.16)$$

This means that the measured spectrum $B'_e(\tilde{\nu})$ is expressed as the convolution of the spectrum of the emission from the source $B_e(\tilde{\nu})$ with $\Phi_{2D}(\tilde{\nu}) = \overline{\Pi_{2D}(x)}$. $\Phi_{2D}(\tilde{\nu})$ is the instrumental function of the interferometer operated at the maximum OPD of D. $\Phi_{2D}(\tilde{\nu})$ is given as

$$\Phi_{2D}(\tilde{\nu}) = (2D)\frac{\sin(2\pi D\tilde{\nu})}{2\pi D\tilde{\nu}} = (2D)\mathrm{sinc}(2\pi D\tilde{\nu}) \qquad (4.17)$$

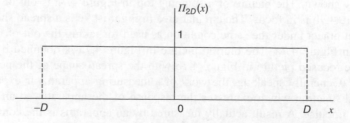

Figure 4.8 *Rectangular function $\Pi_{2D}(x)$.*

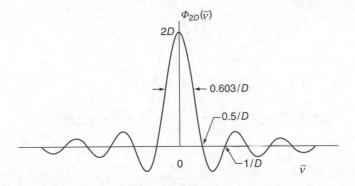

Figure 4.9 *Fourier transform of* $\Pi_{2D}(x)\,[\Phi_{2D}(\tilde{v})]$.

This function is called a *sinc function*, and its shape is shown in Figure 4.9. It is symmetric about $\tilde{v} = 0$ and intersects the $\tilde{v}$ axis at $\pm n/(2D)$ $(n = 1, 2, 3, \ldots)$. The full width at half maximum (FWHM) of the central peak is $0.603/D$.

Let us consider a case where two monochromatic beams of $\tilde{v}_1$ and $\tilde{v}_2$ with equal intensities enter a spectrometer having a fixed instrumental function $\Phi_{2D}(\tilde{v})$. In Figure 4.10, the expected spectra are shown for four different values of separation between $\tilde{v}_1$ and $\tilde{v}_2$. The two peaks corresponding to the lines at $\tilde{v}_1$ and $\tilde{v}_2$ are clearly resolved in the synthesized spectrum in Figure 4.10d, which is depicted for $\tilde{v}_2 - \tilde{v}_1 = 1/D$. In fact, the two peaks are nearly resolved even in the synthesized spectrum in Figure 4.14c, which is for $\tilde{v}_2 - \tilde{v}_1 = 0.7/D$. This separation is close to the FWHM of the central peak of $\Phi_{2D}(\tilde{v})$ $(0.603/D)$ in Figure 4.9. This result supports the usual practice that, for resolving two bands that are closely related, it is required to set the wavenumber resolution of the FT-IR spectrometer in use to a value equal to or smaller than the FWHM of the bands. Usually, the wavenumber resolution $\delta\tilde{v}$ is defined by the relation $\delta\tilde{v} = 1/D$. Then, the wavenumber resolutions 1, 2, and 4 cm^{-1} should be obtained by setting the maximum OPD to 1, 0.5, and 0.25 cm, respectively. If the OPD can be set to a long distance of 100 cm, a high wavenumber resolution of 0.01 cm^{-1} ought to be obtained.

4.4.1.2 Apodization

The instrumental function in Figure 4.9 has many maxima and minima on both sides of the central peak. Spectral measurements with such an instrumental function result in the appearance of artifacts including bands with negative intensities as shown in Figure 4.10. Bands of weak intensities may be lost in actual spectral measurements. Such a situation may mislead the analysis of measured spectra. The side peaks reflect the fact that terminating the measurement of an interferogram at $x = \pm D$ brings discontinuous ends in the interferogram which abruptly becomes zero at $x = \pm D$. The spectrum obtained by the Fourier transform of such an interferogram has ringing components. Apodization is a mathematical operation employed to suppress the artifacts. It uses an apodizing function instead of the rectangular function to make the interferogram continuously decay to zero.

A few typical apodizing functions and their Fourier transforms (instrumental functions) are shown in Figure 4.11, together with the rectangular function and its Fourier transform,

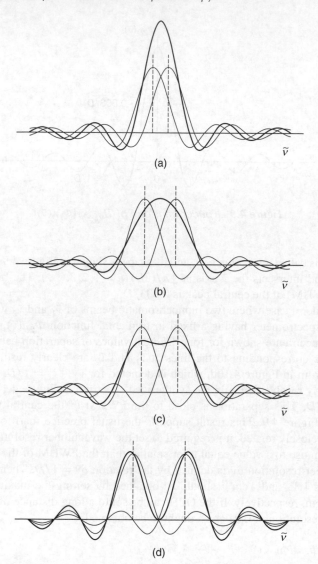

Figure 4.10 *Effect of an instrumental function on two closely located lines. Intervals between the two dashed lines are (a) 0.3/D, (b) 0.6/D, (c) 0.7/D, and (d) 1/D. The thick curve in each spectrum is the sum of the two thin curves corresponding to the two lines.*

the sinc function. It is clear in Figure 4.11B that a higher degree of suppression of the side maxima and minima in the instrumental function is accompanied by an increase in the FWHM, which leads to a decrease in the wavenumber resolution. It is therefore important to choose an apodizing function fit for the purpose of the spectral measurement to be made.

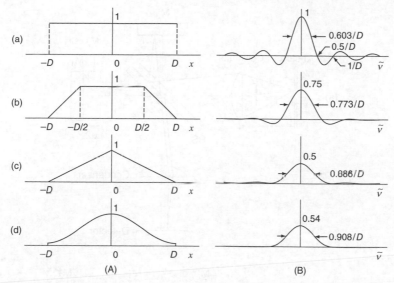

Figure 4.11 *(A) Representative apodizing functions and (B) their instrumental functions. (a) Rectangular function, (b) trapezoid function, (c) triangular function, and (d) Happ–Genzel function.*

4.4.1.3 Effect of Oblique Beams

In the description of the Michelson interferometer in Section 4.3.1, it is postulated that the collimated beam travels parallel to the optical axes to impinge M_1 and M_2 perpendicularly. For considering the OPD, however, collimated beams traveling obliquely to the optical axes also need to be taken into account, because the source actually used is not a point but has a finite size. The existence of oblique beams results in a decrease in wavenumber resolution.

In Figure 4.12a, the optical path of a collimated beam oblique to the optical axis is illustrated. This ray, simply called an *oblique ray*, comes from the upper edge of the circular entrance aperture and makes an angle α to the optical axis after passing through the collimator. The oblique ray is supposed to reach the moving mirror M_1 when it is displaced by $x/2$ from the position of zero OPD.

To facilitate the determination of the OPD for the oblique ray, it is useful to consider a mirror image M_2^* of the fixed mirror M_2, which is depicted at a distance of $-x/2$ from M_1. In Figure 4.12b, the optical path of the oblique ray in the conceived optical configuration consisting of M_1 and M_2^* are illustrated. In this figure, an auxiliary plane $M_2^{*\prime}$ is put at a distance of $x/2$ from M_1 to help determine the OPD. M_2^* and $M_2^{*\prime}$ are symmetrically located with respect to M_1. The ray reflected by M_1 at P_1 is denoted by Ray 1 and that reflected by M_2^* at P_2^* by Ray 2. Point H is the foot of the perpendicular line from P_2 to Ray 1, and $P_2^{*\prime}$ is the mirror image of P_2^* with respect to M_1. Then, the OPD between Rays 1 and 2 is the distance from P_2^* to P_1 to H, which, being equal to the distance from $P_2^{*\prime}$ to H, is expressed as $x\cos\alpha$. Since the OPD for the beam parallel to the optical axis is x, the OPD for the oblique ray is shorter by a factor of $\cos\alpha$. As a result, a lower shift of wavenumber from $\tilde{\nu}$

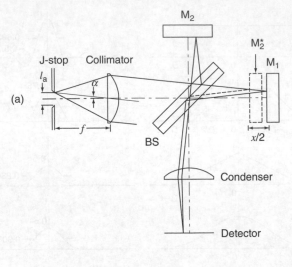

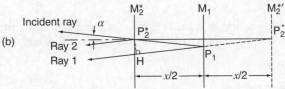

Figure 4.12 *Optical path difference (OPD) for an oblique beam. (a) Oblique beam in the Michelson interferometer and (b) diagram for deriving the OPD for the oblique beam.*

to $\tilde{\nu}\cos\alpha$ occurs for the oblique ray in the Fourier transform of the interferogram. This is equivalent to a decrease in wavenumber resolution.

To avoid a large decrease in wavenumber resolution arising from oblique rays, a circular aperture, called the *Jacquinot stop* (*J-stop*), is placed in the focal plane of the collimator as shown in Figure 4.12a. The optimal diameter l_a of the J-stop is determined in order to make the wavenumber shift due to oblique rays smaller than the wavenumber resolution $\delta\tilde{\nu}$ determined by the OPD for the beam parallel to the optical axis.

If the upper limit of the wavenumber shift due to oblique beams is denoted by $\Delta\tilde{\nu}_{\mathrm{s}}^{\mathrm{up}}$,

$$\tilde{\nu} - \tilde{\nu}\cos\alpha \le \Delta\tilde{\nu}_{\mathrm{s}}^{\mathrm{up}} \tag{4.18}$$

Since α is not a large angle, the following approximate relation holds:

$$\left(\frac{1}{2}\right)\tilde{\nu}(\alpha)^2 \le \Delta\tilde{\nu}_{\mathrm{s}}^{\mathrm{up}} \tag{4.19}$$

If the focal length of the collimator is denoted by f, α is approximately expressed as $l_a/(2f)$. Then, the following relation is obtained:

$$\left(\frac{1}{2}\right)(l_a/2f)^2\tilde{\nu} \le \Delta\tilde{\nu}_{\mathrm{s}}^{\mathrm{up}} \tag{4.20}$$

The value of the left-hand side of this relation is largest for the maximum wavenumber $\tilde{\nu}_{\mathrm{max}}$ in the wavenumber region to be measured by the spectrometer in use. If Relation

(4.20) holds for $\tilde{\nu}_{\max}$, all the other values of $\tilde{\nu}$ also satisfy this Relation. Then, it suffices to take only $\tilde{\nu}_{\max}$ into account. It is customary to set $\Delta\tilde{\nu}_s^{up}$ equal to half of the wavenumber resolution determined by the maximum OPD for the beam parallel to the optical axis; that is, $\Delta\tilde{\nu}_s^{up} = \delta\tilde{\nu}/2 = 1/2D$. Then, l_a is given as

$$l_a = \frac{2f}{(\tilde{\nu}_{\max}D)^{\frac{1}{2}}} = 2f\left(\frac{\delta\tilde{\nu}}{\tilde{\nu}_{\max}}\right)^{\frac{1}{2}} \tag{4.21}$$

If $\tilde{\nu}_{\max} = 4000\,\text{cm}^{-1}$ and $f = 300\,\text{mm}$, the values of l_a for $\delta\tilde{\nu} = 1, 2,$ and $4\,\text{cm}^{-1}$ are calculated, respectively, to be 9.5, 13.4, and 19.0 mm. The J-stop aperture has an area about 100 times larger than the area of a slit used in a dispersive spectrometer with the same wavenumber resolution. This is the second reason why FT-IR spectrometry is superior to dispersive spectrometry. This feature of the FT-IR spectrometer is called the *optical throughput advantage* or *Jacquinot's advantage*.

4.4.2 Measurable Spectral Region

As mentioned in Section 4.3.2, FT-IR spectrometry can measure all the wavenumbers of infrared radiation simultaneously. In practice, however, the measurable spectral data are restricted on account of the digital computer inevitably needed for recording the interferogram and obtaining its Fourier transform. The energy of infrared radiation received by the detector is converted to analog electric signals (A). The analog signals are then converted to digital signals (D), which are discrete in both time and intensity. This process, performed by an analog-to-digital (AD) converter, consists of data sampling and digitization. This sampling operation, taking merely a set of specified discrete signals from a continuous signal, causes some lack of information, which results in restrictions on the measurable spectral region.

For expressing the sampling operation, the Dirac delta comb $\text{Ш}_a(x)$ is useful. The shapes of $\text{Ш}_a(x)$ and its Fourier transform $(1/a)\text{Ш}_{1/a}(\tilde{\nu})$ are illustrated in Figure 4.13. The sampling of the interferogram $F(x)$ is performed as illustrated in Figure 4.14, where points on a continuous interferogram are sampled at intervals of a. This process is expressed as $F(x) \cdot \text{Ш}_a(x)$. The following spectrum $B_e''(\tilde{\nu})$ is obtained by the Fourier transform of $F(x) \cdot \text{Ш}_a(x)$.

$$B_e''(\tilde{\nu}) = \int_{-\infty}^{\infty} F(x)\text{Ш}_a(x)\cos 2\pi\tilde{\nu}x\,dx = \overline{F(x) \cdot \text{Ш}_a(x)} \tag{4.22}$$

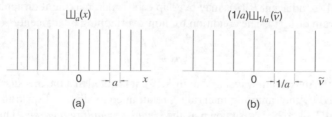

Figure 4.13 *Dirac delta comb and its Fourier transform. (a) $\text{Ш}_a(x)$ and (b) $(1/a)\text{Ш}_{1/a}(\tilde{\nu})$ (arbitrary height).*

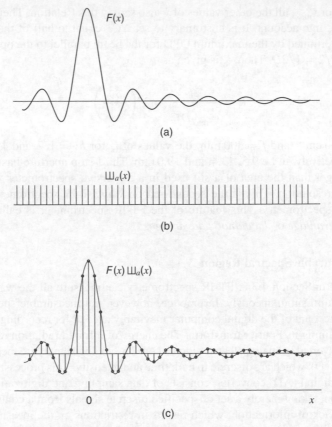

Figure 4.14 *Sampling of an interferogram. (a) Interferogram before sampling, (b) sampling signal, and (c) interferogram after sampling; the sampling points are illustrated by the filled circles.*

By applying the convolution theorem given in Equation (4.13a), Equation (4.22) is rewritten as

$$B''_e(\tilde{v}) = \overline{F(x)} * \overline{\text{Ш}_a(x)} = B_e(\tilde{v}) * \left[\left(\frac{1}{a}\right)\text{Ш}_{1/a}(\tilde{v})\right] \tag{4.23}$$

$B''_e(\tilde{v})$ is illustrated in Figure 4.15. It is noted that the spectrum $B_e(\tilde{v})$ periodically appears at intervals of $1/a$. This phenomenon is called the *aliasing* or *folding*. If the range of $B_e(\tilde{v})$ is wider than $1/a$, adjacent $B_e(\tilde{v})$s may overlap each other, making it difficult to determine the true spectrum of $B_e(\tilde{v})$. The condition for non-overlapping of adjacent $B_e(\tilde{v})$s is given as

$$\tilde{v}_{\text{max}} \leq \frac{1}{2a} \tag{4.24}$$

where $\tilde{v}_{\text{max}}$ is the maximum wavenumber in $B_e(\tilde{v})$. It is noted that the measurable wavenumber is restricted by the sampling interval. A relation equivalent to Equation (4.24), which is expressed as $1/a \geq 2\tilde{v}_{\text{max}}$, is known as the *Nyquist sampling theorem*. This relation dictates that the period of sampling $(1/a)$ must be larger than at least twice the maximum wavenumber of the wavenumber region to be measured.

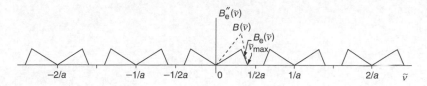

Figure 4.15   *Spectrum obtained by the Fourier transform of the interferogram after sampling.*

In FT-IR spectrometry, the sampling points are determined by monitoring the position of the movable mirror with a laser (usually a He–Ne laser). The wavelength of the He–Ne laser light is 632.991399 nm in vacuum, and this wavelength is denoted by λ_{HN}. Then, if a is $0.5\lambda_{HN}$, λ_{HN}, and $2\lambda_{HN}$, $\tilde{\nu}_{max}$ given by Equation (4.24) is 15 798, 7899, and 3950 cm^{-1}, respectively. If a narrow sampling interval is used, data points increase in number. As a result, a longer time is needed for computing the Fourier transform. A practical sampling interval fit for the purpose of the spectral measurement should be chosen.

4.4.3   Accurate Determination of Wavenumbers

In an FT-IR spectrometer, the wavenumbers of infrared radiation are determined from the OPD, which is obtained by monitoring the translated distance of the movable mirror with a laser. Infrared wavenumbers calibrated by using the accurately determined wavelength of a stabilized laser light are always highly reliable and reproducible. This feature of FT-IR spectrometry is called the *Connes advantage*. This advantage is most effectively utilized for improving S/N ratios by accumulating measured spectra and for determining minute differences between two spectra by the method of spectral subtraction, which is described in Chapter 6.

Further Reading

1. Griffiths, P.R. and de Haseth, J.A. (2007) *Fourier Transform Infrared Spectrometry*, 2nd edn, John Wiley & Sons, Inc., Hoboken, NJ.

5

Hardware and Software in FT-IR Spectrometry

Koji Masutani
Micro Science Inc., Japan

5.1 Introduction

Modern Fourier transform-infrared (FT-IR) spectrometers are highly developed, and a wide variety of spectrometers from small, portable types to larger laboratory-based instruments that can accommodate many kinds of attachments are commercially available. Usually they can be used without difficulty after taking some initial practical training of how to operate them. It may appear unnecessary for an individual user to have enough detailed knowledge of the hardware and hardware-controlling software of an FT-IR spectrometer in order to perform routine work. It may happen, however, that the user gets a new idea that will benefit his/her work if he/she has ample knowledge of the full function and capabilities of the spectrometer in use. In this chapter, the structure and function of a typical FT-IR spectrometer are explained with the intention of providing information useful for understanding the typical specifications and performance capabilities of commercially available FT-IR spectrometers.

5.2 Hardware and Software of an FT-IR Spectrometer

An FT-IR spectrometer may be considered to consist essentially of three parts, namely, an optical system, a computer system, and software for controlling the hardware and processing the recorded data. The former two parts are called the *hardware*. The arrangement of the hardware in an FT-IR spectrometer is illustrated schematically in Figure 5.1.

Introduction to Experimental Infrared Spectroscopy: Fundamentals and Practical Methods,
First Edition. Edited by Mitsuo Tasumi and Akira Sakamoto.
© 2015 John Wiley & Sons, Ltd. Published 2015 by John Wiley & Sons, Ltd.

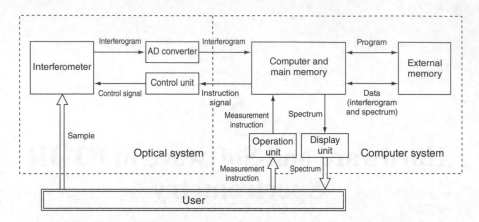

Figure 5.1   *Schematic arrangement of hardware in an FT-IR spectrometer.*

The optical system has the primary role of measuring and sampling an interferogram from a sample placed in the sample compartment. The computer system, together with its software, controls the operations of the optical system, performs numerical calculations including the Fourier transform of an interferogram collected as digitized data, stores data, and provides for user interaction with the system.

In a typical use, the user places a sample in the sample compartment, enters the conditions of measurement into the computer through the operation unit, and gives command instructions to start measuring an interferogram. The computer memorizes the conditions of measurement, adjusts the size of the entrance stop, and so on, in the optical system to the user-given conditions through the control unit, and starts the measurement of an interferogram. After the measurement, the interferogram obtained is stored in the computer via the analog-to-digital (AD) converter and undergoes computations of a Fourier transform and other data processing procedures. As a result, a spectrum is obtained and shown on the display unit.

5.2.1   Optical System

The basic optical system in an FT-IR spectrometer is schematically illustrated in Figure 5.2, and the flow of signals in the optical system is shown in Figure 5.3. The numbering in Figure 5.3 corresponds to that in Figure 5.2. In Figure 5.3① and ④, signals are shown as single-beam spectra to express the process of absorption by the sample.

Infrared radiation from the light source is focused onto the entrance aperture (J-stop), so that any infrared ray with an incident angle α (see Figure 4.12) larger than that determined by the optimal diameter l_a in Equation (4.21) is blocked from advancing to the interferometer.

The infrared radiation incident on the interferometer is modulated to a frequency proportional to the incident infrared wavenumber by the uniform motion of the moving mirror. This modulated frequency is given in Equation (4.9). In Figure 5.3③, the interferogram

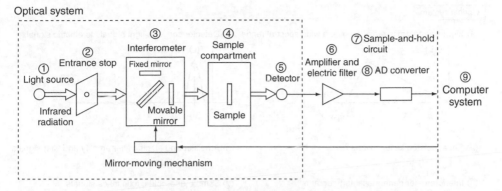

Figure 5.2 *Optical layout of an FT-IR spectrometer.*

$F(x)$, which expresses the modulated signal, is shown. At this stage, the interferogram is a function of time t, but it is given as a function of the optical path difference (OPD) x in Figure 5.3③, because, as given in Equation (4.22), the sampling of an interferogram is performed at constant intervals on x, and the interferogram is converted to a function of x in this process.

The infrared beam emitted from the interferometer is incident on the sample, or its background reference (see later), positioned appropriately in the sample compartment. Radiation energy in the beam is absorbed by the sample at the wavenumbers of its infrared-active vibrations, and the rest of the radiation passes through the sample and is directed toward the detector. In Figure 5.3④, $T(\tilde{v})$ denotes the transmittance of the sample in percentage. The signals actually reaching the detector correspond to $\overline{T(\tilde{v})} * F(x)$ in Figure 5.3⑤. This expression is derived as given below, by referring to the definitions in Equations (4.11) and (4.12) and the convolution theorem in Equation (4.13a).

$$\overline{T(\tilde{v})B(\tilde{v})} = \overline{T(\tilde{v})} * \overline{B(\tilde{v})} = \overline{T(\tilde{v})} * F(\tilde{v}) \tag{5.1}$$

The interferogram in Figure 5.3⑤, which resembles that in Figure 5.3③, should be considered as containing information that represents absorptions by the sample.

In a practical absorption measurement, the same process of measurement is carried out for a sample and an appropriate reference without the sample separately, but only the process for a sample is described here.

The radiation which has passed through the sample is converted to electric signals by the detector, and the signals then enter into the sample-and-hold circuit after passing through the amplifier and electric filter. The sample-and-hold circuit is a part of the AD converter, but it is treated here as an independent component because of its important role. The intensities of signals entered into this circuit are temporarily held here by the sampling instructions given according to the output signals of the laser interferometer monitoring the travel of the movable mirror, until they are processed by the AD converter. As the output signals of the laser interferometer are formed on the basis of the OPD of the interferometer, the

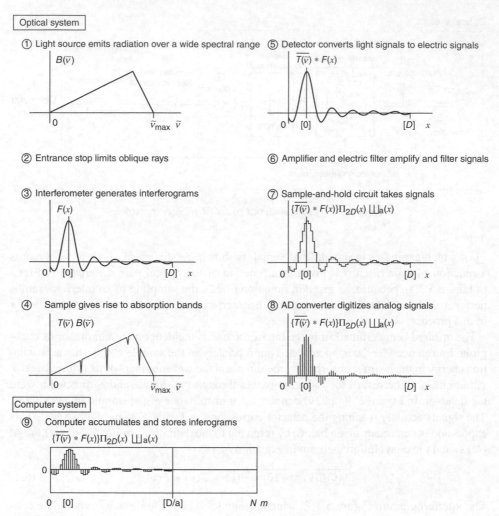

Figure 5.3 *Flow chart for signal processing in an FT-IR spectrometer operation (1). [D] on the x axis indicates the maximum OPD measured from its origin [0].*

interferogram is converted from the function of time to the function of the OPD x. In Figure 5.3⑦, the interferogram after the sampling $\{\overline{T(\tilde{\nu})} * F(x)\}\Pi_{2D}(x)\text{Ш}_a(x)$ is shown. In this expression, $\Pi_{2D}(x)$ is a rectangular function indicating the range of sampling, that is, $|x| \leq D$, and $\text{Ш}_a(x)$ is a Dirac delta comb for the sampling at intervals of a (see Section 4.4.2). As the interferogram should be symmetrical about zero OPD ($x = 0$), the sampling in the range of $0 \leq x \leq D$ is considered to be sufficient. Actually, however, the measured interferogram typically has a shape asymmetrical about zero OPD. To correct this asymmetry, the sampling in the vicinity of $x = 0$ is usually started from a position on the negative side of x as illustrated in Figure 5.3⑦. Correction for the asymmetry of this recorded interferogram will be discussed in Section 5.2.2.3. To simplify the mathematical expression, $\Pi_{2D}(x)$ is used here on the assumption that $F(x)$ is symmetrical about zero OPD.

The AD converter digitizes the magnitudes of the analog signals held in the circuit by comparing them with those of standard signals (successive comparison type to be described in Section 5.2.1.5).

The interferogram measured is read into the computer, accumulated, and stored. This process is repeated until the number of accumulations reaches that instructed by the user. In Figure 5.3⑨, the abscissa axis represents the numbering of sampling points denoted by m.

An explanation of individual elements of the optical system is given in the following parts.

5.2.1.1 Interferometer

An example of an interferometer (Michelson type) used for an FT-IR spectrometer is schematically illustrated in Figure 5.4. The main part of the interferometer consists of a beamsplitter, which divides the incident infrared radiation into two beams, and two mirrors, one fixed and one movable, which reflect the divided two beams back to the beamsplitter. The movable mirror travels along the optical axis, and thus has the role of changing the OPD between the two beams. The output from the interferometer ⓐ goes to the detector.

To change the position of the movable mirror, a driving mechanism consisting of a linear bearing and a linear motor is used. Formerly a gas-bearing was often used. Today, in many designs, it is a common practice to employ a high-precision linear bearing together with a cube-corner retroreflector which replaces the movable plane mirror.

If a gas-bearing is used, the movable mirror with a shaft is floated by gas pressure and changes its position linearly along the moving axis. As the movement is frictionless, its

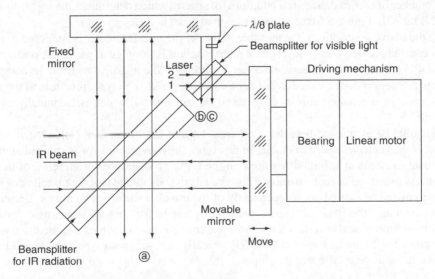

Figure 5.4 *Typical structure of a Michelson interferometer used for FT-IR spectrometry.*

speed can be controlled easily, and the moving axis has little tilt. Either nitrogen gas or dried air is used as the gas for this gas-bearing.

As a linear motor, a voice coil may be used. It is a noncontact linear motor consisting of an electromagnetic coil and a pot-shape magnet. The term *voice coil* originates from its wide use in loud speakers.

In some interferometers, the velocity of the moving mirror can be varied to fit the modulated frequency of the interferogram to the optimum condition of detector response.

To monitor the OPD, a pair of control interferometers using a single laser (usually a He–Ne laser) as the light source are installed. These control interferometers and the main interferometer use some of the optical elements in common. As shown in Figure 5.4, a $\lambda/8$ plate (λ is the wavelength of the laser light used) is inserted into the optical path of laser beam 2, so that a phase difference of $\lambda/4$ exists between the OPDs for laser beams 1 and 2.

The laser interferometer may be regarded as an extremely accurate range finder, based on the monochromaticity of a laser light. Its interferogram, a sinusoidal wave with a wavelength of λ, can be utilized to monitor accurately the position of the movable mirror. This information is used for controlling the speed of the moving mirror as well as for giving instructions on successive positions for sampling the main interferogram (derived from the main interferometer). The output signals ⓑ and ⓒ from the laser interferometers in Figure 5.4 are used as signals sent to the counter which increases or decreases the number of sampling points following the change of the sampling position x (see Figures 5.5 and 5.6).

The process of sampling the main interferogram is shown as a flow chart in Figure 5.5, and the patterns of the signals at steps in the process are illustrated in Figure 5.6. In these two figures, numberings ③–⑥ are commonly associated with the steps so that the signal pattern at each step can be easily identified. The number of sampling points corresponds to the maximum OPD which sets the spectral resolution, and the number of scans is equal to the number of accumulations (co-additions) of spectra which determines the signal-to-noise (S/N) ratio. In Figure 5.6, the sampling interval is set at $\lambda/2$.

In the above descriptions, the movable mirror of the interferometer is supposed to travel at a constant velocity, and the resultant interferogram is sampled at particular positions on the OPD (x) axis placed at equal intervals according to the instructions from the computer. Such an interferometer, called a *continuous-scan* or *rapid-scan* type, produces as the interferogram an alternating current as given in Equation (4.9), which is technically easy to handle.

It should be mentioned here that a different type of interferometer, called the *step-scan type*, is also commercially available. In this type, the movable mirror is stopped stepwise at equal intervals at specified positions on the OPD axis, and the interference of the two beams is measured at each position. This type has the advantage that the sampling of interferogram can be carried out independently of the travel of the movable mirror. Because of this advantage, the step-scan type is particularly useful for photoacoustic, time-resolved, and two-dimensional correlation spectroscopic measurements, which will be discussed in Chapters 14, 20, and 21 respectively. (Historically, the step-scan type had existed before the continuous-scan type was developed.)

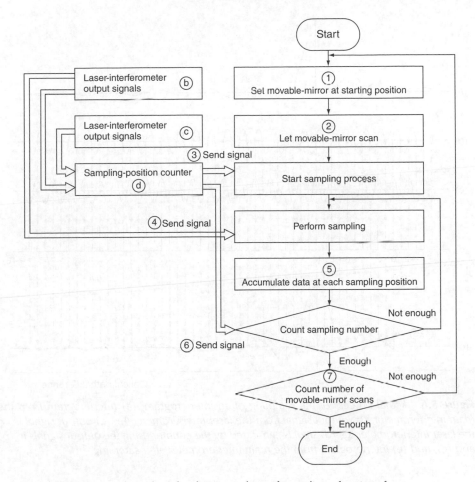

Figure 5.5   *Flow chart for the procedure of sampling of an interferogram.*

Strict conditions are imposed on the motion of the movable mirror. For the movable mirror in the continuous-scan spectrometer, a uniform linear motion without any tilt is necessary. In the case of the step-scan spectrometer, the movable mirror must stop precisely at the instructed positions without any tilt.

The condition that the plane of the movable mirror must always be at a fixed angle to the optical axis and must not have any tilt (deviations from the fixed angle) is an absolute requirement for both the continuous-scan and step-scan spectrometers, because even a small tilt has a detrimental effect on the interference, which results in a weaker interferogram. To prevent the tilt, various devices have been developed. For example: (i) as described earlier, a high-precision gas-bearing is used for the sliding shaft of the movable mirror, (ii) a dynamic alignment mechanism is adopted for the high-speed correction of the tilt, and (iii) as shown in Figure 5.7a, a pair of parallel mirrors which rotate as a united

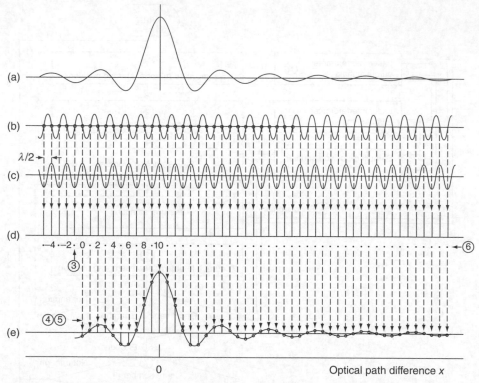

Figure 5.6 *Actual process of the sampling of an interferogram. (a) Interferogram from the main interferometer, (b) output signal from the laser interferometer ⓑ, (c) output signal from the laser interferometer ⓒ, (d) signals generated by the counter using the output signals in (b) and (c), and (e) interferogram from the main interferometer after sampling.*

body may be used. Further, a device called the *cube-corner retroreflector*, the principle of which is schematically illustrated in Figure 5.7b, is also employed for both the fixed and movable mirrors. The cube corner, often also referred to as the *corner cube, corner reflector, retroreflector*, is a reflector consisting of three plane mirrors joined at right angles to each other. The use of the mirror pair or the cube-corner retroreflector has the advantage that, even if they have a tilt against the optical axis, the direction of the reflected beam 2 is always parallel to that of the incident beam 1 (see Figure 5.7a,b). In the case of the mirror pair, beam 2 is reflected back by a mirror which is not shown in Figure 5.7a.

The requirement for the uniform linear motion of the movable mirror in the continuous-scan type is imposed mainly by the frequency-response characteristics of the detector for the main interferometer. The response of the detectors to the beams (indicated by ⓑ and ⓒ in Figure 5.4) of the laser interferometer is usually rapid, so that effects arising from the instability of the motion of the movable mirror which changes the modulated frequency can be disregarded. In contrast, the response characteristic of thermal detectors such as a pyroelectric detector commonly used for detecting the infrared beam from the main interferometer is not as rapid as that of the laser-interferometer detectors. As a result, any

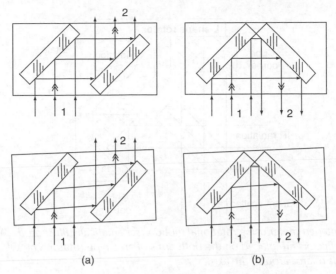

Figure 5.7 *(a) A pair of parallel mirrors and (b) cube-corner retroreflector (only two mirrors are shown). These optical elements are slightly tilted in the lower figures, but reflected rays 2 are always parallel to incident rays 1.*

instability of the motion of the movable mirror alters the measured intensities and phases of the electric signals from the detector. Phase changes shift the sampling positions, and intensity changes distort the shape of the interferogram; either of them leads to unfavorable effects in resultant spectra. The uniform linear motion of the movable mirror furnishes therefore a perfect solution to the problem. It should be mentioned, however, that if a detector such as a semiconductor detector having a flat frequency response as well as a good response to high frequencies is used, the requirement for the uniform linear motion may be relaxed to a certain degree.

In both the continuous-scan and step-scan spectrometers, the extremely precise positioning of the movable mirror is ensured by adopting an efficient feedback mechanism such as the above-mentioned dynamic alignment. Further, some spectrometers are constructed so as to isolate the interferometer base from shocks, including earthquakes. It should be kept in mind that, even if such spectrometers are in use, shocks, vibrations, and tilts given inadvertently to the spectrometer may still have an adverse effect on the results of measurements.

In Figure 5.8, an example of an interferometer which realizes the OPD by rotating a pair of cube-corner retroreflectors is illustrated. As such an interferometer can lessen the effect of shocks and tilts to the spectrometer, restrictions on the place for siting the spectrometer can be relaxed. In this design of spectrometer, however, there is a limit to its spectral resolution, as it is difficult to attain a large OPD. Some spectrometers having such interferometers are, however, portable, and can be used for fieldwork also.

The He–Ne laser is most commonly used for controlling the interferometer, but semiconductor lasers are also used, because they are better suited for making the spectrometer size smaller.

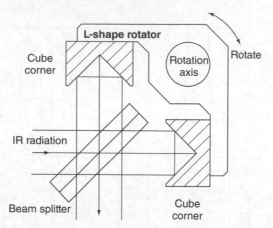

Figure 5.8 *Interferometer using rotational motion (schematically illustrated). A pair of cube–corner retroreflectors are placed on the both arms of an L-type rotator. The OPD is generated by rotating this rotator around the axis.*

Instructions for starting the sampling of interferogram and precise counting of the sampling points are controlled by the output signals Ⓑ and Ⓒ from the two laser interferometers and the counter Ⓓ shown in Figures 5.5 and 5.6.

For the beamsplitter of the main interferometer, various types of materials are used to cover specific wavenumber regions, as a result of the fact that no single material is transparent over the entire infrared region. Representative materials for the beamsplitter are given in Figure 5.9, together with the information on the wavenumber region that each covers.

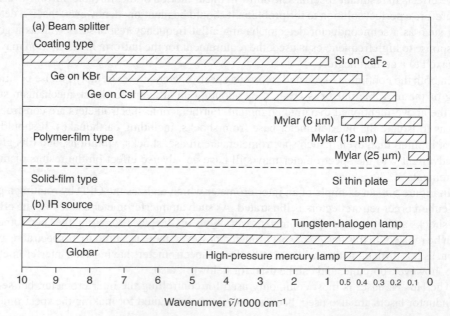

Figure 5.9 *Components of the optical system. (a) Beamsplitter (film thickness in parentheses) and (b) infrared source.*

The beamsplitter of the coating type is a combination of a substrate plate (e.g., a KBr plate) with a thin coating of a material (e.g., Ge) usually vapor-deposited onto one side of the plate and a compensation plate of the same shape. The wavenumber region covered by such a beamsplitter depends on the spectral characteristics of the substrate plate, the coating material, and the thickness of the coating.

The substrate plate has a slightly tapered shape to avoid interference between the beam passing through it and the beam internally reflected by both surfaces of the substrate plate. The compensation plate is used for the purpose of solving the following two problems over the entire infrared region: (i) shifts of the maximum peak of the interferogram at zero OPD (centerburst) and/or increasing asymmetry of the shape of the centerburst arising from the wavenumber dependence of the refractive index of the substrate plate and (ii) a prismatic effect due to the tapered shape of the substrate plate.

It is essential to keep the inside of the optical system sufficiently dry as some of the substrate plates are deliquescent.

The beamsplitter of the polymer-film type used in the far-infrared region is a tightly spread polyester film with high flatness; this is often commonly referred to as a *Mylar*® *beamsplitter* (Mylar is a DuPont trade name for a poly(ethylene terephthalate) film). The wavenumber region covered by such a beamsplitter is determined by the film thickness; the peak of efficient interference shifts toward lower wavenumbers with increasing film thickness. A thin plate of silicon may also be used as the beamsplitter of the solid-film type in the far-infrared region.

5.2.1.2   Source of Radiation

In the mid-infrared region ($4000-400\,\mathrm{cm}^{-1}$), a sintered rod of silicon carbide, usually called a *Globar*, is used as a light source. The thermal emission from this light source has a spectrum close to that of black-body radiation, and can be used over the wavenumber region of $9000-100\,\mathrm{cm}^{-1}$. In Figure 5.9b, three kinds of representative light sources for the infrared region are listed. Like the case of a Globar, thermal emission from a tungsten-halogen lamp is used as an infrared source. Far-infrared radiation from a high-pressure mercury lamp is due to the Bremsstrahlung of electrons as well as thermal emission from its tube, which is made of quartz.

5.2.1.3   Detector

The detector of an FT-IR spectrometer is required to meet the following conditions:

1. It is sensitive to infrared radiation.
2. It detects infrared radiation over a broad wavenumber region.
3. Its photosensitive region has a size comparable with the entrance aperture.
4. It responds to input signals over a broad frequency range.
5. It has a linear response to the intensity of incident infrared radiation.

Condition (3) is required to ensure the optical throughput advantage of FT-IR spectrometry, which was described in Section 4.4.1.3; usually, the area of the photosensitive region has a diameter of $1-2\,\mathrm{mm}$. Condition (4) is concerned with an interferometer of the continuous-scan type described in Section 5.2.1.1; infrared radiation incident on the interferometer has a spectrum over a broad wavenumber region, and as a result, the radiation coming out from the interferometer is modulated to signals over a broad frequency range given by Equation (4.9). Condition (5) requires proportionality between the intensity of the radiation incident on the detector and the magnitude of the electric output from the detector. In a detector which does not meet this condition, the proportionality holds only for weak incident radiation, and an effect of saturation occurs when the incident radiation is intense. The intensity of the radiation incident on the detector reaches a maximum at zero OPD, as given by Equation (4.10) and illustrated in Figure 4.5c. Accordingly, a detector lacking the above proportionality shows the effect of saturation in the range around zero OPD. The spectrum obtained as the Fourier transform of such an interferogram has a negative component spreading over the entire wavenumber region, which would be superimposed on a normal spectrum obtained with a flawless detector. This negative component originates from a decrease in the magnitude of the output signal due to the saturation. It is converted to a positive component by the phase-correction procedure to be described in Section 5.2.2.3. As a result of this procedure, an abnormal spectrum as shown in Figure 3.7b is derived. Such a distorted spectrum cannot be used for quantitative analysis.

As expected from the above description, the simplest way to find the response flaw of a detector is to examine the $400–0 \text{ cm}^{-1}$ region of the measured spectrum, where no spectrum should exist in the case of an FT-IR spectrometer with a Ge/KBr beamsplitter. As for the issue of the detector response, see also Section 3.6.2.

As the flaw of the detector response is noticeable when the incident radiation is intense, it is possible to compensate for this by diminishing the intensity of incident infrared radiation using an optical filter or a dimmer. Some spectrometers have a circuit in the preamplifier of the detector, which corrects the nonlinearity of the detector response. In such a case, it is necessary to adjust the function of the circuit to the response characteristic of the detector in use, as a response characteristic is specific to an individual detector.

The spectral- and frequency-response characteristics of representative detectors are shown, respectively, in Figure 5.10a,b, where the spectral-response data for the frequency of maximum response and the frequency-response data for the wavenumber of maximum spectral response are shown, respectively. The quantity used for the ordinates of these figures is called *specific detectivity*, which is denoted by $D^*(\lambda, f)$ and defined as

$$D^*(\lambda, f) = \frac{\sqrt{A}}{\text{NEP}} \tag{5.2}$$

where λ is the wavelength of infrared radiation ($\lambda = 1/\tilde{\nu}$), f is the modulated frequency ($2v\tilde{\nu}$ in Equation (4.9)), NEP stands for the noise-equivalent power (an input power giving rise to the level of a desired signal equivalent to the background noise level) and A is the area of the photosensitive region of the detector. In many detectors, the value of NEP is proportional to the square root of the area of the photosensitive region. Because, as is clear in Equation (5.2), $D^*(\lambda, f)$ is made independent of the size of the photosensitive region, it is useful for comparing the performances of detectors.

Generally speaking, a measurement performed at a better frequency-response will give a higher S/N ratio in the result obtained. This statement is valid also for FT-IR spectrometry.

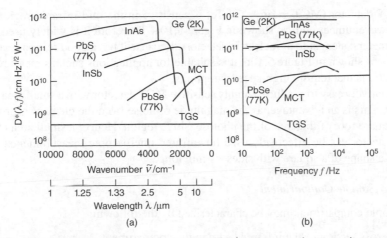

Figure 5.10 *Characteristics of detectors. (a) Specific detectivities of various detectors as functions of infrared wavenumber and wavelength and (b) specific detectivities of various detectors as functions of the modulated frequency.*

In an FT-IR spectrometer, the frequency of the interferogram is proportional to the speed of the moving mirror. This speed should be chosen in order to take advantage of the frequency range over which a good detector response is expected. The condition for comparing the S/N ratios of different measurements is that they must be made over the same period of time. Hence, with increasing speed of the moving mirror, correspondingly more interferograms need to be accumulated.

Pyroelectric detectors, which are commonly used in FT-IR spectrometers at room temperature, utilize the temperature-dependent changes of the polarizabilities of TGS (triglycine sulfate), LATGS (L-alanine doped triglycine sulfate), DTGS (deuterated triglycine sulfate), or DLATGS (deuterated L-alanine doped triglycine sulfate). These detectors, with their photosensitive surfaces blackened to help convert incident radiation to heat, have a good response over a broad infrared region. Their specific detectivities for the conversion of infrared radiation to heat decreases with frequency as exemplified by TGS in Figure 5.10b. This property requires that the speed of the moving mirror should be made slower than $2 \, \text{mm s}^{-1}$. At a mirror speed of $2 \, \text{mm s}^{-1}$, the frequency of the interferogram for the incident radiation of $4000–400 \, \text{cm}^{-1}$ is $1600–160 \, \text{Hz}$.

The pyroelectric crystals must be isolated from the air as they are deliquescent. For this purpose, they are protected by window materials transparent to a specific wavenumber region: that is, KBr or CsI in the mid-infrared region and polyethylene in the far-infrared region.

Detectors based on some photoconductive or photovoltaic semiconductors have very high specific detectivities. These detectors utilize the phenomenon of photoexcitation in which electrons in valence (filled) bands are excited to conduction bands by absorbing the energy of radiation. They function at the temperature of liquid nitrogen (77 K) or a lower temperature. Many semiconductor detectors do not have a good linear response to incident radiation, because the saturation of signal current tends to occur. Among them, the MCT detector (MCT = $Hg_{1-x}Cd_xTe$) is sensitive in the mid-infrared region, even at its low-wavenumber end, as shown in Figure 5.10a. Therefore it is widely used in FT-IR spectrometers, alongside pyroelectric detectors. Because it has a good response at high frequencies as shown in Figure 5.10b, it is suitable for applications such as measurements of fast reactions that require high-speed samplings.

It is relatively easy to produce a highly sensitive MCT detector with a small area detecting element. This is an advantage, as the magnitude of noise from the element is proportional to the square root of the area of the photosensitive region. Hence, a small area element of MCT detector can be effectively used for infrared microscopic measurements; typically, this has a shape of a square with sides <1 mm long.

5.2.1.4   Sample Compartment

The sample compartment may be characterized by the following:

1. The cross section of the infrared beam in the compartment.
2. The type of photometric system.
3. Its isolation from the outside air.

Sample cells and other apparatus to be sited in the sample compartment should be designed so that they fit to the cross section of the infrared beam. If they block any part of the infrared beam, the optical throughput advantage of an FT-IR spectrometer cannot be fully realized. As an image of the entrance stop is formed in the sample compartment, the infrared beam at the focal position has a circular section, the diameter of which can be derived by Equation (4.21) in the same way as employed for calculating the diameter of the entrance stop. For example, in a spectrometer with a condensing mirror having a focal length of 300 mm, the section of the infrared beam has a diameter of about 9.5 mm at a wavenumber resolution of $1 \, cm^{-1}$. The windows of a sample cell and other accessories should be large enough so as not to block the infrared beam. If a sample cell having a window smaller than the beam section is used for quantitative analysis, the cell should be placed at exactly the same position in a series of measurements necessary for the analysis, as the quantity of light blocked by the cell is expected to vary with the position of the cell. It should be noted that the focal position of the beam depends on the spectrometer; that is, it is either at the center of the sample compartment (the center-focus type) or at one side of the compartment (the side-focus type). In some spectrometers, a collimated beam of parallel rays can be directed toward a sample compartment attached to the exterior of a spectrometer for a special experimental purpose.

Only the single-beam type of photometric system of an FT-IR spectrometer is commercially available at present. This requires that two measurements must be made; the first is for a reference single-beam background spectrum recorded without the sample, and the second for the sample placed in the beam. As these two measurements are made under the same optical conditions, accuracy of spectral intensity can be expected. Care must be taken, however, not to change the conditions of measurements during the time interval between the two measurements.

Isolation of the sample compartment from the outside air is very important. Water vapor and carbon dioxide in the air exhibit intense infrared absorptions and it is usually difficult to eliminate them completely from the transmittance (or absorbance) spectrum of a sample by utilizing the single-beam reference spectrum for spectral compensation. The absorptions due to water vapor and carbon dioxide are particularly noticeable and disturbing when spectral measurements are performed at a high wavenumber-resolution, which is easily attainable by FT-IR spectrometry, and also when very low intensity absorption peaks of a sample are studied by utilizing the high sensitivity of FT-IR spectrometry. Before beginning spectral measurements, it is necessary to exchange the air completely from the tightly closed sample compartment by purging it with either dried air or nitrogen. The same operation should be applied also to the box containing the whole optical system, which has a tightly enclosed structure in a modern FT-IR spectrometer.

Various apparatus or accessories are used in the sample compartment to perform a wide range of applications of FT-IR spectrometry. They should be properly placed and carefully aligned to make the best use of the spectrometer's capability. As it is inconvenient and wasteful of time to carry out their alignment every time before starting a new measurement, various ways for securing high reproducibility in their positioning have been developed. For some measurements, control signals from the computer, power source, and cooling water are needed in the sample compartment, and facilities for these are provided in some spectrometers.

5.2.1.5   Amplifier and AD Converter

An interferogram converted into electric signals by the detector is received by the computer via a combination of preamplifier, main amplifier, electric filter, sample-and-hold circuit, and AD converter.

The amplifier increases the interferogram signals without distortion up to an appropriate level of magnitude for input to the AD converter, and the electric filter removes unnecessary signals and noise in the high-wavenumber region, which are folded back in the process of sampling. These components are so designed that their noise levels are smaller than the detector noise, in order to make optimum use of the detector capability.

The AD converter is an electronic circuit that converts signals from analog to digital. Many types of AD converters such as a successive-approximation type and a delta-sigma type are available. The successive-approximation type digitizes the voltage of an input signal by comparing it successively with standard voltages. During this process, the input signal is kept in the sample-and-hold circuit (which maintains the signal magnitudes) to protect the signal from suffering any change. In this type, the dynamic range (power of resolving signals during the AD conversion) is 16 bits at its maximum, and a desired input signal can be taken in at any time by a trigger given by the computer. On the other hand, the delta-sigma type, which is used in many more recent FT-IR spectrometers, is composed of a difference circuit (represented by delta) and an integration circuit (represented by sigma). This type can increase the dynamic range up to 24 bits.

5.2.1.6   Disturbing Factors in FT-IR Measurements

There are a number of factors which disturb FT-IR measurements even if the optical system of the spectrometer itself has no particular problem. Disturbing factors which deserve particular attention include noise from the detector, the dynamic range of the AD converter, the sampling interval of an interferogram, and the spectral background arising from a thermal origin, which is a problem specific to infrared spectrometry.

Noise from the detector is mainly thermal in origin; it arises from the thermal fluctuation of electrons which carry signals. An FT-IR spectrometer uses either a thermal detector or a semiconductor (or quantum) detector. The former has considerable thermal noise as it is used at room temperature. By contrast, the latter, used at low temperatures, most often that of liquid nitrogen, has little thermal noise.

The AD converter is one of the factors limiting the S/N ratio of an FT-IR spectrometer. If a highly sensitive detector like an MCT detector is used, an AD converter with a large dynamic range should be chosen.

As described in Section 4.4.2, the sampling operation for an interferogram causes the folding of signals. This phenomenon occurs also for noise; noise signals generated in the high-frequency region will fold back and overlap each other. For this reason, it is important to remove high-frequency noise signals by an electronic filter before performing the sampling operation. Narrowing the sampling interval is also an effective means of improving the S/N ratio, because the narrowing decreases the number of folding noise signals.

The thermal spectral background may originate from infrared rays emitted by everything in proximity to the spectrometer and its level is temperature dependent. For example, a substance at room temperature gives rise to an emission spectrum having a peak at a wavelength of about 10 µm. Accordingly, the surroundings and environment around the optical system of an FT-IR spectrometer are full of thermal emission, although it is not felt by the human eye. If such an emission enters into the interferometer of an FT-IR spectrometer, it becomes a background that overlaps the measured spectrum. When infrared radiation from the light source is intense enough, the environmental infrared emission does not cause a serious problem in spectral measurements. However, if a sample in a high-temperature furnace is a target of infrared measurements, the high background emission may become an obstacle to the spectral measurements. In such a case, infrared radiation emitted from the sample advances to the interferometer in the direction opposite to the normal, and is reflected back. Such infrared radiation, after passing through the sample, reaches the detector and may give rise to a high background in the measured spectrum. The infrared radiation directly entering the detector from the sample has no effect on the measured spectrum because it has not been modulated by the interferometer. However, if intense radiation exceeding the detector capacity directly enters the detector, the detector will not function properly because of saturation. Careful attention should be given to prevent such a situation.

If a spectral measurement is disturbed, it is important to understand the origin of the disturbance first and take proper action to eliminate it. For example, if the reflectors on the two arms of the interferometer are plane mirrors (not cube-corner retroreflectors), the above case of the infrared radiation emitted by a high-temperature sample may be dealt with in the following way. It is effective to block half of the incident beam from the light source at a point close to the sample position where the incident beam is focused. Then, the infrared radiation from the sample will advance to the interferometer through the remaining half and will be reflected by the interferometer. When it comes back toward the sample, it will be focused on the blocked half symmetrically located around the optical axis, and consequently will not reach the detector.

Other measures such as tilting the sample against the incident beam, roughening the sample surface, and shifting the sample position from the focal point of the incident beam may also be effective in reducing the quantity of the sample-emitted infrared radiation that reaches the detector after being reflected back by the interferometer. If the sample and its surroundings are at high temperatures, it is effective to place an aperture stop in the optical path between the sample position and the detector to let only the radiation coming from the sample pass through it.

5.2.2   Computer System and Software

The optical systems of modern FT-IR interferometers are interfaced to modern computers that have high-speed computing and data-processing capabilities, and which contain a large amount of memory and an operating system with a convenient Windows® function, as well as a wide variety of application-specific software.

The computer system and its relationship with other components of an FT-IR spectrometer are illustrated in Figure 5.11. In this figure, a single computer is depicted as controlling the entire spectrometer. If a multiprocessor system is employed instead, a host computer controls the entire system, and slave processors exclusively carry out their individual functions such as data acquisition and numerical computation.

The memory is composed of main memory and auxiliary storage such as a hard disk. The main memory can exchange information with the central processing unit (CPU) at high speed. The auxiliary storage needs more time for information exchange but usually

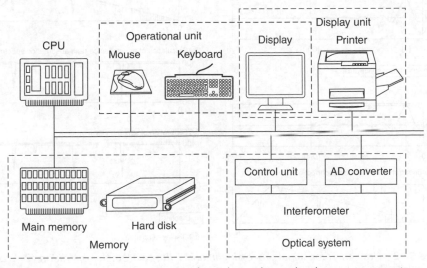

Figure 5.11   *Typical computer system and its relationship with other components in an FT-IR spectrometer setup.*

has a very large amount of memory. Like most application software, programs used in FT-IR spectrometry are kept in the auxiliary storage and are run automatically after they are transferred to the main memory when required. On the other hand, data acquired in spectral measurements are taken into the main memory where various kinds of data processing, including the Fourier-transform computations, may then be undertaken. Then, the measured data are transferred to the auxiliary storage and stored there.

The operation unit has the role of a "window" through which a user communicates with the spectrometer. The unit typically shows on its display a menu of data, instructions, and parameters required at each step of a spectral measurement. Looking at the menu, the user can select and provide instructions on: (i) the parameters for a spectral measurement to be performed, (ii) starting a measurement, and (iii) subsequent necessary data processing.

The user obtains information from the spectrometer according to his/her request, usually on the display and/or a printer. The display is mainly used for monitoring both events still in progress and the results finally obtained, in addition to confirming the parameters selected for the measurement and other instructions. The printer is used to plot measured spectra on charts.

In Figure 5.12, a flow diagram of processing interferogram data acquired by the computer is shown. The numberings from ⑨ to ⑫ in this figure are continued from those in Figure 5.3. The data processing is roughly divided into three steps, that is, (i) acquisition of interferograms, (ii) conversion to a spectrum, and (iii) display of results obtained.

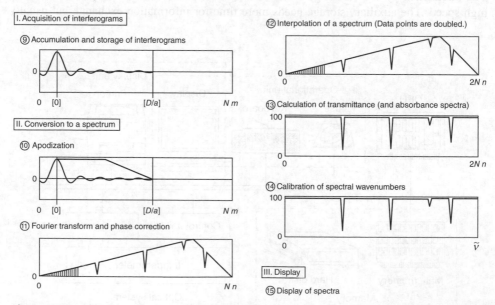

Figure 5.12 *Flow chart for signal processing in an FT-IR spectrometer operation (2) (continued from Figure 5.3).*

5.2.2.1 Accumulation and Storage of Interferograms

Interferograms acquired by the computer are accumulated (co-added and averaged) to increase the S/N ratio. While signals increase in proportion to the number of accumulations (co-additions), noise only increases in proportion to the square root of the number of scans as it is random. Then, the S/N ratio is expected to be improved by a factor of the square root of the number of scans. However, increasing simply the number of scans to improve the S/N ratio requires a longer time for completing a spectral measurement. More specifically, to improve the S/N ratio by a factor of 2^n (i.e., 2, 4, 8, ...), it is necessary to increase the number of accumulations by a factor of $(2^n)^2$ (i.e., 4, 16, 64, ...). This means that a very long time may be needed to obtain a significant improvement of the S/N ratio. Therefore, before performing a spectral measurement, every effort should be made to optimize spectral signals and decrease noise as much as possible; for example, it is advisable to: (i) adjust the optical system so that the strongest intensity of an interferogram reaches the detector, (ii) use a detector having a high detectivity, and (iii) amplify the output signal from the detector to the AD converter to such an extent that it does not saturate.

5.2.2.2 Apodization

Apodization is a mathematical procedure used to overcome the fact that a recorded interferogram is truncated (i.e., does not extend to an infinite distance) and to ensure that the interferogram to be Fourier-transformed terminates smoothly without a step. An explanation of apodization is given in Section 4.4.1. In Figure 5.12⑩, a trapezoidal apodization function is shown overlaid with a measured interferogram to be weighted by this apodization function.

5.2.2.3 Fourier Transform and Phase Correction

An accumulated interferogram is converted to a spectrum by the Fourier cosine transform given in Equation (4.12). Actually, however, the recorded interferogram, which ought to be symmetric with respect to zero OPD, has an asymmetric shape for various reasons. As a result, a true spectrum cannot be obtained by simply performing the Fourier cosine transform of the asymmetric interferogram. The asymmetric shape is caused by multiple factors, including a difference in thickness between the main and compensation plates in coating-type beam splitters, response characteristics of the electronic system, and shifts of the sampling points due to a difference between the OPDs of the main interferometer and the laser interferometer. This asymmetry is attributed to the fact that a deviation occurs in the phase of the ideal interferogram expressed by Equation (4.11). In more concrete terms, $\cos 2\pi\tilde{v}x$ on the right-hand side of Equation (4.11) becomes $\cos[2\pi\tilde{v}x - \delta(\tilde{v})]$, in which $\delta(\tilde{v})$ represents the wavenumber-dependent phase deviation. To deal with this asymmetry, a process called *phase correction* is undertaken. The purpose of this is to remove the phase deviation in order to correct an asymmetric shape of an interferogram to a symmetric one. A brief description of a representative process of the phase correction is as follows.

The asymmetric character of the recorded interferogram is first examined over a narrow range symmetric about (on both sides of) the maximum peak (centerburst). Note that the centerburst is not at zero OPD because of a phase deviation. For the purpose of the above examination, the measurement of the interferogram is started from a position on the negative side of OPD near the centerburst, as can be seen, for example, in Figure 5.12⑨. A complex Fourier transform of the interferogram over this narrow range is calculated to derive both the real and imaginary parts of the spectrum, which are then used to determine the amount of the phase deviation. Because the real part of the complex Fourier transform corresponds to the cosine transform, and the imaginary part to the sine transform, the amount of the phase deviation can be obtained from these. Then, the positive part of the full interferogram is transformed similarly in order to derive the amount of the phase deviation. This amount of phase deviation is corrected for by using that derived earlier. Such a process gives a "true" single-beam spectrum ("target spectrum"), and an example is shown in Figure 5.12⑪.

Another method for phase correction measures an interferogram symmetrically about the centerburst and obtains a spectrum by power Fourier transform. Power Fourier transform performs first the complex Fourier transform of the measured symmetric interferogram, then takes the sum of the squares of the resultant real and imaginary parts, and finally calculates the square root of the sum. By using a symmetrically measured interferogram, this method reduces the difference between the target spectrum and the spectrum calculated by the power Fourier transform of an interferogram having a phase deviation. This method is now widely used, particularly for measurements of low-resolution spectra, in which symmetric interferograms are more easily acquired.

In FT-IR spectrometry, a fast Fourier transform (FFT) algorithm is used to perform a discrete Fourier transform within a short time. The discrete interferogram $\{\overline{T(\tilde{\nu}) * F(x)}\} \varPi_{2D}(x) \text{Ш}_a(x)$ in Figure 5.3⑧ is obtained by sampling an interferogram at intervals of a over the range of $|x| \leq D$, although the negative range of x is not fully depicted in Figure 5.3⑧. If interferogram signals are lacking in the range of $2D$ as seen in Figure 5.3⑧, zero points are added to make up for this in order to enable use of the FFT algorithm. Then, the number of the sampling points N is equal to $2D/a$. By performing the Fourier transform of such an interferogram, a sequence of points is obtained at intervals of $1/2D$, which is the value of wavenumber resolution expected for this case. This procedure is supported by the theory of interpolation to be described in Section 5.2.2.4, which states that interpolation gives appropriate values for all points added by this process between any adjacent points in a spectrum. Thus, the discrete Fourier transform is employed in FT-IR spectrometry to convert an interferogram to a spectrum.

A spectrum obtained by the discrete Fourier transform of a discrete interferogram is expressed as

$$B(n) = \sum_{m=0}^{N-1} F(m) \exp\left(-i\frac{2\pi mn}{N}\right) \quad (n = 0, 1, \ldots, N-1) \tag{5.3}$$

In this equation, to satisfy the number of independent variables (a minimum number of data points to express an obtained spectrum exactly), the following measure is taken: that is, $F(m) = 0$ for $[X/a] < m \leq N - 1$. This means that the interferogram signals in Figure 5.12⑩ are made equal to zero at all points in the range between $[X/a]$ and $N - 1$.

To perform the computation on the right-hand side of Equation (5.3), a value of $F(m)$ is multiplied by the exponential functions N times for one n. Therefore, N^2 times of multiplications are needed in total for all numbers of n. The FFT algorithm is a procedure for accelerating the above calculations by reducing significantly the number of necessary computations. By changing the computational procedure, the same multiplications are achieved at one time. If N is expressed by a binary number (2^K), the number of multiplication times can be reduced to $2N\log_2 N$. For example, the computational time needed for a case of $N = 2^{10}$ becomes as short as about $1/50$ of the time needed before the introduction of the FFT algorithm.

5.2.2.4 Interpolation of a Spectral Curve

Interpolation is a process used for determining precise peak positions and obtaining detailed features in a spectrum by interpolating appropriate data between adjacent recorded points within a discrete spectrum obtained by a discrete Fourier transform. The process introduced here does not estimate values by statistical analysis using the least-squares method, and so on, but rather derives the spectrum which should have been observed, by supplementing the Fourier transform process which deals with a limited number of sampling data values.

The following two methods are available for spectral interpolation.

1. *The method of zero filling.* In this method, spectral interpolation is derived as a result of processing a measured interferogram. As is clear from Equation (5.3), the number of data points in the spectrum obtained by the discrete Fourier transform is equal to the number of data points in the interferogram before the transform. If zero points are added after the measured interferogram as if they are a part of the interferogram ("zero filling"), the intervals between adjacent points in the spectrum after the transform become closer in similarity to the case of a higher-resolution measurement. This process provides a method for performing spectral interpolation. Of course, zero filling does not enhance wavenumber resolution, as it does not add useful spectral information obtainable by measuring an interferogram with a larger OPD.

   As a practical procedure, by performing a zero filling for a measured interferogram having 2^M data points until the total number of points becomes 2^m times the number of the original data points, the resultant interferogram having 2^{M+m} points is transformed. As a result, a process of interpolating one point between each set of adjacent points, before undertaking the process, in the spectrum is repeated m times. In the case of $m = 1$, one point is interpolated in the middle, between each pair of adjacent points in the spectrum before the interpolation. The number of interpolated points becomes three for $m = 2$.

2. *The method of convolution.* In this method, the process of interpolation is directly performed on a spectrum by convolving each data point with a sinc function ($2D \sin 2\pi D\tilde{v}/\pi\tilde{v}$, where D is the maximum OPD). This method has the advantage that interpolation can be performed at any spectral position, but has the limitation that the computation of convolution takes more time.

In Figure 5.12⑫, the case of $m = 1$ is illustrated (cf. Figure 5.12⑪). It should be noted in this figure that the number of data points is increased to $2N$, which is twice the original.

5.2.2.5   Computation of Transmittance and Absorbance Spectra

Up to this point, a spectral measurement consists of a process in which a sample or a reference (no sample) is placed in the sample compartment, and the spectrum of infrared radiation passing through the sample or the reference is measured. From such measurements, $\{T(\tilde{v})B(\tilde{v})\} * \overline{A(x)}$ and $B(\tilde{v}) * \overline{A(x)}$ are determined, respectively, as the spectra of the sample and the reference. In these expressions, $T(\tilde{v})$ is the transmittance, which is set to unity for the reference, and $\overline{A(x)}$ is the instrumental function derived from an apodization function $A(x)$. $B(\tilde{v})$, which was originally defined as the spectrum of the radiation emitted by the light source, includes also the instrumental spectral characteristics of other components such as the interferometer and the detector.

From the two kinds of spectra recorded, the spectrum attributed only to the sample is calculated in the following way.

$$T'(\tilde{v}) = \frac{\{T(\tilde{v})\,B(\tilde{v})\} * \overline{A(x)}}{B(\tilde{v}) * \overline{A(x)}} \tag{5.4}$$

$T'(\tilde{v})$ is usually called the *transmittance spectrum* in FT-IR spectrometry. The corresponding absorbance spectrum is given as

$$A'(\tilde{v}) = -\log_{10} T'(\tilde{v}) \tag{5.5}$$

A natural logarithm is also used to define $A'(\tilde{v})$.

It should be noted that $T'(\tilde{v})$ does not completely agree with $T(\tilde{v})$ which, by definition, means the transmittance spectrum of a sample, because the right-hand side of Equation (5.4) contains the convolution by the instrumental function $\overline{A(x)}$. To obtain a better agreement between $T'(\tilde{v})$ and $T(\tilde{v})$, it is necessary to use an instrumental function having a shape much sharper than the true shape of a band of the sample. It is advisable to find an optimum condition for obtaining a sufficient agreement between $T'(\tilde{v})$ and $T(\tilde{v})$, by examining spectral changes that may occur when measurements are made at different wavenumber resolutions.

5.2.2.6   Calibration of Spectral Wavenumbers

At the end of a spectral measurement, calibration of the spectral abscissa is needed to allow for the following two factors, namely, the size of the entrance aperture and the refractive index of air. The latter factor is disregarded in this chapter, because its effect is negligibly small at a wavenumber resolution employed for most practical analyses. The effect arising from the size of the entrance aperture is calibrated in the following way.

$$\tilde{v} = \tilde{v}_0 \left(1 + \frac{a}{2D}\right) \tag{5.6}$$

where $\tilde{v}_0$ is the wavenumber before calibration, a is the interval of the interferogram sampling, and D is the maximum OPD. In deriving this equation, the optimum diameter of the entrance aperture given in Equation (4.21) is assumed, and a half of the wavenumber shift for the outermost oblique ray is adopted. This equation implies that the wavenumber shift to be calibrated becomes smaller with the decreasing size of the entrance aperture. The entrance aperture is made smaller when the maximum wavenumber in the measurable wavenumber region (which is inversely proportional to a) is made larger and also when the wavenumber resolution (which is defined as $1/D$) is made higher (smaller in number).

Further Reading

1. Griffiths, P.R. and de Haseth, J.A. (2007) *Fourier Transform Infrared Spectrometry*, 2nd edn, John Wiley & Sons, Inc., Hoboken, NJ.
2. Jackson, R.S. (2002) Continuous scanning interferometer for mid-infrared spectrometry, In: *Handbook of Vibrational Spectroscopy*, Vol. 1 (eds J.M. Chalmers and P.R. Griffiths), John Wiley & Sons, Ltd, Chichester, pp. 264–282.
3. Manning, C.J. (2002) Instrumentation for step-scan FT-IR modulation spectrometry, In: *Handbook of Vibrational Spectroscopy*, Vol. 1 (eds J.M. Chalmers and P.R. Griffiths), John Wiley & Sons, Ltd, Chichester, pp. 283–297.

6

Computer Processing of Measured Infrared Spectra

Shukichi Ochiai

S. T. Japan, Inc., Japan

6.1 Introduction

The as-measured infrared spectrum of a sample usually contains, in addition to the genuine spectrum of the sample, what is not required for spectral analysis: for example, noise signals arising from the detector and electronic circuits of the spectrometer, a tilted background arising from a coarse surface of the sample, bands due to water vapor and carbon dioxide existing inside the spectrometer, and so on. If these unnecessary features can be removed from the as-measured spectrum, the spectral features intrinsic to the sample will be seen clearly and they will become more useful for their intended purpose.

Computational methods are available for processing an as-measured spectrum in order to extract the useful information contained or hidden within it. For example, the method of absorbance difference spectroscopy may be utilized to obtain a solute spectrum from a solution spectrum by subtracting the pure solvent spectrum.

In measured infrared spectra of many samples, especially organic materials, a large number of bands are usually present. Many overlap each other to differing extents, and weak bands are sometimes buried beneath intense bands. By looking at such spectra, it is often difficult to determine precisely the true number of existing bands and their intensities. To solve these difficulties, at least to a certain extent, the methods of difference spectroscopy, derivative spectroscopy, Fourier self-deconvolution (FSD), and band decomposition (curve fitting) have been developed.

Data processing programs for the above-mentioned purposes are often supplied with commercial Fourier transform-infrared (FT-IR) spectrometers as integral parts of their software packages. They are also available in commercial software packages

Introduction to Experimental Infrared Spectroscopy: Fundamentals and Practical Methods,
First Edition. Edited by Mitsuo Tasumi and Akira Sakamoto.
© 2015 John Wiley & Sons, Ltd. Published 2015 by John Wiley & Sons, Ltd.

sold independently. Representative methods of data processing are outlined in the following sections.

6.2 Computer Processing of Measured Spectra

6.2.1 Spectral Smoothing

Noise signals in a spectrum arising from the detector and electronic circuits of an FT-IR spectrometer can be reduced by accumulating and averaging essentially identical interferograms obtained by repeated measurements; higher signal-to-noise (S/N) ratios are expected to be obtained by this method as described in Section 5.2.2.1. However, this method is not always applicable; for instance, when the sample under study is unstable, it is not practicable to repeat measurements over a long time period. In such a case, spectral noise may be apparently reduced by the method of spectral smoothing.

Spectral smoothing is based on the method of moving average. The Savitzky–Golay smoothing algorithm [1] is most commonly used for this purpose. This algorithm assumes that the spectral curve over a certain region can be approximately expressed by a polynomial and determines the weighting coefficient of the polynomial by fitting measured data points to the polynomial curve by the method of least-squares. The measured data points are chosen over a defined range successively at equal intervals. Savitzky and Golay have reported a method for deriving the weighting coefficients and given a table of the weighting coefficients. Smoothing can be achieved by taking a moving average at points in a spectrum, that is, an average at a desired spectral point is calculated by using the intensities of m spectral points before and after the desired point (($2m + 1$) points in total) to be multiplied by their respective weighting coefficients, and this averaging procedure is repeated by successively moving the desired point to neighboring points.

The weighting coefficient is largest for the desired point, becoming successively smaller for points more distant from the desired point. The smoothing effect becomes greater with increasing m value but at the same time the spectral shape becomes more distorted from the true one. This distortion is more apparent for a sharp band; its peak intensity is lowered and its bandwidth is broadened. In Figure 6.1, an example of spectral smoothing is shown for the OH stretching region of kaolin (China clay). It is clearly seen in this figure that noise decreases with increasing m value and that band shapes change at the same time, with a consequent apparent loss in spectral resolution. When spectral smoothing is performed, it is important to keep the distortion within a range allowable for the purpose by comparing the spectra before and after the operation of smoothing.

6.2.2 Removal of Bands due to Water Vapor and Carbon Dioxide

As described in Section 3.6.3, water vapor and carbon dioxide existing in the spectrometer in use should be minimized or removed as they have strong infrared absorption bands. The strong absorption bands of water vapor in the region of $1700–1500\,\text{cm}^{-1}$ are particularly troublesome because they overlap with absorption bands, which characterize many organic molecules in general, and hinder analysis of their spectra. The effect of water vapor can be removed by purging the entire optical system of the spectrometer with either dried air or nitrogen gas but it is often difficult to eliminate the absorption bands of water vapor

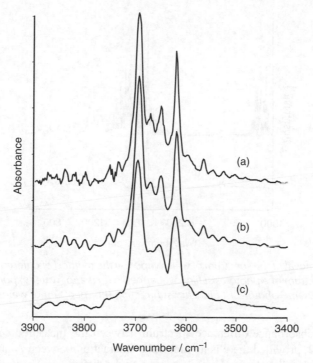

Figure 6.1   *Smoothing of the infrared spectrum of kaolin. (a) Measured spectrum, (b) spectrum after 9-point smoothing, and (c) spectrum after 17-point smoothing.*

completely. Then, the method of difference spectrum (described in Section 6.2.3) can be applied in order to subtract out the remaining absorption bands of water vapor. The infrared absorption spectrum of water vapor in the ambient air is first measured without placing any sample in the unpurged sample compartment. This spectrum is then used to eliminate the water-vapor bands overlapping with the spectrum of a sample by applying the same idea given in Equation (3.4). The result obtained by this procedure is shown in Figure 6.2.

If the water-vapor bands have considerable intensities, the above procedure may result in an increase in spectral noise. Some spectrometers have built-in software for automatically removing bands due to water vapor and carbon dioxide from measured spectra. This software is usually convenient for obtaining apparently "clean" spectra, but may not function satisfactorily if considerable amounts of water vapor and carbon dioxide exist inside the spectrometer. It is always advisable to purge the entire optical system of the spectrometer thoroughly with either dried air or nitrogen gas in order to make residual water-vapor bands as weak as possible. Purging with nitrogen gas is required to remove carbon dioxide as well as water vapor.

6.2.3   Difference Spectrum

When the spectra of two substances, A and B, are known to be overlapping in a measured spectrum, spectrum A, which is the target of a spectral measurement, can be obtained by

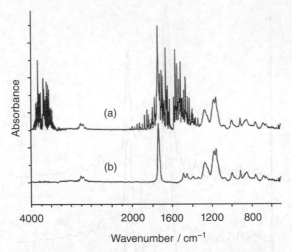

Figure 6.2 *Removal of the spectrum of water vapor by the method of difference spectroscopy. (a) Infrared spectrum of a photoresist film overlapped with that of water vapor and (b) infrared spectrum of the same photoresist film after subtracting the spectrum of water vapor.*

computing an absorbance difference spectrum between the measured spectrum (A + B) and spectrum B, provided there is no molecular interaction between A and B. Of course, spectrum B needs to be obtained before computing the difference. The procedure of taking a difference spectrum can be applied to determining the spectrum of a solute in a spectrum of a solution by subtracting the spectrum due to the solvent from the solution spectrum, as discussed in Section 3.3. In this case, it is necessary to adjust appropriately the magnitude of a coefficient to be multiplied to the solvent spectrum (the coefficient k_d in Equation (3.4)). The same procedure may be applicable to the detection of an extraneous low-concentration substance in a sample, which gives rise to only very weak absorption bands. By computing a difference spectrum, it is often possible to resolve bands of such a substance that are hidden in the overall spectrum dominated by many intense bands of a major component(s). Taking a difference spectrum is a procedure frequently utilized in FT-IR spectral analysis. In Figure 6.3, a difference spectrum between spectra obtained from two kinds of polyethylene samples with different degrees of branching is shown. The intensity of the band at $1378 \, \text{cm}^{-1}$ (attributed to the CH_3 groups at branched chain-ends) is used to estimate the degree of branching in polyethylene.

6.2.4   Derivative Spectrum

The first and second derivatives of an absorbance spectrum $A(\tilde{v})$ with respect to wavenumber $\tilde{v}$ can provide useful information for spectral analysis. In a first-derivative spectrum, local maxima, local minima, and inflection points in the original spectrum are more easily detected. As the first-derivative spectrum has an effect of baseline correction, bands of first-derivative shape are more clearly observable than the corresponding bands in the original

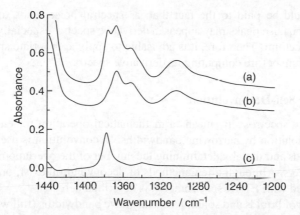

Figure 6.3 *Another example of difference spectroscopy. Infrared spectra of films of polyethylene samples with (a) a high degree of branching and (b) a small degree of branching, (c) difference between spectra in (a) and (b).*

spectrum; that is, peaks of overlapping bands, shoulders, and weak bands become more discernible. In a second-derivative spectrum, absorption peaks in the original spectrum become local minima, which can be easily detected as the baseline in the second-derivative spectrum is flat (i.e., baseline slope is removed). The peak intensity of a local minimum in a second-derivative spectrum can be used for a quantitative analysis. In Figure 6.4, the infrared absorption spectrum of *n*-butyl stearate is shown, together with its first- and second-derivative spectra.

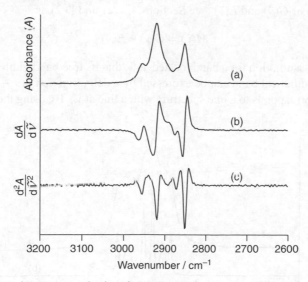

Figure 6.4 *Infrared spectrum of* n-*butyl stearate in the CH stretching region and its derivative spectra. (a) Measured spectrum, (b) first-derivative spectrum, and (c) second-derivative spectrum.*

Attention should be paid to the fact that, as spectral noise tends to be amplified in derivative spectra, false peaks may appear in derivative spectra, especially from a poor S/N ratio original spectrum. Therefore, it is advisable to apply appropriate spectral smoothing to a noisy spectrum before computing its derivative spectra.

6.2.5 Fourier Self-Deconvolution

Deconvolution in spectroscopy means a mathematical operation for enhancing apparent wavenumber resolution by narrowing bandwidths. Deconvolution is useful for separating overlapping bands and thereby determining the number of the overlapping bands and their peak wavenumbers. Although a few methods of deconvolution exist, only FSD, which is closely associated with FT-IR spectrometry, is described here.

Infrared bands of liquids and solutions usually have bandwidths (full widths at half maximum, FWHMs) of more than $10\,\mathrm{cm}^{-1}$. If a few such bands are closely overlapping within a spectral region, it is not always possible to resolve them by measuring the spectrum even at a wavenumber resolution higher (smaller in number) than the intrinsic FWHMs of individual bands. FSD of such a spectrum helps decompose the overlapping bands by making the shapes of component bands artificially sharper. An example which demonstrates the effectiveness of FSD is shown in Figure 6.5, where the overlapping broad bands are clearly decomposed into six component bands.

The process of FSD is illustrated stepwise in Figure 6.6, where, in accordance with a common practice, a Lorentz profile is assumed for the bandshape in the condensed phase. An observed bandshape $M(\tilde{v})$ is generally broader than its true shape $E(\tilde{v})$ for various reasons. If a function $G(\tilde{v})$ is used to express the broadening of a bandshape, $M(\tilde{v})$ is expressed as a convolution of $G(\tilde{v})$ and $E(\tilde{v})$ (see Sections 4.4.1.1 and D.2) as

$$M(\tilde{v}) = G(\tilde{v}) * E(\tilde{v}) \tag{6.1}$$

Let us make an assumption for a band located at $\tilde{v}_0$ that its true bandwidth is infinitely narrow. Such a hypothetical band can be expressed by a delta function $\delta(\tilde{v} - \tilde{v}_0)$ (see Section D.3.2), which corresponds to a line spectrum with a line at $\tilde{v}_0$. By using the Lorentz profile

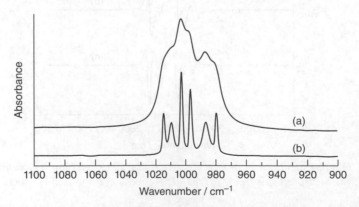

Figure 6.5 *An example of spectral deconvolution. (a) Spectrum synthesized from six bands and (b) spectrum obtained by the FSD of the spectrum in (a).*

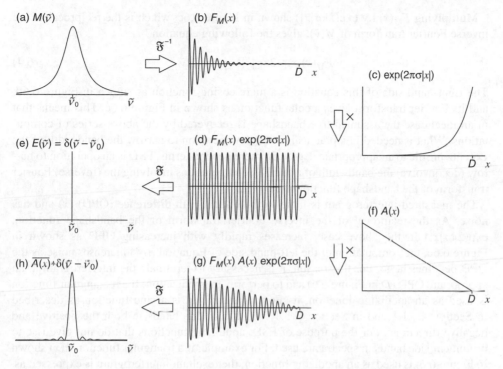

Figure 6.6   *(a–h) Flow diagram of the FSD procedure. See text for details.*

function $\sigma/(\sigma^2 + \tilde{v}^2)$ for $G(\tilde{v})$, Equation (6.1) can be rewritten as

$$M(\tilde{v}) = \frac{\sigma/\pi}{\sigma^2 + \tilde{v}^2} * \delta(\tilde{v} - \tilde{v}_0) = \frac{\sigma/\pi}{\sigma^2 + (\tilde{v} - \tilde{v}_0)^2} \qquad (6.2)$$

The formula transformation from the middle to the right-hand side of this equation is explained in Section D.4.3. The Lorentz profile has a FWHM of 2σ, and its area is equal to π but it is normalized to unity in Equation (6.2). Thus, $M(\tilde{v})$ in Equation (6.2) expresses a Lorentz profile with its peak wavenumber at $\tilde{v}_0$ shown in Figure 6.6a.

The FSD process for $M(\tilde{v})$ in Equation (6.2) is as follows. $M(\tilde{v})$ ought to be obtained by the Fourier transform of an interferogram. To reverse this process, that is, to obtain the interferogram from $M(\tilde{v})$, an inverse Fourier transform should be performed on $M(\tilde{v})$. The relation between the Fourier transform and its inverse transform is described in Section D.1.1. In Figure 6.6, their operations are denoted by $\mathfrak{F}$ and $\mathfrak{F}^{-1}$, respectively. $F_M(x)$, which denotes the interferogram to be obtained by the inverse Fourier transform of $M(\tilde{v})$, is given as

$$F_M(x) = \exp(-2\pi\sigma|x|) \cos 2\pi\tilde{v}_0 x \qquad (6.3)$$

To derive this equation, the convolution theorem described in Section 4.4.1.1 and in Section D.2 is utilized, and the descriptions in Sections D.3.4 and D.3.5 will also help derive this equation. $F_M(x)$ is depicted in Figure 6.6b.

Multiplying $F_M(x)$ by $\exp(2\pi\sigma|x|)$ shown in Figure 6.6c, which is the reciprocal of the inverse Fourier transform of $M(\tilde{\nu})$, gives the following equation.

$$F_M(x)\exp(2\pi\sigma|x|) = \cos 2\pi\tilde{\nu}_0 x \qquad (6.4)$$

The right-hand side of this equation is a mere cosine function as shown in Figure 6.6d, and its Fourier transform gives a delta function as shown in Figure 6.6e. This means that in an ideal case the assumed true bandshape is recovered by the above series of computations. What is needed in a practical case is an operation to narrow the bandwidth of the Lorentz profile to an appropriate degree. In more general terms, FSD is an operation to narrow (deconvolve) the bandwidth of a band by computations involving the (inverse) Fourier transform of the bandshape function itself (FSD).

The measured interferogram is finite in the optical path difference (OPD) (x) and has noise. As the reciprocal of the inverse Fourier transform of the bandshape function, $\exp(2\pi\sigma|x|)$ in the above case, increases rapidly with increasing OPD as shown in Figure 6.6c, the computation of the left-hand side of Equation (6.4) increases noise as the OPD becomes large. For this reason, it is necessary to terminate the interferogram at an appropriate OPD (D in Figure 6.6) and to perform an apodization. If a rectangular function is used as an apodizing function, its Fourier transform is a sinc function as described in Section 4.4.1.1 and in Section D.3.1, which has side bands in both the positive and negative directions. For the purpose of FSD, apodizing functions that do not give rise to prominent side bands in spectra are used. For example, if a triangular function $A(x)$ shown in Figure 6.6f is used as an apodizing function, the resultant interferogram is expressed as

$$F_M(x)A(x)\exp(2\pi\sigma|x|) = A(x)\cos 2\pi\tilde{\nu}_0 x \qquad (6.5)$$

The interferogram corresponding to the right-hand side of this equation is depicted in Figure 6.6g. The Fourier transform of this interferogram is expressed as

$$\overline{A(x)\cos 2\pi\tilde{\nu}_0 x} = \overline{A(x)} * \delta(\tilde{\nu} - \tilde{\nu}_0) \qquad (6.6)$$

Because, as described in Section D.3.6, the real part of $\overline{A(x)}$ is proportional to a sinc function squared, the right-hand side of Equation (6.6) corresponds to the spectrum shown in Figure 6.6h. In the FSD procedure described already, the FWHM of $M(\tilde{\nu})$ is reduced to the FWHM of the central band of $\overline{A(x)}$, which is obtained by the Fourier transform of an apodizing function terminating the interferogram at D. In other words, FSD is a special form of apodization.

In the actual process of performing FSD, it is necessary to specify the FWHM (2σ), the effective range of OPD (D), and an apodizing function that can suppress the side bands $[A(x)]$. If many bands with different bandwidths exist in the spectrum under study, the minimum bandwidth should be used to avoid excessive deconvolution, which will be accompanied by side bands. If a large value is chosen for D, the resultant bandwidth becomes narrow and the wavenumber resolution appears to be higher, but noise levels increase at the same time. As the S/N ratio depends on the apodizing function, it is advisable to test various apodizing functions.

As an example of FSD, the spectra of n-butyl stearate in the CH stretching region before and after FSD are shown in Figure 6.7.

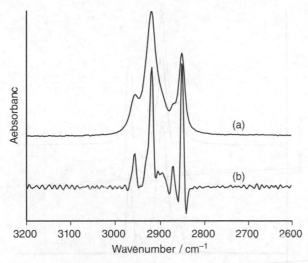

Figure 6.7 *FSD of the infrared spectrum of n-butyl stearate in the CH stretching region. (a) Measured spectrum and (b) spectrum obtained by the FSD of the spectrum in (a).*

6.2.6 Band Decomposition (Curve Fitting)

If bands are overlapping over a spectral region, information on component bands may be obtained by the method of band decomposition (often called *curve fitting*). In this method, it is assumed that the bandshape of each component band can be expressed by one of the functions expressing a Lorentz profile, a Gaussian profile, or a linear combination of these two profiles, and computations are performed to minimize the difference between the measured spectrum and the spectrum synthesized from component bands. The peak wavenumber, intensity, and FWHM of each component band are optimized in this process by a nonlinear least-squares method. To start the computation, it is necessary to specify a bandshape function and to set initial values of the above data for each component band. To estimate an initial set of values, it helps to calculate the derivative and FSD spectra and see the results obtained. To reach a satisfactory agreement between the measured and synthesized spectra, it is advisable to change the initial set of values in various ways and repeat the computation. This may help avoid cases where reasonable solutions cannot be obtained. In Figure 6.8, a result of band decomposition is shown.

6.2.7 Other Software for Processing Measured Spectral Data

Many commercial FT-IR spectrometers have a variety of built-in software routines used for purposes other than those described above; for example, calculations of the refractive and absorption indices from a reflection spectrum by using the Kramers–Kronig relations (see Chapter 8), "baseline correction," "peak picking," "zap," (i.e., drawing a straight line over a desired wavenumber region in a spectrum), and so on.

As described in detail in Chapter 8, calculations based on the Kramers–Kronig relations give the real and imaginary parts of a complex refractive index ($n(\tilde{v})$ and $k(\tilde{v})$; see Section 1.2.4) from a reflection spectrum measured by the method of specular reflection from a

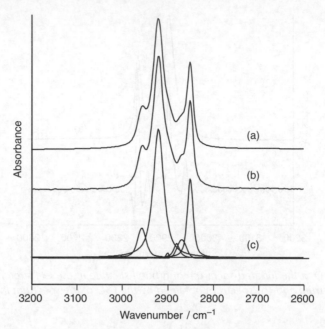

Figure 6.8 *Band decomposition for the infrared spectrum of n-butyl stearate in the CH stretching region. (a) Measured spectrum, (b) spectrum synthesized from the six decomposed bands in (c), and (c) decomposed six bands.*

flat surface of a target substance. It should be emphasized that this method is particularly important for analyzing inorganic substances in general, including ceramics and glasses [2]. As a representative example obtained by this method, the results for Si_3N_4 are shown in Figure 6.9.

"Baseline correction" is a procedure used to make the spectral baseline flat (parallel to the abscissa axis), when a measured spectral baseline is curved and/or has a slope for reasons which are not always clearly known. This correction helps clarify spectral features, and makes it easier to compare band intensities between different spectra. This correction should be applied to absorbance spectra (not to transmittance spectra).

"Peak picking" is a routine that automatically determines the wavenumber positions of absorption peaks in a measured spectrum. However, it is advisable to examine the results obtained by this software, as they may not coincide completely with the wavenumbers read by the user. The peak-picking routines often cite band positions to far greater precision (e.g., three decimal places), but such numerical values should not be considered to be correct in view of the spectral resolution of the recorded spectrum.

"Zap" is a cosmetic procedure that obliterates a spectral curve between two selected wavenumber positions, and draws a straight line instead. It may be used to show a measured spectrum in an apparently better form by removing unwanted (sometimes strong) absorptions due to carbon dioxide, and so on, without losing necessary information.

Algorithms of these various pieces of software vary among the different manufacturers of FT-IR spectrometers. Users are advised to utilize them after consulting the instruction manuals provided by individual manufacturers.

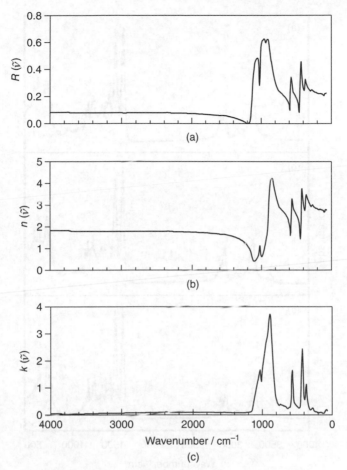

Figure 6.9 *An example of the application of the Kramers–Kronig relations. (a) Reflectance spectrum measured for Si_3N_4, (b) calculated $n(\tilde{v})$ spectrum, and (c) calculated $k(\tilde{v})$ spectrum. (Source: Reproduced from Ref. [2] with permission from the Society of Applied Spectroscopy, 2013).*

6.2.8 Spectral Search

Various infrared spectral databases or libraries are available, which contain collections of the infrared spectra of a number of specific chemical species. Spectral search or data retrieval is a technique enabling one to identify a material of unknown origin by comparing its spectrum with library spectra, or to make a guess at the chemical structure of the unknown material from the similarity of its spectrum to some library spectra.

Two kinds of algorithms are available for spectral search; one is to take the difference between the spectrum of the unknown material and library spectra and to select those library spectra for which the residual sum of squares are small, while the other is to calculate the correlation coefficient between the spectrum of the unknown material and library spectra and to select those library spectra which have large correlation coefficients.

In Figure 6.10, the infrared spectra of an unknown sample (a polymer film) and two library spectra, which were the first-place and second-place results (hits) of a spectral

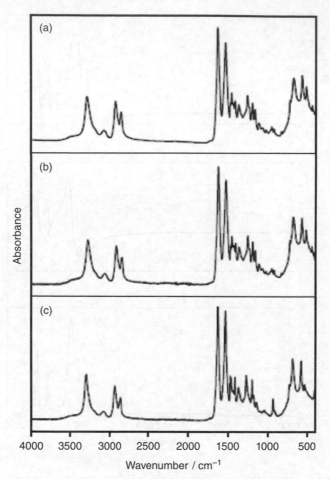

Figure 6.10   *An example of the result of spectral search. (a) Spectrum of an "unknown" material, (b) closest-match search spectrum: nylon 6, and (c) second-best-match search spectrum: nylon 6.6.*

search, are shown. These three spectra are very similar to each other, but a close examination indicates that the spectrum of the unknown sample in Figure 6.10a and the first-place spectrum in Figure 6.10b are very similar in detail. Accordingly, the unknown sample is identified as nylon 6. Many similar materials are found as the result of spectral search, including nylon 6.6 (second place) and nylon 6.9 (third place). Detailed visual comparisons of full spectra as shown in Figure 6.10 are therefore important in identifying an unknown sample without error.

In some databases, in addition to compound names, the CAS (Chemical Abstracts Service) Registry Numbers and some physical constants are also stored; these additional data are sometimes useful to aid identification of unknown materials.

Infrared spectral databases consisting of spectra of a great number of chemical compounds are commercially available. When the purpose of spectral search is clearly defined, databases sorted by the class of compounds (polymers, for example) and a collection of spectra measured by a specific method (for example, attenuated total reflectance (ATR) (see Chapter 13)) may be more useful than databases for a more general purpose.

References

1. Savitzky, A. and Golay, J.E. (1964) Smoothing and differentiation of data by simplified least squares procedures. *Anal. Chem.*, **36**, 1627–1639.
2. Yamamoto, K. and Masui, A. (1995) Complex refractive index determination of bulk materials from infrared reflection spectra. *Appl. Spectrosc.*, **49**, 639–644.

Further Reading

DeNoyer, L.K. and Dodd, G. (2002) Smoothing and derivatives in spectroscopy, In: *Handbook of Vibrational Spectroscopy*, Vol. **3** (eds J.M. Chalmers and P.R. Griffiths), John Wiley & Sons, Ltd, Chichester, pp. 2173–2184.

Griffiths, P.R. and de Haseth, J.A. (2007) *Fourier Transform Infrared Spectrometry*, 2nd edn, John Wiley & Sons, Inc., Hoboken, NJ.

Saarinen, P.E. and Kauppinen, J.K. (2002) Resolution enhancement approaches, In: *Handbook of Vibrational Spectroscopy*, Vol. **3** (eds J.M. Chalmers and P.R. Griffiths), John Wiley & Sons, Ltd, Chichester, pp. 2185–2214.

7

Chemometrics in Infrared Spectroscopic Analysis

Takeshi Hasegawa
Institute for Chemical Research, Kyoto University, Japan

7.1 Introduction

Quantitative infrared spectroscopic analysis is based on Beer's law that directly relates the concentration of an analyte (target of analysis) in a sample solution with the intensity (in absorbance) of an absorption band of the analyte [1]. As Beer's law, which can be derived from Maxwell's equations, is physically established, a reliable model for quantitative analysis can be built on it.

Beer's law is particularly applicable to a situation in which the solution for analysis contains a single solute (analyte), and the shape of a key absorption band due to the analyte remains unchanged with changes in the analyte concentration. In such a case, quantitative spectroscopic analysis can be performed by using either the integrated intensity of a key band or its peak maximum intensity. This type of analysis may be called the *single-band method*.

It is usually easy to apply the single-band method to ultraviolet–visible absorption spectra, as only one or two bands of the analyte are observed. In ultraviolet–visible absorption measurements, it is possible to use a solvent that has no absorption band within this spectral region. By contrast, it is not always so easy to apply the single-band method to infrared absorption spectra, which usually consist of a number of bands arising not only from the solute but also from the solvent. If a sample contains two or more dissolved substances as solutes, the observed spectrum becomes even more complex. Many bands may overlap other bands at least partially. In such a case, a key band for quantitative analysis should be selected with care and its intensity should be measured in an appropriate manner as described in Section 3.4.2. It is necessary to make the signal-to-noise ratio of

Introduction to Experimental Infrared Spectroscopy: Fundamentals and Practical Methods,
First Edition. Edited by Mitsuo Tasumi and Akira Sakamoto.
© 2015 John Wiley & Sons, Ltd. Published 2015 by John Wiley & Sons, Ltd.

a recorded spectrum high enough for good, reproducible quantitative measurements to be undertaken.

In addition to these problems in applying the single-band method to quantitative infrared spectroscopic analysis, the single-band method is not suitable for determining the molar ratios of two or more substances existing in a sample. The single-band method, which depends only on the selected key band, does not utilize all the other bands in the observed infrared spectrum. Thus, it is reasonable to seek an alternative method that makes the optimum use of an entire infrared absorption spectrum for quantitative analysis.

Chemometrics [2–6], a suite of the multivariate data analyses techniques, has been developed to overcome the limitations of the single-band method. These techniques utilize all the infrared absorption bands over a wide wavenumber region as multivariate data for quantitative analysis, and can handle multicomponent samples simultaneously. In other words, the methods of chemometrics in quantitative infrared spectroscopic analysis are mathematical procedures to apply the concept of Beer's law to various problems in order to extract from them as much useful information as possible. In this chapter, the term *spectroscopic calibration* or just *calibration* is used to represent such procedures.

7.2   Role of the Molar Absorption Coefficient

Beer's law [1] states that the absorbance of a key band of an analyte is proportional to its concentration. Thus, the law is expressed as

$$A = \varepsilon c d \tag{7.1}$$

In this equation, A (absorbance) is a dimensionless number by definition. The other quantities are usually expressed in the following units: ε (molar decadic absorption coefficient) in cubic decimeter per mole per centimeter (or liter per mole per centimeter), c (concentration) in cubic decimeter per mole (or mole per liter), and d (cell thickness or pathlength) in centimeter.

In quantitative analysis by the single-band method, the concentration c of an analyte is determined by measuring the absorbance A of a key band of the analyte. For this purpose, the relationship between c and A should be determined in advance from sample solutions with known values of c, as illustrated in the example calibration plot shown in Figure 7.1. The result in Figure 7.1 indicates that a linear relationship exists between c and A. This means that A is proportional to c, and Equation (7.1) is applicable to the sample. The molar absorption coefficient ε can be calculated from the gradient of the linear relationship obtained (usually called the *calibration line*).

A calibration line should be obtained in the following way.

1. At least three different standard sample solutions with known analyte concentrations are prepared, and their infrared absorption spectra are measured; the analyte concentrations in these standard solutions must cover the range expected for the test samples. A key absorption band of the analyte is selected, and its peak maximum intensity (in absorbance) is measured for each standard sample and plotted against its concentration.
2. A calibration line is obtained for the plots by the least-squares method.

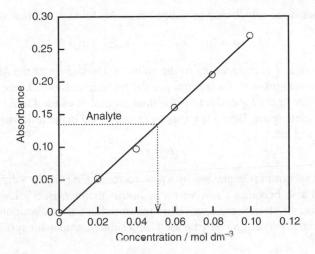

Figure 7.1 *Calibration and analysis using a single band.*

The concept of determining the molar absorption coefficient ε from the calibration line can be transferred to chemometrics. Although it is not practicable in a multivariate space to draw something like a calibration line in Figure 7.1, a multivariate parameter corresponding to ε can be defined.

Equation (7.1) relates the absorbance at a fixed wavenumber of a key band to the concentration of the analyte. As both the absorbance and molar absorption coefficient depend on wavenumber, they should be expressed as functions of wavenumber in the following way:

$$A(\tilde{v}) = \varepsilon(\tilde{v})cd \tag{7.2}$$

As $A(\tilde{v})$ represents an infrared absorption spectrum, $\varepsilon(\tilde{v})$ can also be regarded as the infrared spectrum of a solution with a concentration of $1 \ mol \ dm^{-3}$ and a pathlength of 1 cm.

With a change in concentration, $A(\tilde{v})$ changes in magnitude but is still proportional to $\varepsilon(\tilde{v})$; in other words, the spectral shape of $A(\tilde{v})$ is kept unchanged with changing concentration. Such a spectral behavior corresponds to a case in which the analyte consists of a single chemical component. If a mixture of solutes is present in a sample, $\varepsilon(\tilde{v})$ represents a constituent spectrum corresponding to the spectrum of one chemical component. Note that the term *component* used here has a meaning different from "species." If an analyte solution contains two chemical species which form an aggregate and the aggregate yields a characteristic spectrum, the aggregate is regarded as one chemical component. Thus, a chemical component is an entity based on its spectrum.

7.3 Re-Expression of a Spectrum in Multivariate Space

$A(\tilde{v})$ is an expression of an infrared absorption spectrum as a function of wavenumber. Actually, however, a spectrum obtained from a Fourier transform infrared (FT-IR) spectrometer is not a continuous function but a collection of absorbance values at M discrete wavenumber positions, which may be defined as an M-dimensional row vector a.

$$a = [a_1 \quad a_2 \quad a_3 \cdots a_{M-1} \quad a_M] \tag{7.3}$$

where a_M is a variate corresponding to the value of absorbance at the Mth wavenumber position. This re-expression of a spectrum in the multidimensional space (also called the *multivariate space* or the *hyperspace*) is the fundamental of chemometrics [2, 4].

In the multivariate space, Beer's law may be re-expressed in the following form by fixing the pathlength to 1 cm.

$$a = c\varepsilon \tag{7.4}$$

As a measured spectrum is expressed by a row vector a, the corresponding molar absorption coefficient also becomes a row vector ε. In this manner, Beer's law is expanded to involve a number of absorbance data values measured at various wavenumber positions. This formulation, however, can only be used for a single-component system.

7.4   Beer's Law for a Multicomponent System: CLS (Classical Least Squares) Regression

Let us extend the idea that any spectrum can be expressed by a row vector to a multicomponent system. Consider a solution sample containing a mixture of two chemical components "a" and "b," each of which gives rise to an independent infrared spectrum. Their spectra for a concentration of $1 \, mol \, dm^{-3}$ and a pathlength of 1 cm are expressed, respectively, by the following row vectors.

$$k_a = \begin{bmatrix} k_{a1} & k_{a2} & k_{a3} \cdots k_{aM-1} & k_{aM} \end{bmatrix} \tag{7.5}$$

$$k_b = \begin{bmatrix} k_{b1} & k_{b2} & k_{b3} \cdots k_{bM-1} & k_{bM} \end{bmatrix} \tag{7.6}$$

In these equations, in the field of chemometrics, ε in the single-component spectrum is replaced by k.

If the two components are not chemically interacting with each other, the spectrum of a mixture of the two components should be a superposition (linear combination) of the spectra of the two components, which is equivalent to the sum of k_a and k_b multiplied by the respective concentrations of components "a" and "b" (c_a and c_b). Spectra of mixtures of two components with changing relative concentrations are schematically shown in Figure 7.2, where each "channel" on the abscissa axis represents a wavenumber position.

The linear combination can be expressed by a product of a vector and a matrix in the following form:

$$[a_1 \quad a_2 \quad a_3 \cdots a_{M-1} \quad a_M] = [c_a \quad c_b] \begin{bmatrix} k_{a1} & k_{a2} & k_{a3} \cdots k_{aM-1} & k_{aM} \\ k_{b1} & k_{b2} & k_{b3} \cdots k_{bM-1} & k_{bM} \end{bmatrix}$$

$$\Longleftrightarrow a = [c_a \quad c_b] \begin{bmatrix} k_a \\ k_b \end{bmatrix} \tag{7.7}$$

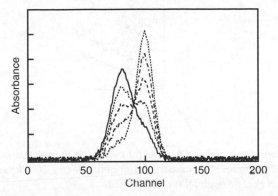

Figure 7.2 *Simulated spectra obtained as linear combinations of two bands centered at Channels 80 and 100.*

When the number of samples with changing relative concentrations of the two components is increased up to N, the collected N spectra ($a_1 \sim a_N$) are represented as follows:

$$\begin{bmatrix} a_1 \\ a_2 \\ \vdots \\ a_N \end{bmatrix} = \begin{bmatrix} c_{a1} & c_{b1} \\ c_{a2} & c_{b2} \\ & \vdots \\ c_{aN} & c_{bN} \end{bmatrix} \begin{bmatrix} k_a \\ k_b \end{bmatrix} \iff A = CK \qquad (7.8)$$

As is self-explanatory in Equation (7.8), A, C, and K are, respectively, matrices with N rows and M columns, N rows and two columns, and two rows and M columns.

The above matrix formulation can be visualized by using vectors in a multivariate space as shown in Figure 7.3. The two continuous line arrows in this figure correspond to the vectors k_a and k_b, and the five dots (points) represent five observed spectra. The position of each point, which corresponds to a row vector in the A matrix, is determined by adding the two k vectors multiplied by different concentrations given in the C matrix. The five points are therefore in a plane spanned by the two k vectors. This means that the number of dimensions needed for spanning the space to contain all the points is equal to the number of chemical components. This concept will be used positively later for considering PCA (principal component analysis).

By expressing spectral information and component data as vectors, all the scalar parameters in Beer's law are now replaced by a matrix equation $A = CK$. This relation can accommodate any number of chemical components by altering the matrix size. The matrix formulation is thus the basis for quantitative infrared spectroscopic analysis. In practice, however, this equation does not hold strictly, because most of the observed data contain uncertain factors such as noise. To remove the uncertain factors from A, a matrix composed of the uncertain factors is introduced. This matrix, which is denoted by R, is called the *residual matrix* (or *error matrix*). Then, the equation is rewritten as

$$A = CK + R \qquad (7.9)$$

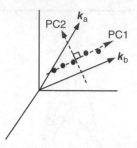

Figure 7.3   *Variation of two-component spectra plotted in the multivariate space and vectors in CLS and PCA. Filled circles indicate two-component spectra, and k_a and k_b depict vectors in CLS, and PC1 and PC2 orthogonal vectors in PCA.*

What is meant by Equation (7.9) is a modeling of the observed spectra (A) by using the concentrations (C) and the component spectra (K). The target of modeling in Equation (7.9) is the matrix of the observed spectra (A). As this equation is an expansion of the classical form of Beer's law, it is called the *classical least-squares* (*CLS*) regression equation. The term *regression* means a process of analysis for predicting optimum values.

When the two matrices A and C are experimentally determined, the unknown matrix K can be calculated without regard to R by the following equation [2–6]:

$$K = (C^T C)^{-1} C^T A \tag{7.10}$$

where superscripts T and −1 denote, respectively, transposed and inverse matrices. As K in the multivariate space corresponds to ε of Beer's law in the single-band method, the calculation of K by Equation (7.10) in the multivariate space corresponds to the determination of ε by drawing a calibration line as shown in Figure 7.1. The result of calculation by Equation (7.10) is called the *compromise solution* or the *least-squares solution*, because it is mathematically proved [4] that the result calculated by Equation (7.10) is equivalent to the least-squares solution for sets of quantities underlying the points (corresponding to A) plotted in the multivariate space. This is the reason why the name of this equation has the term *least-squares*.

As an example of a system to which Equation (7.10) can be applied, the case shown in Figure 7.2 was examined. The calculation was performed by taking the spectra shown in Figure 7.2 for A and the known concentrations of components 1 and 2 for C. The results are shown in Figure 7.4, where the calculated spectra (K) are given as pred #1 and pred #2. In this figure, the observed spectra of components 1 and 2 are also shown as comp #1 and comp #2. It is clear that pred #1 and pred #2 accurately reproduce comp #1 and comp #2, respectively. This example shows that the CLS regression calculation is effective for a case where the number of components is definitely known. This calculation may be less effective if a reliable estimate of the number of components cannot be made. A method for dealing with such a case will be discussed in a later section.

Once K has been determined, the unknown concentrations of the analytes (C_u) in samples to be analyzed can be calculated by using K and the spectra observed from the samples (A_u) by the following equation:

$$C_u = A_u K^T (K K^T)^{-1} \tag{7.11}$$

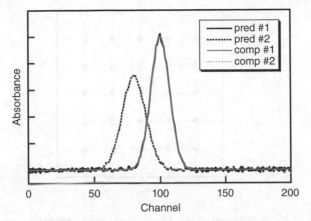

Figure 7.4 *CLS-predicted **K** and component spectra. Solid and dotted curves in black correspond, respectively, to the spectra predicted for components 1 and 2, and solid and dotted curves in gray show, respectively, the spectra of components 1 and 2.*

It should be noted that this method can handle any number of samples, any number of chemical components in the samples, and any number of wavenumber positions in any spectral region. The concentrations of all components in each sample are simultaneously determined. This is the great advantage of applying this method to spectroscopic calibration.

7.5 Experimental Design

In applying the single-band method to quantitative analysis, it should be kept in mind that the analyte concentration must be within the range for which the calibration line was obtained; in other words, an extrapolation of the calibration line beyond this range should be avoided. If this is called the experimental design in using the single-band method, what sort of experimental design is needed to apply the CLS regression method to a multicomponent system?

Let us take the simplest case, a system consisting of two components denoted by "a" and "b." If the concentrations of the two components are expressed by C_a and C_b, respectively, a set of C_a and C_b may be represented by a point in a two-dimensional space as depicted in Figure 7.5. Then, a set of unknown concentrations in a sample to be analyzed, C_a^u and C_b^u, may take a position indicated by a cross (×) in Figure 7.5. To deal with such a sample, samples having the sets of concentrations at positions surrounding the cross, for example, at the lattice points indicated by dots (●) in Figure 7.5 should be prepared, and their spectra should be measured and used as A in Equation (7.10) with the sets of concentrations at the dots to be used as C. In this process, it is important to have an experimental design in such a way that, as shown in Figure 7.5, the position of the cross is completely surrounded by the dots for which the spectral data for A were obtained. It is not necessary to place the dots regularly on the lattice points so long as they completely surround the cross.

The number of dots required (corresponding to samples with known concentrations of chemical components) increases rapidly with the number of chemical components present.

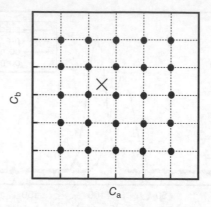

Figure 7.5   *Experimental design for a two-component system. See text for details.*

If the number of chemical components is denoted by n and, for each chemical component, samples with m different concentrations are prepared, the number of dots required becomes m^n. This means that, when the CLS regression method is used for a multicomponent system, the task for deriving A by measuring spectra may become very difficult to perform.

7.6   ILS Regression Using the Inverse Beer's Law

To calculate matrix K by Equation (7.10), the matrix size of C and its elements should be determined in advance. The dimension of the column in C is equal to the number of independent chemical components, which is equal to the dimension of the row in K. The size of matrix A is automatically determined when measured spectra are stored in the computer used for the data analysis.

The analytical accuracy of the CLS regression method is known to be high when the number of chemical components (or the number of independent component spectra) is estimated appropriately [4]. This is confirmed by the results given in Figure 7.4. However, the reliability of this method greatly decreases when the dimension of the column in C is less than the number of chemical components that actually exist in a sample. Such a case occurs if interactions between known chemical components unexpectedly yield new chemical components in the sample [2]. This is a drawback, which is a characteristic of the CLS regression method. If the situation is described in more general terms, a reliable result cannot be expected from an attempt to model a set of quantities having more information by using another set of quantities having less information.

To overcome the difficulty of the CLS regression method, an idea of exchanging A and C was developed; that is, C having less information is modeled with A having more information. This inverse modeling is expressed as follows:

$$C = AP_{\text{ILS}} + R \qquad (7.12)$$

The relation expressing the concentration of a solute c by its absorbance A may be called the *inverse Beer's law*. As the new regression method given by Equation (7.12) is based on the inverse Beer's law [2, 3, 5], it is called the *inverse least-squares* (*ILS*) method or the *multiple linear regression* (*MLR*) method [3, 5, 6].

Prior to applying the ILS method to samples to be analyzed, matrix P_{ILS} should be determined by the following equation, which is similar to Equation (7.10):

$$P_{ILS} = (A^T A)^{-1} A^T C \tag{7.13}$$

Once P_{ILS} is obtained, concentrations in C are calculated by using Equation (7.12).

In practice, however, the computation in Equation (7.13) cannot be performed in a straightforward manner [2–6], because A is usually a rectangular matrix, that is, the dimension of rows (N) is smaller than that of columns (M). It is well known that in such a case the determinant of $A^T A$ becomes zero and its inverse matrix cannot be computed. This "singularity" problem is a matter specific to ILS [2]. To avoid this problem, a condition $M \le N$ should be satisfied by reducing the number of wavenumber positions used.

The reduction of wavenumber positions has its drawback in that information recorded in an observed spectrum cannot then be used in full for analysis. This is a big disadvantage in multivariate analysis. The reduction of wavenumber positions has some arbitrariness and this arbitrariness may affect the results of a calculation. If the number of samples (N) is increased and becomes equal to the number of wavenumber points (M), high accuracy is expected for the results of calculation.

The fact that the accuracy of quantitative analysis depends on the number of samples, which cannot be easily increased, is a disadvantage of the ILS method. The opposing requirements, that is, the necessity of reducing M and increasing N, place this method in an intrinsic dilemma. Actually, however, the ILS method is frequently employed, particularly for quantitative analysis using near-infrared spectra. In some cases, this method gives better results than the PCA method [7], as discussed in the next section.

7.7   Principal Component Analysis (PCA)

PCA [2–8] is a method developed to overcome the disadvantage of the ILS method. The PCA method may be regarded as an extension of CLS regression [4], as PCA is a linear algebraic technique applied to multidimensional space.

Equation (7.9), which is the basis of CLS regression, can be rewritten as a sum of products of a column vector c_j and a row vector k_j.

$$
\begin{aligned}
A &= CK + R \\
&= \begin{bmatrix} c_{11} & c_{21} & \cdots & c_{r1} \\ \vdots & \vdots & \ddots & \vdots \\ c_{1N} & c_{2N} & \cdots & c_{rN} \end{bmatrix} \begin{bmatrix} k_{11} & \cdots & k_{1M} \\ k_{21} & \cdots & k_{2M} \\ \vdots & \ddots & \vdots \\ k_{r1} & \cdots & k_{rM} \end{bmatrix} + R \\
&= \begin{bmatrix} c_{11} \\ \vdots \\ c_{1N} \end{bmatrix} [k_{11} \ \cdots \ k_{1M}] + \begin{bmatrix} c_{21} \\ \vdots \\ c_{2N} \end{bmatrix} [k_{21} \ \cdots \ k_{2M}] + \cdots + \begin{bmatrix} c_{r1} \\ \vdots \\ c_{rN} \end{bmatrix} [k_{r1} \ \cdots \ k_{rM}] + R \\
&= \sum_{j=1}^{r} c_j k_j + R \tag{7.14}
\end{aligned}
$$

The sum of products $\sum_{j=1}^{r} c_j k_j$ in Equation (7.14) may be regarded as an expansion of A by k_j. In this expansion, c_j is now defined as a column vector which expresses the concentration profile of a chemical component, while k_j has the same meaning as k_a and k_b in Equations (7.7) and (7.8). Equation (7.14) shows that the CLS regression method may be characterized by an expansion with the two chemically meaningful vectors.

The meaning of Equation (7.14), which is based on the chemically meaningful vectors, is easy to understand. The problem with this equation lies in the fact that r should be given in advance as the number of chemical components known to exist. Unexpected components such as molecular aggregates and hydrated species cannot be taken into account. If such unexpected components actually exist, the CLS regression method gives poor performance as mentioned in Section 7.6.

The PCA method borrows from Equation (7.14) the idea of expanding A into the products of two vectors, and expands A as the products of two mutually orthogonal vectors, t_j and p_j, as

$$A = \sum_{j=1}^{N} t_j p_j \tag{7.15}$$

This equation has the same form as Equation (7.14), but they are different in the properties of the quantities used. The quantities in Equation (7.15) have the following properties.

1. Vectors p_j are orthogonal to each other: $p_j^{\mathrm{T}} \cdot p_j = \delta_{ij}$.
2. Neither t_j nor p_j is related to a particular chemical quantity, except for some special cases [9].
3. The upper limit of j is N instead of r, if $N < M$.

Matrix A represents points in multidimensional space, and vectors p_j provide the new orthogonal axes spanning most effectively the space in which the points of A are plotted. Vector p_j is called the *loading vector* and its elements are called *loadings*. The loading vectors are obtained by solving the eigenvalue/eigenvector problem for $A^{\mathrm{T}}A$ [2–6].

$$A^{\mathrm{T}}Ap_j = \lambda_j p_j \tag{7.16}$$

As $A^{\mathrm{T}}A$ is a square matrix having M rows and M columns, this equation has $M(M > N)$ solutions, but only the N solutions having larger eigenvalues than the rest are useful.

Vectors t_j, coefficients for p_j, indicate positions of the N points on the new axes. Vector t_j is called the *score vector* and its elements are called *scores*. As scores are projections of the N points on the loading vectors, they can be calculated by taking the inner product of A and transposed p_j as

$$t_j = Ap_j^{\mathrm{T}} \tag{7.17}$$

In the PCA method, the variance of the N points in the multidimensional space spanned by the orthogonal coordinates is determined in this manner.

The concept of PCA is schematically illustrated in Figure 7.3 in comparison with CLS regression. In the latter method, five spectra of a two-component system are indicated by five points that are modeled by linear combinations of two vectors, k_a and k_b. As a result, the five points are on the plane spanned by the two axes, k_a and k_b, which are related with the molar absorption coefficients of the two chemical components. In PCA, by contrast, the five points are expressed by two loading vectors, PC1 and PC2, which have no chemical meanings. PC1 expresses the largest variance of the five points [4]. PC2, which is orthogonal to PC1, expresses the residual variance which cannot be expressed by PC1. If the variance of the five points is confined in the two-dimensional plane, the five points are perfectly expressed by the two loading vectors. This leads to the conclusion that there are only two components in the samples. The PC1 and PC2 vectors represent the principal components in this case.

The number of loading vectors may be increased up to N as given in Equation (7.15). By using N loading vectors, all spectral features including noise would be modeled thoroughly, and R would no longer be needed. However, the use of N loading vectors is excessive in that the modeling of noise is meaningless from the viewpoint of chemistry. To avoid such a case, an appropriate number of loading vectors (denoted by b in the later discussion) should be determined by using an analytical technique such as the eigenvalue plot [2–6], the empirical Malinowski's IND function [8, 10], or the spectral reconstruction method [5].

The eigenvalues λ_j, which are obtained together with the loading vectors p_j by solving Equation (7.16), show the variance of the N points along the loading vectors [2, 4]. This means that an eigenvalue reflects the degree of spectral variation. If the spectral variation is due to noise only, the degree of variation would be small, and the corresponding eigenvalue would also be small. This characteristic of an eigenvalue is useful in establishing a chemically meaningful spectral variation; if the eigenvalues are arranged in order of their magnitudes and a significant decrease in magnitude occurs on going from one eigenvalue to the next, the loading vectors for the first group of higher eigenvalues (b in number) should be used for the PCA method. These considerations may be expressed in the following way:

$$A = \sum_{j=1}^{b} t_j p_j + R \tag{7.18}$$

The first b terms on the right-hand side of Equation (7.18) are called the *basis factors*, and the remaining $(N - b)$ terms, which are now included in R, are called the *noise factors*. In this way, the variance of the N points can be discussed in the b-dimensional space.

As the dimension size b is often larger than three, it is difficult to visualize the b-dimensional space. Then, in order to have an image of the N points in the b-dimensional space, they are projected onto a two-dimensional plane formed by two selected loading vectors. This projected image is called the *score score plot*, which is often used for visually understanding similarities and differences between spectral plots.

It is worth pointing out that reconstruction of spectra using the basis factors in Equation (7.18) is useful for noise reduction. This noise-reduction technique is much better than spectral smoothing, as chemical information contained in original spectra is not lost, if the number of basis factors is appropriately determined.

Reduction of noise by the PCA method is expected to give good performance when it is employed for generating an image obtained by measurements made using an infrared microscope-spectrometer equipped with a focal-plane array detector. Collection of spectra needed for infrared imaging over a short timescale usually results in noisy spectra. Even if a measurement is performed with a relatively low spatial resolution, at least 32×32 ($=1024$) pixels are used, and the number of spectra obtained readily exceeds 1000. If the sample area under study contains five independent chemical components, the number of basis factors b can now be set to 5. Then, the remaining $1019(= 1024 - 5)$ terms are discarded in R as noise factors, and effective reduction of noise is expected to be realized. In fact, significant results have been obtained in this manner [11, 12]. The benefit of the PCA noise reduction technique is understood by seeing the greatly improved imaging [11].

The PCA method can model observed spectra without knowing the number of chemical components in the samples prior to the spectroscopic calibration. This characteristic makes it possible to model the spectra for any unexpected chemical components, a situation which often makes a CLS calibration an unreliable method.

As the loading vectors, which are obtained by a mathematical procedure, are orthogonal to each other, most of the loadings have no chemical meanings except for a special case [9]. Similarly, scores have no explicit chemical meanings either. Thus, loadings and scores are sometimes called *latent variables* [2–6].

Scores indicate new positions of the points (observed spectra) on the coordinate axes corresponding to the loading vectors. It should be noted that the variation of spectral quantity is fully recorded in the scores. The number of scores on each loading vector is exactly the same as the number of points in the multivariate space. This plays an important role in principal component regression (PCR), described in the next section.

7.8   Principal Component Regression (PCR)

To apply the ILS method to spectroscopic calibration properly, the A matrix corresponding to spectra must have a square or portrait form. To fulfill this requirement, it is necessary to reduce greatly the number of wavenumber points selected from the measured spectra. However, the information required for spectroscopic calibration need not be taken from the measured spectra; it can also be obtained from the PCA scores, which have information equivalent to the recorded spectra. The number of scores is equal to that of the points corresponding to the measured spectra in the multivariate space. Therefore, A with a size of $N \times M$ can be replaced by a matrix of scores with a reduced size of $N \times b$, the rank of which never exceeds N. In this manner, the singularity problem inherent in the ILS method can be overcome. Note that no reduction of wavenumber points is needed and full use of the measured spectra is made, as the scores are obtained by Equations (7.16) and (7.17).

The ILS method taking scores as the modeling target is called the *principal component regression (PCR) method*. The PCR method is formulated by expressing A with the PCA scores T as follows:

$$A = \sum_{j=1}^{b} t_j p_j + R_{\text{PCA}} = TP + R_{\text{PCA}} \tag{7.19a}$$

$$C = TP_{\text{ILS}} + R \tag{7.19b}$$

Equations (7.19a) and (7.19b) represent the PCA and PCR methods, respectively.

An example of application of the PCR method is given in Figures 7.6 and 7.7 [13]. In Figure 7.6, the infrared spectra of pullulan (an edible polysaccharide) measured at various temperatures are shown. The intensity changes of the bands at about 1640 and 1420 cm^{-1}

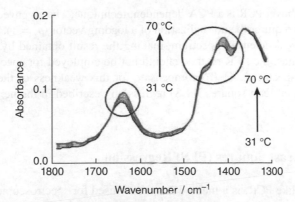

Figure 7.6 *Temperature-dependent infrared spectra of a cast film of pullulan. (Source: Adapted from Ref. [13] with permission from Elsevier.)*

seem to reflect, respectively, variations in the content of water and its interaction with the polysaccharide chain. The spectra are used as A in Equation (7.19a) to obtain the PCA scores T. In addition to recording the infrared spectra, shrinkage in the film length was measured by using an apparatus for thermo-mechanical analysis (TMA) to obtain C in Equation (7.19b). In this case, C is a row vector. Then, P_{ILS} is calculated by the same equation as a compromise solution to build a PCR model.

The interrelationship between the shrinkage values predicted from the infrared spectra and the data obtained by TMA is shown in Figure 7.7 [13]. The good linearity seen in this figure indicates that chemometrics is a powerful tool for extracting molecular information from an infrared spectral region, which may be used for predicting a macroscopic property such as film shrinkage.

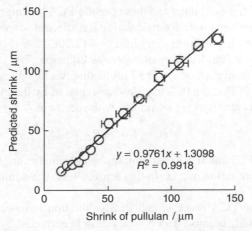

$$y = 0.9761x + 1.3098$$
$$R^2 = 0.9918$$

Figure 7.7 *Relationship between the shrink values predicted from infrared spectra and those measured by TMA. (Source: Adapted from Ref. [13] with permission from Elsevier.)*

As described above, PCR is a PCA-dependent technique. The eigenvector calculation in the PCA process requires the normalization of a loading vector ($p_j = 1$). This may emphasize noise in a weak-intensity spectrum, making the result obtained by the PCR method unreliable [7]. Thus, the PCR method should not be employed for spectroscopic calibration with noisy spectra [2, 4]. To compensate for this weakness of the PCR method to some extent, partial least squares (PLS) regression described in the next section is often employed.

7.9   Partial Least Squares (PLS) Regression

PLS regression, like PCR, is a method frequently used for spectroscopic calibration. The fundamental concept of PLS regression is expressed by the following two equations.

$$A = t_1 p_1 + t_2 p_2 + t_3 p_3 + \cdots + R_A \qquad (7.20a)$$

$$C = u_1 q_1 + u_2 q_2 + u_3 q_3 + \cdots + R_C \qquad (7.20b)$$

In Equation (7.20a), p_j and t_j are, respectively, the PLS loading and score vectors, which are different from the loading and score vectors of the PCA method. In Equation (7.20b), q_j and u_j are the loading and score vectors for the concentration matrix (C).

An essential characteristic of the PLS method is that both A and C are modeled, so that two residual terms R_A and R_C are independently generated [2, 4]. As a result, the experimental error in adjusting experimentally the sample concentration is separated from the error in spectral measurements. This independent modeling, however, disregards the correlation between A and C. To correct this shortcoming, a rather complicated algorithm using a matrix correlating t_j with u_j has been developed.

In short, the PLS method also employs for modeling mutually orthogonal vectors like those in the PCA method. A proxy vector called the *weight loading vector* (w_j) is introduced in place of the PLS loading vector, and it is calculated by a procedure similar to PCA through calculations which are essentially the same as CLS. In this procedure, loadings similar to those in PCA are obtained and these pseudo-PCA loadings converge at w_j. These weight loadings w_j are used only for calculating the spectral scores t_j, which are further used for obtaining the PLS loadings p_j by Equation (7.20a).

Although Equation (7.20a) has the same form as Equation (7.18), their calculation procedures are different from each other. The PLS loading vectors do not have strict orthogonality. PLS becomes PCR if the PLS loadings are replaced by the PCA loadings.

It is noted that concentrations are correlated with spectra as

$$C = t_1 q_1 + R_C \qquad (7.21)$$

which indicates that PLS and ILS have common fundamental concepts in that their modeling target is the concentration matrix. In this sense, PLS has a common feature with PCR also [2, 4].

PLS has no explicit PCA process, and PLS calibration is more useful for analyzing noisy spectra than PCR, because a PLS calibration is corrected by using the concentration information. Therefore, PLS is recognized as the best of the spectroscopic calibration techniques. However, it should be pointed out from a theoretical viewpoint that results from

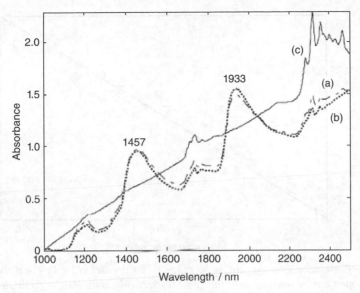

Figure 7.8 *Near-infrared diffuse reflection spectra of (a) cheese wrapped in a polyethylene film, (b) cheese without a wrapping film, and (c) a polyethylene film. (Source: Adapted from Ref. [14] with permission from Elsevier.)*

PLS may be inferior to those from PCR when spectral intensity varies nonlinearly against concentration due to a systematic error [2]. The PLS method is expected to be superior for handling spectra with random noise, whereas the PCR method should be applied to spectra with systematic noise. Thus, it is desirable to decide, prior to performing a spectroscopic calibration, which of the two methods should be employed.

An example of application of the PLS method to a practical problem is given here. A spectroscopic prediction of quantities of fat, protein, and sodium chloride in cheese slices was performed by utilizing their near-infrared spectra [14]. Diffuse reflection spectra in the near-infrared region were measured for cheese slices wrapped with polyethylene (PE) films and also for the same slices without PE-film wrapping. Representative spectra obtained are shown in Figure 7.8, together with the spectrum of a PE film used for wrapping the cheese slice. The authors of Ref. [14] claim that the broad bands at 1457 and 1933 nm are attributable to water contained in cheese, and weaker features in the regions of 1700–1900 nm and around 2300 nm are associated with fat and protein and indirectly with water interacting with protein and sodium chloride. Therefore these bands are useful for calibration and prediction of these ingredients.

The slopes and drifts of the spectral baseline as seen in Figure 7.8, which arise from scattering of near-infrared light, disturb the accuracy of analysis of minute spectral changes. These disturbing factors can be removed by applying the second-derivative technique to the spectra [15]. In the study in Ref. [14], spectra were pretreated by the second-derivative technique and the resultant second-derivative spectra were analyzed by the PLS method. A total of 51 samples of cheese slices were divided into two groups; 41 samples were employed for developing calibration models (calibration set) and the remaining 10 samples were used for validating the developed models (prediction set).

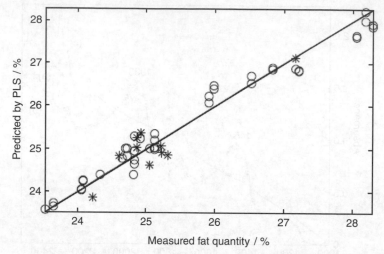

Figure 7.9 *Correlation between the percentage of fat predicted by PLS and the measured value. The open circles and asterisks indicate, respectively, data for the calibration and prediction sets. (Source: Adapted from Ref. [14] with permission from Elsevier.)*

In Figure 7.9, the quantities of fat obtained by the PLS method for the calibration and prediction sets are compared with the values determined by a more conventional, non-spectroscopic method of quantitative analysis. The PLS calculations were performed by using four PLS loadings. In spite of negative factors such as near-infrared absorptions being generally weak and overlapping each other considerably, the predicted quantities are close to the actual values. Similarly, good agreements between the PLS-calculated quantities and actual values were obtained for protein and water combined with sodium chloride, indicating the power of chemometrics in performing multicomponent calibration with accuracy.

The PLS method formulated by Equations (7.20a) and (7.20b) is often called *PLS2*. If the concentration data of a single component are put into the concentration matrix (C), the matrix becomes a vector (c). This case is called *PLS1*, where the concentration loading becomes a scalar quantity (q_j) and the spectral scores are equal to the concentration scores, that is, $t_j = u_j$. Then, the fundamental equations of PLS1 are given as:

$$A = t_1 p_1 + t_2 p_2 + t_3 p_3 + \cdots + R_A \tag{7.22a}$$

$$c = t_1 q_1 + t_2 q_2 + t_3 q_3 + \cdots + R_C \tag{7.22b}$$

The relation of $t_j = u_j$, which is intrinsic to PLS1, greatly reduces the amount of calculations, removing the convergence calculation required in PLS2. As PLS1 and PLS2 are comparable in their calibration performance, PLS1 is often preferred because of its low cost of computation. Although, in some textbooks, the PLS approach is explained by using the PLS1 model, the concept of PLS is explicitly embodied in the formulation of PLS2.

References

1. Griffiths, P.R. (2002) Beer's law, In: *Handbook of Vibrational Spectroscopy*, Vol. **3** (eds J. M. Chalmers and P. R. Griffiths), John Wiley & Sons, Ltd, Chichester, pp. 2225–2234.
2. Hasegawa, T. (2002) Principal component regression and partial least squares modeling, In: *Handbook of Vibrational Spectroscopy*, Vol. **3** (eds J. M. Chalmers and P. R. Griffiths), John Wiley & Sons, Ltd, Chichester, pp. 2293–2312.
3. Gemperline, P. (2006) *Practical Guide to Chemometrics*, 2nd edn, Taylor & Francis Group, Boca Raton, FL.
4. Kramer, R. (1998) *Chemometric Techniques for Quantitative Analysis*, Marcel Dekker, New York.
5. Brereton, R.G. (2003) *Chemometrics Data Analysis for the Laboratory and Chemical Plant*, John Wiley & Sons, Ltd, Chichester.
6. Martens, H. and Martens, M. (2001) *Multivariate Analysis of Quality: An Introduction*, John Wiley & Sons, Ltd, Chichester.
7. Hasegawa, T. (2006) Spectral simulation study on the influence of the principal component step on principal component regression. *Appl. Spectrosc.*, **60**, 95–98.
8. Malinowski, E.R. (2002) *Factor Analysis in Chemistry*, 3rd edn, John Wiley & Sons, Inc., New York.
9. Hasegawa, Y. (1999) Detection of minute chemical species by principal-component analysis. *Anal. Chem.*, **71**, 3085–3091.
10. Malinowski, E.R. (1977) Determination of the number of factors and the experimental error in a data matrix. *Anal. Chem.*, **49**, 612–617.
11. Sakabe, T., Yamazaki, S. and Hasegawa, T. (2010) Analysis of cross-section structure of a polymer wrapping film using infrared attenuated total reflection imaging technique with an aid of chemometrics. *J. Phys. Chem. B*, **114**, 6878–6885.
12. Salzer, R. and Siesler, H.W. (2009) *Infrared and Raman Spectroscopic Imaging*, Wiley-VCH Verlag GmbH, Weinheim, pp. 65–112.
13. Sakata, Y. and Otsuka, M. (2009) Evaluation of relationship between molecular behaviour and mechanical strength of pullulan films. *Int. J. Pharm.*, **374**, 33–38.
14. Pi, F., Shinzawa, H., Ozaki, Y. and Han, D. (2009) Non-destructive determination of components in processed cheese slice wrapped with a polyethylene film using near-infrared spectroscopy and chemometrics. *Int. Diary J.*, **19**, 624–629.
15. Griffiths, P.R. and de Haseth, J.A. (2007) *Fourier Transform Infrared Spectrometry*, 2nd edn, John Wiley & Sons, Inc., Hoboken, NJ, pp. 237–240.

Part II
Practical Methods of Measurements

8

Reflection Measurements at Normal Incidence

Takeshi Hasegawa
Institute for Chemical Research, Kyoto University, Japan

8.1 Introduction

What technique should be employed for measuring infrared spectra from thick samples for which a transmission measurement does not work? Examples of target samples in this category are crystals and polymers (including rubber) having flat surfaces. To analyze such samples, reflection measurements should be considered. To record infrared reflection spectra from such samples, two representative techniques are available, namely, specular reflection (reflection at normal or near-normal incidence) and attenuated total reflection (ATR). This chapter deals with external reflection at normal incidence, which has been used for a long time for measuring mid-infrared spectra from optically thick materials with flat surfaces. ATR will be discussed in Chapter 13.

In addition to this chapter and Chapter 13, reflection-related measurements are described in Chapter 9 (external reflection measurements for thin films and interfaces), Chapter 10 (reflection–absorption measurements for thin films on metal substrates), and Chapter 12 (diffuse reflection measurements).

8.2 Measurements of Reflection at Normal Incidence

The optical arrangement of equipment for reflection measurements at normal incidence is schematically illustrated in Figure 8.1. Measurements of the background (reflection from

Introduction to Experimental Infrared Spectroscopy: Fundamentals and Practical Methods,
First Edition. Edited by Mitsuo Tasumi and Akira Sakamoto.

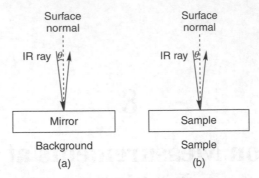

Figure 8.1 *An example of the equipment used for specular-reflection measurements. Arrangements for (a) background measurement and (b) sample measurement.*

the mirror) and the reflection from the sample, shown respectively in Figures 8.1a and 8.1b, are performed with the modulated infrared beam from the interferometer of a Fourier transform-infrared (FT-IR) spectrometer. The reflected beam is directed to the detector of the spectrometer. An aluminum-deposited glass plate may be used as the mirror for measuring the single-beam background corresponding to the instrumental function of the equipment. In the equipment set-up, the mirror is interchanged with the sample, so that the measurements of the background and the reflection from the sample surface are performed with the same optical pathlength.

The angle of incidence θ is defined as that between the direction of the incident infrared beam and the surface normal indicated by the dotted line in Figure 8.1. Thus, $\theta = 0°$ corresponds to normal incidence. In practical measurements, however, it is impossible to realize the condition of $\theta = 0°$. Then, θ should be set as small as possible, usually $<15°$ as will be mentioned again later. As the polarization of light has no meaning in such geometry of measurement, unpolarized light is usually employed for reflection measurements at normal incidence.

The result recorded in this way is usually displayed as a reflectance spectrum $R(\tilde{\nu})$, which is defined as

$$R(\tilde{\nu}) = \frac{I^{\mathrm{sam}}(\tilde{\nu})}{I^{\mathrm{BG}}(\tilde{\nu})} \tag{8.1}$$

where $I^{\mathrm{sam}}(\tilde{\nu})$ is the single-beam spectrum (a function of wavenumber $\tilde{\nu}$) of the reflection from the sample, and $I^{\mathrm{BG}}(\tilde{\nu})$ is that of the background. It should be mentioned that this equation is not always applicable; this is particularly the case for low-reflectance samples or, in more general terms, results from low-reflectance measurements. How to modify Equation (8.1) for such cases will be described later.

To enhance the qualities of infrared spectra obtained by reflection measurements, a double-reflection type equipment, which is schematically shown in Figure 8.2, can be used. In this equipment, the background measurement is performed using the V-shape optical arrangement which has a small mirror (aluminum-deposited glass plate) positioned at the bottom of the V-shape as shown in Figure 8.2a. The angle of incidence in this optical arrangement is designed to be 12°, which is appropriate for specular-reflection measurements. The optical stage involving the mirror can be rotated by 180° around its axis indicated by the filled circle in Figure 8.2, so that the V-shape optical arrangement can be converted into the W-shape optical arrangement shown in Figure 8.2b. When the

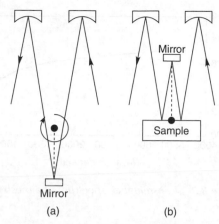

Figure 8.2 *An Example of the equipment used for double-reflection measurements. Arrangements for (a) background measurement and (b) sample measurement. A rotator is turned around the axis at • to switch arrangements between (a) and (b).*

sample surface is at this position in the W-shape arrangement, the optical pathlength in the W-shape arrangement is exactly the same as that in the V-shape arrangement. In this way, accurate double-reflection measurements are possible by using this equipment.

The reflectance R of a dielectric material (nonmetallic material) is generally low. In particular, reflectance for a double-reflection measurement R^2 is often much smaller than unity. This makes the quality of the resultant infrared spectrum unsatisfactory. Even in a single-reflection measurement, the situation is similar if the reflectance from the sample is low.

To obtain infrared spectra of tolerable signal-to-noise (S/N) ratios, a high-sensitivity detector such as an MCT (mercury cadmium telluride) detector (see Sections 3.5 and 5.2.1.3) should be used for measuring the weak signals of reflection from the sample. In contrast, the background signals from the small mirror at the bottom of the V-shape arrangement are much stronger than the signals from the sample. If the background signals should directly reach the MCT detector, the detector might not function properly because of saturation. To avoid this problem, a mesh filter should be inserted into the optical path before or after the mirror to reduce the strong intensity of the background signals. However, the function of the filter depends on the wavenumber region, and the instrumental function of the filter (filter function for short) should be determined prior to the reflection measurements.

The filter function can be obtained by treating the filter like a sample in a transmission measurement. The transmission spectrum of the filter can be measured with the equipment in Figure 8.2a, but an MCT detector should not be used for this purpose. Instead, a triglycine sulfate (TGS) or deuterated triglycine sulfate (DTGS) detector (see Section 5.2.1.3) should be used. If this detector is not available, a method of using two filters with an MCT detector should be tested (see the statement in parentheses in the next paragraph).

The filter function is given as $T_{\text{filter}}(\tilde{v}) = I_{\text{filter}}(\tilde{v})/I_{\text{air}}(\tilde{v})$, where $I_{\text{filter}}(\tilde{v})$ denotes the single-beam reflection spectrum with the filter, and $I_{\text{air}}(\tilde{v})$ is that without the filter. An example of the filter function is shown in Figure 8.3. The background spectrum $I^{\text{BG}}(\tilde{v})$ in Equation (8.1) is related to the background spectrum for the specular reflection with the filter $I^{\text{BG}}_{\text{filter}}(\tilde{v})$ as $I^{\text{BG}}_{\text{filter}}(\tilde{v}) \equiv I^{\text{BG}}(\tilde{v})\, T_{\text{filter}}(\tilde{v})$. (If two filters are used with an MCT detector, $I_{\text{filter}}(\tilde{v})$ should

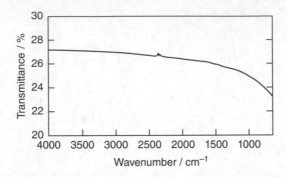

Figure 8.3   *Transmittance spectrum of a mesh filter.*

correspond to the single-beam reflection spectrum with two filters, and $I_{\mathrm{air}}(\tilde{\nu})$ to that with one filter.)

When a double-reflection measurement is performed, the observed reflectance is the square of the reflectance to be determined $R(\tilde{\nu})$, which is given as

$$R(\tilde{\nu}) = \left[\frac{I^{\mathrm{sam}}(\tilde{\nu})}{I^{\mathrm{BG}}(\tilde{\nu})}\right]^{\frac{1}{m}} = \left[\frac{I^{\mathrm{sam}}(\tilde{\nu})}{I^{\mathrm{BG}}_{\mathrm{filter}}(\tilde{\nu})} \cdot T_{\mathrm{filter}}(\tilde{\nu})\right]^{\frac{1}{m}} \tag{8.2}$$

where m is the number of reflections from the sample.

An example of the result of a double-reflection measurement is shown in Figure 8.4. The sample used is a thick film of low-density polyethylene (LDPE). The three quantities in

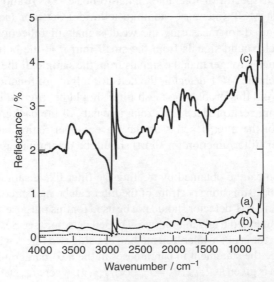

Figure 8.4   *(a) Double-reflection spectrum of a thick LDPE film (thickness about 1 mm), (b) double-reflection spectrum obtained by correcting the spectrum in (a) with the filter function, and (c) single-reflection spectrum calculated from the spectrum in (b).*

Equation (8.2), namely, $A = I^{sam}(\tilde{v})/I^{BG}(\tilde{v})$, $B = A \cdot T_{filter}(\tilde{v})$, and $C = (B)^{1/2} = R(\tilde{v})$, are shown, respectively, in Figure 8.4a–c. The reflectance in Figure 8.4a is very low because of the double-reflection, but $R(\tilde{v})$ finally determined and shown in Figure 8.4c has a tolerable quality.

As is clear in the above example, the reflection spectrum has bands having derivative shapes which are not observed in normal transmission spectra. This is characteristic of near-normal incidence external reflection spectra because they are greatly influenced by the real part (n) of the complex refractive index $\hat{n}$ ($= n + ik$) (see Section 1.2.4) [1]. To derive an absorption spectrum from a reflectance spectrum, the procedure of analysis described in the next section should be followed.

8.3 Analysis of Specular-Reflection Spectra

According to classical electromagnetic theory, Fresnel's amplitude reflection coefficient $\hat{r}$ for specular reflection (reflection at normal incidence) from a sample (i.e., air/sample interface) is given as [1, 2]

$$\hat{r} = \frac{n - ik - 1}{n + ik + 1} \equiv |\hat{r}|e^{i\delta} = \sqrt{R}e^{i\delta} \qquad (8.3)$$

where n and k are, respectively, the real and imaginary parts of the complex refractive index $\hat{n}$. The amplitude reflection coefficient $\hat{r}$, which is the ratio of amplitudes of the electric fields before and after reflection, is related to the reflectance R as $R = |\hat{r}|^2$. The angle δ is called the *canonical phase shift* [3], which measures the change of phase angle on reflection. Equation (8.3) can be separated into two equations corresponding to the real and imaginary parts, from which the following two relations are obtained.

$$n = \frac{1 - R}{1 - 2\sqrt{R}\cos\delta + R} \qquad (8.4a)$$

$$k = \frac{2\sqrt{R}\sin\delta}{1 - 2\sqrt{R}\cos\delta + R} \qquad (8.4b)$$

The imaginary part of the complex refractive index k, which is called the *absorption index*, is a quantity directly related with absorbance (see Section 1.2.4). In Equations (8.4a) and (8.4b), k and δ as well as R are functions of wavenumber $\tilde{v}$, and $k(\tilde{v})$ is essentially the same as an absorbance spectrum from a transmission measurement.

As normal incidence is assumed for deriving Equation (8.3), the angle of incidence in actual measurements should ideally be $<15°$. If both R and δ can be obtained from the reflection spectrum measured under this condition, the $k(\tilde{v})$ spectrum corresponding to a absorbance spectrum would be calculated by using Equation (8.4b). Although $\delta(\tilde{v})$ cannot be directly obtained in a specular-reflection measurement, it is possible to calculate it from $R(\tilde{v})$ by using the Kramers–Kronig (KK) relations [4].

The KK relations are derived from mathematical analyses of response of a dielectric material to a dynamic external stimulus. The KK relations indicate that the real and imaginary parts of a complex variable like the complex refractive index are dependent on

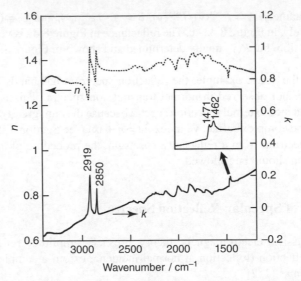

Figure 8.5 *The $n(\tilde{\nu})$ and $k(\tilde{\nu})$ spectra calculated from the single-reflection spectrum in Figure 8.4c. (- - -): $n(\tilde{\nu})$ and (—): $k(\tilde{\nu})$*

each other. A few formulations exist for the KK relations; the expression for deriving $\delta(\tilde{\nu})$ from $R(\tilde{\nu})$ is given below [4].

$$\delta(\tilde{\nu}) = -\frac{\tilde{\nu}}{\pi} \int_0^\infty \frac{\ln[R(\tilde{\nu}')/R(\tilde{\nu})]}{\tilde{\nu}'^2 - \tilde{\nu}^2} d\tilde{\nu}' \qquad (8.5)$$

This equation may be rewritten in the following form [4, 5], which has no singular point and is more convenient for practical calculations.

$$\delta(\tilde{\nu}) = -\frac{2}{\pi} \int_0^\infty \sin 2\pi c \tilde{\nu} t dt \int_0^\infty \ln [R(\tilde{\nu}')]^{\frac{1}{2}} \cos 2\pi c \tilde{\nu}' t d\tilde{\nu}' \qquad (8.6)$$

It should be noted that the integral range extends from zero to infinity in every KK formulation. This means that the observed reflection spectrum must also cover an infinitely wide wavenumber region. In practice, this is not attainable in the strictest sense. However, provided the absorption intensity is not extremely large and the observed reflection spectrum covers a wide enough wavenumber region, the above KK relations are still useful for calculating $\delta(\tilde{\nu})$ to a close approximation.

The above method for calculating $\delta(\tilde{\nu})$ was applied to the $R(\tilde{\nu})$ spectrum in Figure 8.4c. Then, the $n(\tilde{\nu})$ and $k(\tilde{\nu})$ spectra were calculated by the use of Equations (8.4a) and (8.4b), and the results obtained are shown in Figure 8.5. The $k(\tilde{\nu})$ spectrum has the shape of an absorption spectrum, from which band positions can be accurately determined. The bands observed at 2919 and 2850 cm^{-1} are, respectively, due to the CH$_2$ antisymmetric and symmetric stretching vibrations characteristic of the all-trans planar zigzag conformation of the n-alkane chain. The doublet bands at 1471 and 1462 cm^{-1} are assigned to the CH$_2$

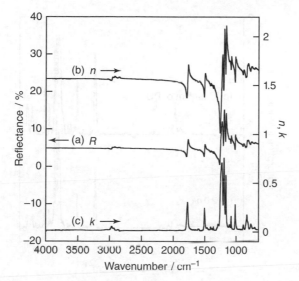

Figure 8.6 *(a) Reflection spectrum measured from the surface of a compact disk, (b) calculated n(ṽ) spectrum, and (c) calculated k(ṽ) spectrum.*

scissoring vibration, which is known to split into two bands due to the crystal-field effect in the orthorhombic subcell of polyethylene.

The above-mentioned bands are found at the same positions in the transmission spectrum of polyethylene, indicating that the recorded specular-reflection spectrum can indeed be converted, by using the KK relations, into a spectrum that is essentially the same as the absorbance spectrum derived from a transmission measurement. However, several bands observed in the region of 2000–1700 cm^{-1} in the $k(\tilde{\nu})$ spectrum of Figure 8.5 are not expected for polyethylene. A more careful examination is needed to clarify their origin.

On the other hand, the $n(\tilde{\nu})$ spectrum in Figure 8.5 is not as reliable from a quantitative viewpoint; a value of n smaller than 1.4 in a relatively flat region in Figure 8.5 is apparently too small, as the refractive indices of many organic compounds are known to be closer to 1.5 [6]. This discrepancy must be due to inaccurate integral calculation, which should be performed for a range from zero to infinity.

Another example of a reflection spectrum measured from the flat surface of polycarbonate (PC) which has the same shape as a compact disk (CD) is presented in Figure 8.6a. The results of analysis using the KK relations are also shown in the same figure. The calculated dispersion curve, the $n(\tilde{\nu})$ spectrum in Figure 8.6b, is apparently similar to the reflectance spectrum in Figure 8.6a, and the $k(\tilde{\nu})$ spectrum in Figure 8.6c has the familiar shape of an absorbance spectrum. Although the $k(\tilde{\nu})$ spectrum is often thought to be the final result, a further analysis may give a more useful spectrum, when a direct comparison with a transmittance spectrum is needed, for example, when the spectrum is used for a library search.

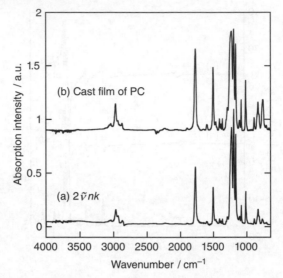

Figure 8.7   (a) Absorption spectrum calculated by using the $n(\tilde{\nu})$ and $k(\tilde{\nu})$ spectra in Figure 8.6 and (b) absorbance spectrum of a cast film of polycarbonate (PC) deposited onto a germanium substrate.

According to classical electrodynamics, the absorbance of a thin film determined by a transmission measurement is proportional to $\tilde{\nu}\mathrm{Im}[\hat{\varepsilon}_r(\tilde{\nu})] = 2\tilde{\nu}\, n(\tilde{\nu})\, k(\tilde{\nu})$, where $\hat{\varepsilon}_r(\tilde{\nu})$ is the relative permittivity, and the proportionality constant in this case has the dimension of length [1]. The spectrum of $2\tilde{\nu}n(\tilde{\nu})k(\tilde{\nu})$ calculated by using the values of $n(\tilde{\nu})$ and $k(\tilde{\nu})$ in Figure 8.6 is shown in Figure 8.7a. This spectrum resembles closely that of the absorbance spectrum of a cast film of PC on a germanium substrate, which is shown in Figure 8.7b. As this example shows, materials analysis using specular reflection and the KK relations is powerful, particularly when the analyte is strongly absorbing.

Generally speaking, it is possible to convert reflection spectra into the $n(\tilde{\nu})$ and $k(\tilde{\nu})$ spectra by using the KK relations. However, when integration involving a reflectance spectrum having weak signals over a wide wavenumber region is performed, the magnitude of $n(\tilde{\nu})$ and the intensity of $k(\tilde{\nu})$ calculated may not be reliable enough. In such a case, it is advisable to use only the $k(\tilde{\nu})$ spectrum for determining band positions, which are more reliable than intensities.

References

1. Tolstoy, V.P., Chernyshova, I.V. and Skryshevsky, V.A. (2003) *Handbook of Infrared Spectroscopy of Ultrathin Films*, John Wiley & Sons, Inc., Hoboken, NJ.
2. Hecht, E. (1990) *Optics*, 2nd edn, Addison-Wesley, Reading, PA, p. 95.
3. Toll, J.S. (1956) Causality and the dispersion relation: logical foundations. *Phys. Rev.*, **104**, 1760–1770.

4. Bardwell, J.A. and Dignam, M.J. (1985) Extensions of the Kramers–Kronig transformation that cover a wide range of practical spectroscopic applications. *J. Chem. Phys.*, **83**, 5468–5478.
5. Peterson, C.W. and Knight, B.W. (1973) Causality calculations in the time domain: An efficient alternative to the Kramers–Kronig method. *J. Opt. Soc. Am.*, **63**, 1238–1242.
6. Yamamoto, K. and Masui, A. (1995) Complex refractive index determination of bulk materials from infrared reflection spectra. *Appl. Spectrosc.*, **49**, 639–644.

9

External-Reflection Spectrometry for Thin Films and Surfaces

Takeshi Hasegawa
Institute for Chemical Research, Kyoto University, Japan

9.1 Introduction

If a material could be made extremely thin, for example, to the level of a single layer
of molecules, this thin layer would transmit almost all of the infrared radiation, so that
its infrared transmission spectrum could be measured. In fact, it is possible to measure a
mid-infrared transmission spectrum from a thin soap film. It is usually practically difficult,
however, to maintain such a thin film without it being supported by a substrate. For a thin
film supported on a substrate, its infrared spectrum is often obtained by utilizing a reflection geometry. Two reflection methods are available for measuring infrared spectra from
substrate-supported thin films, depending on the dielectric properties of the substrates used.
External-reflection (ER) spectrometry, which is the subject of this chapter, is a technique
for extracting useful information from thin films on dielectric (or nonmetallic) substrates,
while reflection–absorption (RA) spectrometry, described in Chapter 10, is effective for
thin films on metallic substrates [1]. In addition to these two reflection methods, attenuated
total-reflection (ATR) spectrometry, described in Chapter 13 and emission spectroscopy,
described in Chapter 15 may also be useful in some specific cases.

9.2 Characteristics of ER Spectrometry in Comparison with Transmission Measurement

In an infrared transmission measurement, the intensity of an absorption band is expressed
by Lambert–Beer's law (see Section 1.2.3). This law states that: (i) the intensity of an

Introduction to Experimental Infrared Spectroscopy: Fundamentals and Practical Methods,
First Edition. Edited by Mitsuo Tasumi and Akira Sakamoto.
© 2015 John Wiley & Sons, Ltd. Published 2015 by John Wiley & Sons, Ltd.

absorption band is proportional to the thickness of a sample and (ii) the intensity of a band of a solute in a solution sample is proportional to the concentration of the solute. The second statement of this law is useful for quantitative analysis of the concentration of a solute by a transmission measurement. Lambert–Beer's law ignores reflection of infrared radiation occurring at the surface of the liquid cell containing the sample, but this approximation does not cause a serious problem in the results of quantitative analysis.

It is important to point out that, when a thin film on a substrate is the target of analysis, the situation is different from the transmission measurement mentioned above, because reflection from the substrate surface has a great effect and cannot be ignored. In fact, it is difficult to obtain a reliable absorption spectrum of a thin film on a substrate by measuring its transmission spectrum and subtracting from it the transmission spectrum of the substrate alone. The absorption spectrum of the thin film is influenced by the optical property of the substrate. In such a situation, it is therefore necessary to resort to the measurement of reflection from the surface of a thin film by ER spectrometry and to analyze the result by a method based on electromagnetic principles.

An important difference exists between a solution and a thin film on a substrate as a sample for infrared measurement. In a solution, chemical species randomly change their positions and orientations, whereas molecules in a thin film on a substrate cannot move freely and have fixed orientations as a result of interactions between molecules and with the substrate. The molecular orientation is associated with a quantity which is called *permittivity* in electromagnetics or refractive index in optics. Both permittivity and refractive index are complex quantities that have components representing their anisotropic properties. It is necessary to consider these quantities in order to deal with the spectroscopic properties of an oriented thin film.

The theoretical treatment based on electromagnetics for analyzing reflection spectra from a thin film on a substrate has already been established [2, 3] and explaining it in detail is beyond the scope of this chapter, because it involves somewhat complicated mathematical formulations. In this chapter, emphasis is placed on the surface selection rule that is derived from theoretical considerations and is important for the practical use of ER spectrometry.

9.3   Infrared ER Spectra and the Definition of Reflection–Absorbance

Here, the target of measurement by infrared ER spectrometry is a thin film deposited on a flat surface of a dielectric material. Multilayered adsorbed species (adsorbates) may also be a target, but, if a quantitative analysis of their observed spectra is to be performed, each layer in the adsorbate film should have a flat and parallel structure. For this reason, the discussion in this chapter assumes that thin films like a Langmuir–Blodgett (LB) film [4], which has an ideally flat layer, are the target of study. A "dielectric" substrate is made of a nonmetallic material and includes semiconductors; the surface of water is also regarded as a dielectric substrate. Monolayers spread onto the surface of water are often analyzed by ER spectrometry.

Reflection at an interface between two phases having refractive indices n_1 and n_2 ($n_1 < n_2$) is categorized into two cases, depending on the optical geometry of the reflection measurement. The optical geometry of ER spectrometry is schematically illustrated in Figure 9.1a. The beam of incident infrared radiation, making an "angle of incidence"

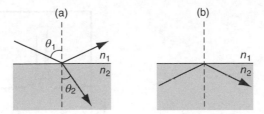

Figure 9.1 *Reflection at an interface ($n_1 < n_2$). (a) External reflection and (b) Internal reflection.*

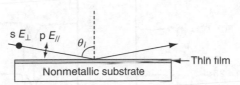

Figure 9.2 *Measurement geometry of polarized ER spectrometry for a thin film deposited on a nonmetallic substrate: s indicates s-polarization and p indicates p-polarization.*

(or "incident angle") θ_1 with the interface normal (indicated by the dashed line) passes through the phase having a lower refractive index n_1 and is reflected at the interface. The reflection occurring in this geometry is called *external reflection*. By contrast, the case shown in Figure 9.1b, in which the incident radiation passing through the phase having a higher refractive index n_2 is reflected at the interface, is called *internal reflection*.

In a discussion of ER at an interface, the polarization and the incident angle of radiation are usually defined as shown in Figure 9.2. The plane formed by the incident and reflected radiation beams is called the *plane of incidence* or *incident plane*. The polarization of radiation with an electric-field vector perpendicular to the incident plane is called *s-polarization* ("s" comes from the German word "senkrecht"), while that with an electric-field vector parallel to the plane is called *p-polarization*.

In Figure 9.3 [5], the infrared ER spectra of a nine-monolayer LB film of cadmium stearate deposited onto a GaAs wafer are shown. These spectra were obtained by a typical ER measurement, in which the infrared beam was incident at various angles through the air ($n_1 - 1.0$) onto the film ($n_2 \approx 1.5$) deposited on the GaAs wafer ($n_3 = 3.28$), and the reflected beam was directed to the detector. The spectra in Figure 9.3a,b were obtained, respectively, by s- and p-polarization measurements, and by changing the angle of incidence for each polarization measurement.

The ordinate represents a quantity called the *reflection–absorbance* denoted by A_R, which is defined as

$$A_R \equiv -\log_{10} \frac{R^{\text{sam}}}{R^{\text{BG}}} \qquad (9.1)$$

where R^{sam} and R^{BG} refer, respectively, to the "reflectance" from the thin film on the substrate and that from the substrate alone (BG, background) (see Equation (9.3) for the strict

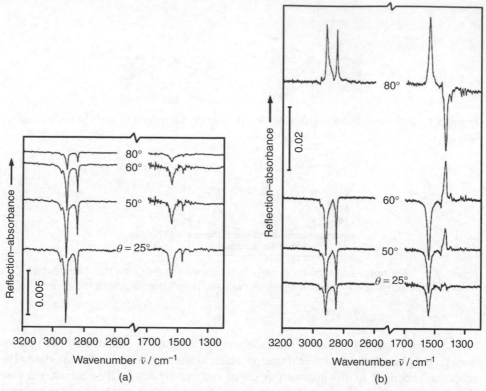

Figure 9.3 *Polarized infrared ER spectra of a nine-monolayer cadmium stearate LB film deposited on a GaAs wafer. (a) s-polarization and (b) p-polarization. (Source: Reproduced from Ref. [5] with permission from [5]. Copyright (1993) American Chemical Society).*

definition of the "reflectance"). The above definition of A_R is similar in form to the definition of the absorbance A for a transmission measurement, which is expressed as

$$A \equiv -\log_{10}\frac{I^{sam}}{I^{BG}} = -\log_{10}T \tag{9.2}$$

where I^{sam} refers to the infrared intensity measured with a sample in the sample compartment and I^{BG} is that without a sample (background intensity), and T is called the *transmittance*. A difference between A and A_R is that A is definable by directly using measured intensities, but A_R is defined via reflectance measurements, in which the infrared intensity I does not explicitly appear as given in Equation (9.1). The reflectance R is a dimensionless quantity defined as

$$R \equiv \frac{I_{refl}}{I_{inc}} \tag{9.3}$$

where I_{inc} and I_{refl} are the intensities of the incident and reflected radiation, respectively. However, as described later, the measurement of I_{inc} is not needed.

To measure the intensity I_{inc} incident on the sample position, any instrument accessory should be removed from the sample compartment and a single-beam spectrum should be measured as the background spectrum. On the other hand, to measure the reflection from a sample, a reflection attachment should be installed in the sample compartment. This difference in the measuring conditions means that the following two different instrumental functions should be taken into account: Φ for the background measurement and Θ for the measurement of reflection from the sample. Then, the measured single-beam spectra for the two cases are expressed as $I_{inc}\Phi$ and $I_{refl}\Theta$. As a result, the actually measured reflectance $I_{refl}\Theta/I_{inc}\Phi$ is not equal to R defined in Equation (9.3). This is characteristic of an ER measurement, which has aspects different from a transmission measurement.

When the light source and the detector are in good operating condition, it is reasonable to consider that $I_{inc}\Phi$ is unchanged during the measurements for the background and a sample, and can therefore be treated as a constant.

Measurements of ER spectra by using a reflection attachment should be made for the background (without any sample) and for a sample. The observed intensities for these two cases are denoted, respectively, by $I_{refl}^{BG}\Theta$ and $I_{refl}^{sam}\Theta$. Then, the ratio of the reflectances of the sample and the background is as follows.

$$\frac{R^{sam}}{R^{BG}} = \frac{I_{refl}^{sam}\Theta}{I_{inc}\Phi} \cdot \frac{I_{inc}\Phi}{I_{refl}^{BG}\Theta} = \frac{I_{refl}^{sam}}{I_{refl}^{BG}} \tag{9.4}$$

This equation shows that it is unnecessary to measure I_{inc} as $I_{inc}\Phi$ cancels out, and that neither of the instrumental functions Φ and Θ is required in deriving the right-hand side of this equation. Consequently, R^{sam}/R^{BG} is simply equal to the ratio of the reflected intensities $I_{refl}^{sam}/I_{refl}^{BG}$, which corresponds to the transmittance (usually expressed in percentage) of a transmission measurement.

As described above, the reflection–absorbance A_R is analogous to the absorbance A in transmission spectrometry. Nonetheless, A_R is a quantity characteristic of reflection spectrometry. The most significant characteristic of A_R is that it can be negative. In a transmission measurement, the transmittance always decreases after passing through the absorbing sample; that is, $T = I^{sam}/I^{BG} < 1$. By contrast, in ER measurements, the reflectance of the sample on the substrate may be greater than that of the substrate alone ($R^{sam}/R^{BG} > 1$). This increase in the reflectance results in a negative value of A_R.

It is noted in Figure 9.3 that all the s-polarization spectra have negative peaks, whereas the p-polarization spectra show both positive and negative peaks. It is only in the p-polarization measurements that A_R changes its sign at a specific angle of incidence; that is, as is observed in Figure 9.3, with the increase of θ from 60° to 80°, three bands in the region of 3000–1500 cm^{-1} change their A_R signs from negative to positive, while a band at around 1400 cm^{-1} changes its A_R sign from positive to negative. Such complicated characteristics of infrared ER spectra can be understood by analyses based on electromagnetic properties, which can be summarized as the "surface selection rule" to be described later in Section 9.5.

9.4   Reflectance on a Dielectric Material and Infrared ER Spectra

Infrared ER spectrometry depends on the reflection characteristics of differently polarized radiation at an interface. Figure 9.4 shows the calculated dependencies of the reflectances

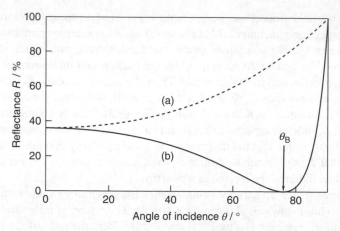

Figure 9.4 *(a,b) Dependencies of reflectances of polarized radiation on the angle of incidence at the interface between the air ($n_1 = 1.0$) and a Ge substrate ($n_1 = 4.0$). See text for details.*

of s- and p-polarized radiations on the angle of incidence θ at a two-phase interface (air/germanium). Let us consider first the case of $\theta = 90°$ (parallel incidence). In this case, the incident radiation advances without being reflected, giving the nominal reflectance of 100% for both s- and p-polarizations. On the other hand, in the case of $\theta = 0°$ (normal incidence), the reflectance should be <100%, but both polarizations should yield the same reflectance, as the two polarizations cannot be discriminated.

Between the two limiting angles, the shape of the reflectance curve greatly depends on the polarization. The reflectance for s-polarization rises monotonically with increasing θ until it becomes 100% at $\theta = 90°$. The reflectance for the p-polarization case decreases to zero, and then rapidly increases up to 100% at $\theta = 90°$. The angle at which the p-polarization exhibits no reflection is called the *Brewster angle* and is denoted by θ_B. As shown in Figure 9.3b, p-polarization spectra exhibit, at an angle somewhere between 60° and 80°, the inversion of the signs of A_R values for the observed bands, and it is most probable that this occurs at the Brewster angle. The results shown in Figure 9.3 indicate that a strong correlation seems to exist between the reflection at the two-phase interface (air/substrate) and the results of ER measurements for the three-phase sample (air/thin film/substrate). In other words, the substrate (background) has a significant influence on the ER spectra even after taking the ratio of R^{sam}/R^{BG}. This influence is clearly noticeable when the film is thin enough or comparable with the wavelength of the incident infrared radiation.

9.5   Surface Selection Rule of the Infrared ER Spectrometry

The following discussion of infrared ER spectrometry is carried out using a theoretical framework of the "three-phase system" consisting of air/film/substrate. The substrate should have a polished flat surface on the film side and a coarse surface on the other side. Otherwise, the system would have to be dealt with as a four-phase system consisting of air/film/substrate/air, which is totally different from a three-phase system.

As the substrate for the three-phase system, a semiconductor wafer is useful, as most semiconductor wafers in use have polished flat surfaces on one side only. When an infrared beam is incident on its polished flat surface, at least a part of the beam will pass through the substrate and be scattered at the coarse surface on the other side. In such a situation, the influence of the coarse surface can be ignored, and the system may be considered as consisting of three phases. Water can also be a suitable substrate for a three-phase system. A water trough used for ER measurements is usually deep enough for absorbing the infrared beam entering the water phase, and the reflection from the bottom of the trough container is negligible. As a result, it is sufficient to take into account only the reflections at the air/film/water interfaces.

Let us consider the reflection of an electromagnetic wave incident on an interface between two optically isotropic phases. When an electromagnetic wave is reflected at an interface, both the electric and magnetic vectors of the electromagnetic wave must be continuous across the interface. The boundary condition for the magnetic fields can be converted to another form using the electric fields by applying Maxwell's equations [6]. The amplitude ratio of the electric fields of the incident and reflected waves (or the amplitude reflection coefficient) denoted by r, is given for the s- and p-polarizations, respectively, as [7]

$$\hat{r}_s = \frac{\hat{n}_1 \cos \theta_1 - \hat{n}_2 \cos \theta_2}{\hat{n}_1 \cos \theta_1 + \hat{n}_2 \cos \theta_2} \tag{9.5a}$$

$$\hat{r}_p = \frac{\hat{n}_1 \cos \theta_2 - \hat{n}_2 \cos \theta_1}{\hat{n}_1 \cos \theta_2 + \hat{n}_2 \cos \theta_1} \tag{9.5b}$$

The hat (^) attached over r and n means that these are complex quantities, and $\hat{n}$ is the complex refractive index given as

$$\hat{n} = n + ik \tag{9.6}$$

where the real part n corresponds to the ordinary refractive index, and the imaginary part k is called the *absorption index* [8, 9], which is closely associated with the absorption of the radiation by the sample (see Section 1.2.4). If the infrared radiation is not absorbed by the sample over a wavenumber region, k is nonexistent (value 0) over that region and only the real part n has a value. The reason that the refractive index should be treated as a complex quantity is due to the fact that the electric dipole induced by the radiation has a phase shift with respect to the electric field oscillation of the radiation.

In Equations (9.5a) and (9.5b), θ_1 and θ_2 are, respectively, the angles of incidence and refraction measured from the surface normal, and Snell's law holds between them if phases 1 and 2 above and below the interface are both optically isotropic. As is well known, Snell's law is given as

$$n_1 \sin \theta_1 = n_2 \sin \theta_2 \tag{9.7}$$

The reflectance R is an observable quantity, which is related to the amplitude reflection coefficient $\hat{r}$ by the following equation

$$R = |\hat{r}|^2 \tag{9.8}$$

In the case of the reflection at the air/germanium interface, the reflectances for the s- and p-polarizations at the interface are calculated as functions of the angle of incidence by using the refractive indices of the air ($n_1 = 1.0$) and germanium ($n_2 = 4.0$). The results of calculations shown in Figure 9.4 can be regarded as R^{BG}.

To calculate the reflectance from a sample R^{sam} for simulating its ER spectrum, the following two conditions should be taken into account: (i) the sample consists of stratified layers with a thin film between infinitely thick outer layers and (ii) the sample has an anisotropic optical property. By considering the continuities of the electric and magnetic fields at the interface, condition (i) was theorized as the transfer-matrix method by Abelès [2, 3] and condition (ii) has been discussed by Berreman [10] and Yeh [2]. As the transfer-matrix method uses a matrix denoted by M, the method is also known as the *M-matrix method*. If the anisotropic permittivity and the thickness of the thin-film layer are known, it will be possible to calculate the ER spectrum of the sample exactly and use the calculated results for the discussion of molecular orientation in the thin film.

The permittivity denoted by ε is a quantity that relates the electric displacement D to the electric field E as follows:

$$D = \varepsilon E \tag{9.9}$$

The electric displacement is expressed in another form using the electric polarization P as

$$D = \varepsilon_0 E + P \tag{9.10}$$

where ε_0 is called the *electric constant* or the *permittivity of vacuum*.

As the electric polarization may have a direction different from that of the electric field applied to the material, and D and E are vector quantities, the permittivity must be a second-order tensor (matrix) expressed by ε. As a result, Equation (9.9) should be rewritten as

$$D = \varepsilon E \tag{9.11}$$

Determination of all the tensor elements in an arbitrary coordinate system is a difficult task. Therefore, the optic axis is usually assumed to be perpendicular to the surface of the thin film. This makes the complex permittivity tensor ε_2 (suffix 2 refers to the thin film) a diagonal matrix in a uniaxial approximation ($\varepsilon_x = \varepsilon_y \neq \varepsilon_z$). Then, ε_2 is given as

$$\varepsilon_2 = \begin{pmatrix} \varepsilon_x & 0 & 0 \\ 0 & \varepsilon_y & 0 \\ 0 & 0 & \varepsilon_z \end{pmatrix} \tag{9.12}$$

It is a common practice to use relative permittivity $\varepsilon_r \equiv \varepsilon / \varepsilon_0$ instead of ε, and to treat ε_r as a complex quantity $\hat{\varepsilon}_r = \varepsilon_r' + i\varepsilon_r''$. As $\hat{\varepsilon}_r = \hat{n}^2$, the following relations hold: $\varepsilon_r' = n^2 - k^2$ and $\varepsilon_r'' = 2nk$ [9].

Let us consider the three-phase system consisting of air/film/substrate shown in Figure 9.5. Phases 1, 2, and 3 (air, thin film, and substrate) are indicated by suffixes 1, 2, and 3, respectively. If the thickness of the thin film d_2 is much smaller than the wavelength λ of the incident infrared radiation ($d_2/\lambda \ll 1$), and the substrate has no absorption ($k_3 = 0$ and $\hat{n}_3 = n_3$), the exact formulas of the reflection–absorbances derived by the transfer-matrix method can lead to approximate forms by applying the Taylor expansions with respect to d_2/λ [6].

The above procedure, which is known as *Hansen's approximation* [1, 11], gives the reflection–absorbances for measurements by using incident radiations with the s- and

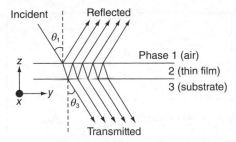

Figure 9.5 *Reflection and transmission of incident radiation at a three-phase system placed in rectangular coordinates* x, y, *and* z.

p-polarizations. The s-polarized radiation interacts with the thin film through absorption in the x-direction, by k_{2x}, while the p-polarized radiation interacts through absorptions in both the y- and z-directions, by k_{2y} and k_{2z}, respectively. The reflection–absorbances for these cases, denoted, respectively, by A_{Rsx}, A_{Rpy}, and A_{Rpz}, are given as

$$A_{Rsx} = -\frac{4}{\ln 10}\left(\frac{\cos\theta_1}{n_3^2 - 1}\right)n_2\alpha_{2x}d_2 \tag{9.13a}$$

$$A_{Rpy} = -\frac{4}{\ln 10}\left(\frac{\cos\theta_1}{\xi_3^2/n_3^4 - \cos^2\theta_1}\right)\left(-\frac{\xi_3^2}{n_3^4}\right)n_2\alpha_{2y}d_2 \tag{9.13b}$$

$$A_{Rpz} = -\frac{4}{\ln 10}\left(\frac{\cos\theta_1}{\xi_3^2/n_3^4 - \cos^2\theta_1}\right)\frac{\sin^2\theta_1}{(n_2^2 + k_2^2)^2}n_2\alpha_{2z}d_2 \tag{9.13c}$$

In the above equations, $\alpha_2 = 4\pi k_2/\lambda$ and $\xi_3 = n_3\cos\theta_3$.

If the transition dipole moment of a group vibration is oriented parallel to the thin-film surface, k_2 has no component in the z direction. Then, the result of the ER measurement for the s-polarization will correspond to A_{Rsx}, and that for the p-polarization will correspond to A_{Rpy}. If the transition dipole moment is oriented perpendicularly to the surface, k_2 only has a z component. In this case, the result of the ER measurement only for the p-polarization is meaningful and will be described by A_{Rpz}. For the s-polarization measurement, A_{Rsz} is always zero, regardless of the angle of incidence.

The dependencies of A_{Rsx}, A_{Rpy}, and A_{Rpz} on the angle of incidence θ_1 can be calculated by using Equations (9.13a), (9.13b), and (9.13c), respectively. The results of such calculations are shown in Figure 9.6 for a band at 2850 cm^{-1} (due to the CH$_2$ symmetric stretching vibration) of a nine-monolayer LB film of cadmium stearate deposited onto a GaAs wafer. In this example, the following parameters were used: $n_1 = 1.0$ (air), $n_2 = 1.5$ (the LB film), and $n_3 = 3.28$ (GaAs); $k_2 = 0.2$ (a value for the CH$_2$ symmetric stretching band of a bulk sample of cadmium stearate [12]); $d_2 = 22.5$ nm. Phases 1 and 3 were assumed to be semi-infinite in extent.

The calculated curves in Figure 9.6 show that the signs of A_{Rpy} and A_{Rpz} are reversed across the Brewster angle, while A_{Rsx} is always negative for any angle of incidence. In fact, the measured A_R values in Figure 9.3 agree with the calculated curves satisfactorily [13].

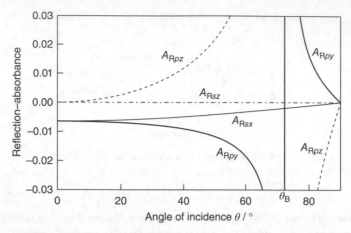

Figure 9.6 *Dependencies of reflection–absorbances on the angle of incidence calculated by Hansen's approximate equations (see text for details).*

The theoretical considerations and experimental findings described above can be summarized as the "surface selection rule" [1] in the following way:

1. In s-polarization ER spectra of a thin film (phase 2) of a three-phase system, the reflection–absorbance A_R is always negative for any band regardless of the angle of incidence. Only those vibrational modes having transition dipoles parallel to the surface (parallel modes) give rise to bands whose negative values of A_R depend on the angle of incidence, while the vibrational modes with transition dipoles perpendicular to the surface (perpendicular modes) cannot be observed.

2. In p-polarization ER spectra, the A_R sign of a band depends on the direction of the transition dipole of the vibrational mode giving rise to the band. If the angle of incidence is smaller than the Brewster angle, parallel and perpendicular modes give, respectively, negative and positive values of A_R. When the angle of incidence exceeds the Brewster angle, this relationship for parallel and perpendicular modes is reversed.

In s-polarization measurements, it is desirable to set the angle of incidence at about 20° to obtain a reflection spectrum with a reasonably high S/N ratio, because the absolute value of A_{Rsx} is larger for a smaller angle of incidence (Figure 9.6), and the value of R is moderately large for a small angle of incidence (Figure 9.4). In p-polarization measurements, on the other hand, the absolute values of both A_{Rpy} and A_{Rpz} are large when the angle of incidence is close to Brewster's angle (Figure 9.6). However, as $R = 0$ at the Brewster angle, p-polarization measurements performed near the Brewster angle should be avoided [14]. Practically it is difficult to determine the angle of incidence precisely, because the infrared beam incident on the sample for ER measurements is not completely collimated.

9.6   Molecular Orientation Analysis Using Infrared ER Spectra

The procedure for deriving molecular orientation present in a thin film by analyzing infrared ER spectra is outlined in this section.

For a sample like a liquid that is optically isotropic, the absorption index can be expressed by a single parameter k, which is denoted here as k_{bulk}. For a sample such as a thin solid film, which is usually optically anisotropic, deposited onto a substrate, it is necessary to take into account three absorption indices in the x, y, and z directions: k_x, k_y, and k_z, respectively. These are related with k_{bulk} as

$$\frac{k_x + k_y + k_z}{3} = k_{bulk} \tag{9.14}$$

If the thin film is a "uniaxially oriented system" and the transition dipole of a molecular vibration in the thin film is "perfectly parallel" to the film surface, $k_x = k_y$ and $k_z = 0$. Then, the following relation holds:

$$k_y = \frac{3}{2}k_{bulk} \tag{9.15a}$$

If the transition dipole is "perfectly perpendicular" to the film surface, $k_x = k_y = 0$, and the following relation holds:

$$k_z = 3k_{bulk} \tag{9.15b}$$

In this manner, k_y ($= k_x$) and k_z can be estimated by using k_{bulk}. They are related with the orientation angle ϕ of a long-chain molecule like cadmium stearate deposited on GaAs (i.e., ϕ is the angle between the molecular axis and the surface normal) in the following way:

$$k_y = \frac{3}{2}k_{bulk} \sin^2\phi \tag{9.16a}$$

$$k_z = 3k_{bulk} \cos^2\phi \tag{9.16b}$$

These values can be used for evaluating the A_R values in Equations (9.13a)–(9.13c). Particularly, A_{Rp} ($= A_{Rpy} + A_{Rpz}$) is useful for evaluating ϕ from the comparison of the calculated value and the observed infrared ER spectrum [12].

In Figure 9.7, the dependence of A_{Rp} on ϕ calculated by Equations (9.13b) and (9.13c) using k_y and k_z obtained by Equations (9.16a) and (9.15b) is shown, together with the value observed for the CH_2 symmetric stretching band at $2850\,cm^{-1}$ [11]. From the comparison of the calculated and observed results, the orientation angle of the transition dipole of the CH_2 symmetric stretching mode has been evaluated to be $83°$ ($\equiv \alpha$) from the surface normal. In the same way, the transition dipole of the CH_2 antisymmetric stretching mode at $2916\,cm^{-1}$ has been found to have an angle of $78°$ ($\equiv \beta$) from the surface normal. The hydrocarbon chain in cadmium stearate has an all-trans zigzag conformation, and the tilt angle of the chain axis denoted by γ is orthogonal to both α and β. Then, the following "direction cosine relationship" holds:

$$\cos^2\alpha + \cos^2\beta + \cos^2\gamma = 1 \tag{9.17}$$

By using this equation, γ has been calculated to be $14°$.

Hansen's thin-film approximation is not applicable to films thicker than 100 nm. For such thick films, an exact analysis based on the transfer-matrix method is needed. If the substrate absorbs infrared radiation ($k_3 \neq 0$), Equations (9.13a)–(9.13c) cannot be used for simulating ER spectra of thin films on such a substrate. This means that these equations are not useful for analyzing the RA spectra of thin films on metals, which are to be discussed in Chapter 10, as metals are infrared-absorbing materials.

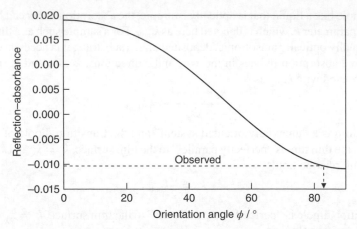

Figure 9.7 *Calculated dependence of reflection–absorbance on the orientation angle and comparison with the observed value.*

Details of application of the transfer-matrix method to anisotropic systems have been discussed in Refs [12, 14]. The approaches in these references may appear different from each other as they use different processes for solving Maxwell's equations, but both of them have obtained results which would help understand infrared ER spectra.

To estimate necessary k values, Equations (9.16a) and (9.16b) can be employed, but it is not possible to estimate the n values in a similar way. Refractive indices should be taken from appropriate literature, or should be determined by spectral simulations using permittivity functions [15].

Infrared ER spectrometry has no mechanism of intensity enhancement, in contrast to other methods such as RA spectrometry to be described in Chapter 10 and surface-enhanced infrared absorption (SEIRA), mentioned in Chapter 13. Nonetheless, infrared ER spectrometry provides a unique technique for utilizing s- and p-polarized radiations for obtaining information governed by the surface selection rule on the transition dipoles of molecular vibrations. Theoretical analysis of the information obtained by this technique has the possibility for elucidating molecular orientations in thin films on dielectric substrates and molecular interactions in a wide variety of materials, including liquid crystals.

References

1. Tolstoy, V.P., Chernyshova, I.V. and Skryshevsky, V.A. (2003) *Handbook of Infrared Spectroscopy of Ultrathin Films*, John Wiley & Sons, Inc., Hoboken, NJ.
2. Yeh, P. (1998) *Optical Waves in Layered Media*, John Wiley & Sons, Inc., Hoboken, NJ.
3. Born, M. and Wolf, E. (2009) *Principles of Optics*, 7th (expanded) edn, Cambridge University Press, Cambridge.
4. MacRitchie, F. (1990) *Chemistry at Interfaces*, Academic Press, San Diego, CA.
5. Hasegawa, T., Umemura, J. and Takenaka, T. (1993) Infrared external reflection study of molecular orientation in thin Langmuir–Blodgett films. *J. Phys. Chem.*, **97**, 9009–9012.

6. Itoh, Y., Kasuya, A. and Hasegawa, T. (2009) Analytical understanding of multiple-angle incidence resolution spectrometry based on classical electromagnetic theory. *J. Phys. Chem. A*, **113**, 7810–7817.

7. Hecht, E. (1990) *Optics*, 2nd edn, Addison-Wesley, Reading.

8. Bertie, J.E. (2002) Glossary of terms used in vibrational spectroscopy, In: *Handbook of Vibrational Spectroscopy*, Vol. **5** (eds J.M. Chalmers and P.R. Griffiths), John Wiley & Sons, Ltd, Chichester, pp. 3745–3794.

9. International Union of Pure and Applied Chemistry, Physical and Biophysical Chemistry Division (2007) *Quantities, Units and Symbols in Physical Chemistry*, 3rd edn, RSC Publishing, London.

10. Berreman, D.W. (1972) Optics in stratified and anisotropic media 4 × 4-matrix formation. *J. Opt. Soc. Am.*, **62**, 502–510.

11. Hansen, W.N. (1970) Reflection spectroscopy of adsorbed layers. *Symp. Faraday Soc.*, **4**, 27–35.

12. Hasegawa, T., Takeda, S., Kawaguchi, A. and Umemura, J. (1995) Quantitative analysis of uniaxial molecular orientation in Langmuir–Blodgett films by infrared reflection spectroscopy. *Langmuir*, **11**, 1236–1243.

13. Blaudez, D., Buffeteau, T., Desbat, B., Fournier, P., Ritcey, A.-M. and Pézolet, M. (1998) Infrared reflection-absorption spectroscopy of thin organic films on nonmetallic substrates: optical angle of incidence. *J. Phys. Chem. B*, **102**, 99–105.

14. Parikh, A.N. and Allara, D.L. (1992) Quantitative determination of molecular structure in multilayered thin films of biaxial and lower symmetry from photon spectroscopies. I. Reflection infrared vibrational spectroscopy. *J. Chem. Phys.*, **96**, 927–945.

15. Hasegawa, T., Nishijo, J., Umemura, J. and Theiss, W. (2001) Simultaneous evaluation of molecular-orientation and optical parameters in ultrathin films by oscillators-model simulation and infrared external reflection spectrometry. *J. Phys. Chem. B*, **105**, 11178–11185.

10

Reflection–Absorption Spectroscopy of Thin Layers on Metal Surfaces

Koji Masutani[1] and Shukichi Ochiai[2]
[1]*Micro Science, Inc., Japan*
[2]*S. T. Japan, Inc., Japan*

10.1   Introduction

In studies related to understanding the properties of adsorption, adhesion, corrosion, anti-corrosion, and so on, associated with thin films on metal surfaces, it is essential to know the chemical structure of the thin films, including those formed from materials such as paint and other thin films (of the order of 100 nm or less) adsorbed onto metal surfaces. Although mid-infrared spectroscopy might be expected to give useful information on such studies, the transmission measurement method is useless for such purposes because infrared radiation does not pass through metals. Therefore, one has to resort to using a reflection method. However, the specular-reflection method (see Chapter 8), in which the infrared beam is typically set to irradiate a sample at an angle of incidence close to the normal to the sample surface, is not useful because the reflection signal from the sample is usually far too weak to give a good-quality spectrum from which significant information can be extracted. However, in 1966 (before Fourier-transform infrared (FT-IR) spectrometers began to be widely used), Greenler [1] predicted, using theoretical calculations, that when an infrared beam is incident on a monolayer sample adsorbed onto a metal surface at a near-grazing angle and the beam reflected from it is examined, the radiation polarized parallel to the plane of incidence may be strongly absorbed by the monolayer sample. Since that time, this phenomenon has become well recognized and has been widely studied and utilized, both theoretically and experimentally, by many researchers, and it is now well established that this phenomenon really occurs [2, 3] and such measurements have

Introduction to Experimental Infrared Spectroscopy: Fundamentals and Practical Methods,
First Edition. Edited by Mitsuo Tasumi and Akira Sakamoto.
© 2015 John Wiley & Sons, Ltd. Published 2015 by John Wiley & Sons, Ltd.

proved to be analytically useful. This method of measuring the infrared absorption spectra of thin layers on metal surfaces by grazing-angle incidence of the infrared beam is called *reflection–absorption spectroscopy* (abbreviated as *RAS*); it is also sometimes referred to as *IRRAS* (*infrared reflection–absorption spectroscopy*) or *RAIRS* (*reflection–absorption infrared spectroscopy*).

10.2   Reflection–Absorption Spectroscopy

10.2.1   Principles

Let us consider the behavior of a linearly polarized beam of infrared radiation incident on a metal surface, as illustrated in Figure 10.1. The plane containing the normal N to the metal surface and the incident beam is called the *plane of incidence*, and the electromagnetic wave vibrating in this plane is called the *parallel polarized wave* ("p-wave") indicated by subscript p, while that vibrating perpendicularly to the same plane is called the *perpendicularly polarized wave* ("s-wave") indicated by the subscript s; the subscript "s" originates from a German word "senkrecht" (meaning perpendicular).

Reflection from a surface is generally described by the Fresnel equations, which are well-known in optics [4–6]. If the surface absorbs the incident radiation, the complex refractive index (see Section 1.2.4) of the surface material needs to be taken into account in these equations in order to evaluate the reflection from the surface. In this subsection, the results of such a treatment are summarized without going into mathematical details.

In Figure 10.1, the electric vectors of the incident and reflected rays on the metal surface are shown; as illustrated, the electric vector of the reflected ray for the s-wave always has a phase shift of about −180° from that of the incident ray (effectively creating a node at the metal surface), whereas the phase shift for the reflected p-wave depends on the angle of incidence as described below. The definition of the phase shift follows established practice [1, 4–6]. In Figure 10.2, the amplitude of the reflection coefficients (ρ_s, ρ_p) and phase shifts (τ_s, τ_p) calculated by the Fresnel equations for the radiation reflected from a metal surface are shown. The definition of the phase shifts in this figure is the same as those shown in Figure 10.1. For this calculation, the complex refractive-index values of gold ($n = 15.9$

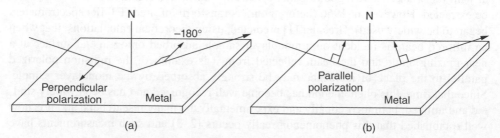

Figure 10.1 *Grazing-incidence reflection of an infrared ray on a metal surface. N, the normal to the metal surface. (a) Infrared ray polarized perpendicular to the plane of incidence (s-wave) showing a phase shift of −180° when reflected. (b) Infrared ray polarized parallel to the plane of incidence (p-wave). See text for details of the phase shift of the p-wave when reflected.*

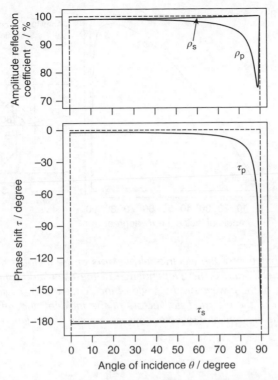

Figure 10.2 *Dependence of the calculated amplitude reflection coefficients (ρ_s and ρ_p) and phase shifts (τ_s and τ_p) on the angle of incidence θ.*

and $k = 53.4$) were used, and the wavelength of the incident radiation λ was assumed to be 10 µm.

In Figure 10.2, it can be seen that, for the case of parallel polarization, the phase shift of the reflected ray is close to zero unless the angle of incidence is very large. When this occurs, the electric-field vectors of the incident and reflected rays combine (sum) with each other at the metal surface to form an enhanced electric field in the direction perpendicular to the metal surface. The calculated amplitude of the strengthened electric field for parallel polarization is shown in Figure 10.3a. This calculation was made using the same conditions as for calculating Figure 10.2 and by using unity for the amplitude of the incident radiation and ρ_p given in Figure 10.2 for that of the reflected ray. The amplitude of this electric field is seen to increase with the angle of incidence θ, increasing from zero until it reaches its maximum when θ is about 80°, as the components of electric-field vectors perpendicular to the metal surface increase with θ. When θ exceeds 80°, due to the rapid decrease in the phase shift of the reflected ray τ_p toward $-180°$ (see Figure 10.2), the amplitude of the combined electric-field begins to rapidly decrease.

Figure 10.3b illustrates how the incident and reflected rays are combined at $\theta = 80°$ in the plane of incidence. At this angle of incidence, the reflected ray is delayed from the incident ray by a phase shift of 11.4°; that is, the incident ray vector occurs after its maximum

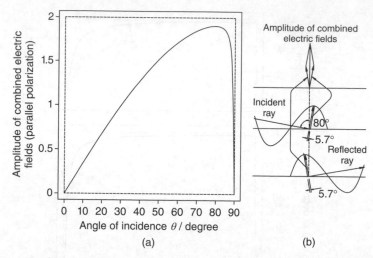

Figure 10.3 *Combination of the electric-field vectors of the incident and reflected rays of the p-wave on a metal surface. (a) Dependence of the calculated amplitude of the combined electric fields of the p-wave on the angle of incidence θ and (b) schematic illustration of the combination of the electric-field vectors of the incident and reflected rays, and an example showing the incident and reflected rays at an angle of incidence of 80° with a phase shift of 11.4°.*

amplitude by a phase of 5.7°(= 11.4°/2), and the reflected ray vector occurs before its maximum amplitude by a phase of 5.7°.

For the case of perpendicular polarization, on the other hand, the phase shift between the incident and reflected rays is always close to −180°, and their electric vectors cancel each other. As a result, little perpendicular electric field remains at the metal surface, and essentially a node exists at the surface.

When the infrared beam polarized parallel to the plane of incidence irradiates a thin absorbing layer adsorbed onto a metal surface, the thin layer interacts with the enhanced electric field at the metal surface and gives rise to infrared absorptions [1, 2]. Analysis of the process of absorption and reflection for such a system is considerably complicated [7, 8] and beyond the scope of this book, but the following simple equation has been derived [9] by making some assumptions on the results given in Refs [7, 8].

$$\frac{\Delta R}{R_0} = \frac{4n_1^3 \sin^2 \theta}{n_2^3 \cos \theta} \alpha d \tag{10.1}$$

where $\Delta R = R_0 - R_d$ (R_0 and R_d refer, respectively, to the intensity of reflection from the bare metal surface and that from a thin layer with thickness d adsorbed onto the metal surface); n_1 and n_2 are, respectively, the refractive index of air and that of the thin layer; θ is the angle of incidence; and α is the absorption coefficient of the thin layer.

Although Equation (10.1) is only a rough approximation, it has a significance in that it gives an insight into the characteristic of $\Delta R/R_0$ which has a similarity to the absorption observed in a transmission measurement. By denoting the transmittance of the thin layer with thickness d by I_d and that without the thin layer by I_0, the relation $\Delta I/I_0 = \alpha d$ ($\Delta I = I_0 - I_d$) is obtained by applying the Lambert–Beer law (see Section 1.2.3) to the thin film with $\alpha d \ll 1$. This means that, if the angle of incidence θ is close to 90°, $\Delta R/R_0$ obtained from the reflection measurement gives an absorption intensity larger than that obtained by the transmission measurement by a factor of $4n_1^3\sin^2\theta/n_2^3 \cos\theta$. $4\sin^2\theta$ in the numerator of this factor is associated with the intensity of the enhanced electric-field formed at the metal surface, and $1/\cos\theta$ expresses the dependence on the angle of incidence θ of the path length of the infrared beam in the thin layer. (In Equation (10.1), for $\theta = 90°$, $\Delta R/R_0$ increases to infinity, but this is of course an error arising from the approximate nature of this equation.) Although the angle of incidence giving rise to the maximum absolute value of $\Delta R/R_0$ depends on the metal and the wavelength of the infrared radiation, it is usually in a range of 85–89° (except for metals of low reflectance); it is higher for a metal of higher reflectance. In practice, the angle of incidence used is usually about 80° because of restrictions due to the sample size and the shape of the optical accessory used for the measurement. In the region where no absorption of the sample exists, $\Delta R/R_0 = 0$; the intensity of reflection does not change from that of the metal surface. Figure 10.4 shows the dependence of $(\Delta R/R_0)_\perp$ for the s-wave and $(\Delta R/R_0)_{//}$ for the p-wave on the angle of incidence calculated for the 1717 cm^{-1} band of a 1 nm-thick liquid film of acetone on gold [2]. In this figure, it is clear that $(\Delta R/R_0)_{//}$ depends greatly on the angle of incidence θ with its maximum occurring at $\theta = 88°$, whereas $(\Delta R/R_0)_\perp$ remains almost zero irrespective of the angle of incidence.

The characteristics of the reflection–absorption measurement may be summarized as follows:

1. Only the infrared radiation polarized parallel to the plane of incidence, the amplitude of which is increased by the combined electric-fields of the incident and reflected rays, interacts with the thin layer on the metal surface. The infrared radiation polarized perpendicular to the plane of incidence, the amplitude of which is almost canceled by the

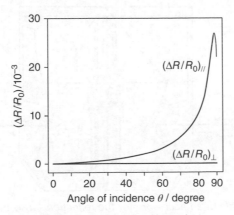

Figure 10.4 *Dependence of $\Delta R/R_0$ on the angle of incidence θ, calculated for the maximum-intensity band of 1 nm-thick liquid film of acetone on gold. (Source: Reproduced with permission from [2]. Copyright Elsevier (1985).)*

combined electric-fields of the incident and reflected rays, does not interact with the thin layer on the metal surface.

2. Among the vibrational modes of a molecule (or a part of it) adsorbed onto the metal surface, only those modes having transition dipole moments perpendicular to the metal surface can interact with the infrared ray polarized parallel to the plane of incidence, thereby giving rise to characteristic infrared absorption bands. The vibrational modes which have transition dipole moments parallel to the metal surface cannot be observed in the reflection–absorption infrared spectrum. These reflection–absorption phenomena are sometimes referred to as the *surface selection rule*.

3. When the angle of incidence of the infrared beam polarized parallel to the plane of incidence is close to 90°, absorption bands arising from the thin layer have their maximum intensities.

10.2.2   Method of Measurement

A polarizer and an appropriate accessory for reflection measurement are needed in order to measure a grazing-incidence reflection–absorption spectrum. Descriptions of a polarizer used in the mid-infrared region were given in Section 2.5.4. An example of an accessory for this type of reflection measurement is illustrated in Figure 10.5. The incident polarized infrared radiation is collimated by parabolic mirror 3 and focused by parabolic mirror 5 onto the sample S. The radiation reflected by the sample passes through an optical system symmetric to that for the incident radiation, and then reaches the detector (which is not shown in Figure 10.5). By sliding mount 6′ (on which mirrors 4 and 4′ are placed) on rail 6, it is possible to alter the angle of the collimated beam going to parabolic mirror 5, and thus be able to vary the angle of incidence to the sample S.

The intensity of a reflection–absorption spectrum is expressed by $\Delta R/R_0$ (called the *absorption factor* by Greenler [1]), or more frequently by the "reflection–absorbance"

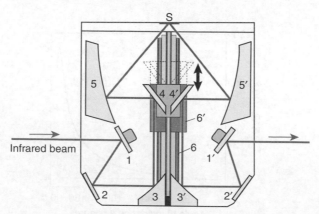

Figure 10.5   *An example of an accessory for the reflection–absorption measurement with a capability of varying the angle of incidence. S, sample; 1, 1′, 2, 2′, 4, and 4′, plane mirrors; 3, 3′, 5, and 5′, parabolic mirrors; 6 and 6′, a mount (on which mirrors 4 and 4′ are placed) and a slide rail. (Source: Adapted with permission from the catalog of Pike Technologies © 2013.)*

(sometimes simply called *absorbance*) expressed by $-\log(R_S/R_R)$ or $\log R_R/R_S$ where R_S and R_R refer, respectively, to the intensity of reflection measured from the sample and that from the reference material. The reflection–absorbance is analogous in form to the absorbance in a transmission measurement but it should be used for quantitative purposes with due care (see Section 10.2.3 below).

It is advisable to make reflection–absorption measurements at two or three different angles of incidence because the optimum angle of incidence varies with the type of substrate metal. As the reference material, the same metal as used for the substrate for the thin film is most appropriate. If this is not available, a stainless-steel plate or a plane mirror may be used but this may then cause some details of the measured sample spectrum to be obscured. It is very important to purge the inside of the spectrometer with dried air or nitrogen in order to remove water vapor and carbon dioxide as completely as possible, so that the measurements for the sample and reference material can be made under exactly the same conditions.

As the angle of incidence is usually as large as 80°, the result of such a reflection measurement may be affected by any slight curvature or unevenness of the surface of the substrate metal. This is an important consideration, especially when measuring a monolayer sample, for which the surface of the substrate metal must be sufficiently flat.

10.2.3 Examples of Measurement

A comparison of the spectra measured by the reflection–absorption method and that measured by the transmission method was made by using machine oil as a test sample; the results are shown in Figure 10.6. The sample oil was thinly spread onto a chromium-coated glass substrate without any particular measure to control the orientation of component molecules of the sample. The spectrum obtained by the reflection–absorption measurement shown in (a) of Figure 10.6 agrees satisfactorily with the absorbance spectrum shown

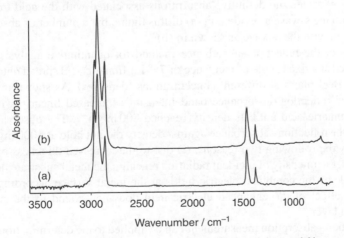

Figure 10.6 *(a) Reflection–absorbance spectrum of a thin machine-oil film on a chromium–coated glass substrate and (b) absorbance spectrum of the same sample measured from a thin film of the oil by the transmission method. Spectra have been offset for clarity.*

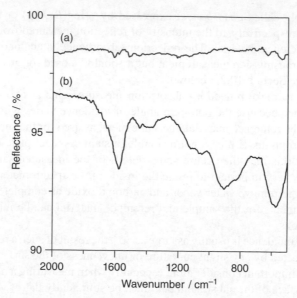

Figure 10.7   *Reflectance spectra of an iron plate processed by chromic acid. (a) Reflectance spectrum for perpendicular polarization and (b) reflectance spectrum for parallel polarization. Spectra have been offset for clarity.*

in (b) obtained by the conventional transmission measurement recorded from a thin film of the oil.

In a second example of RAS, reflection–absorption measurements were made from an iron plate treated with chromic acid by using either the s-wave or the p-wave, and the results are compared in Figure 10.7. In this case, an untreated iron plate was used as the reference material. As expected, no definitive absorptions associated with the acid treatment were observed with the s-wave as evident in (a) of this figure, but a number of absorption bands were observed with the p-wave as shown in (b).

Reliability of the reflection–absorbance method for quantitative analysis applications was examined at a fixed angle of incidence of 75° for fumaric acid spread onto the surfaces of polished steel plates at different concentrations (densities). As shown in Figure 10.8, the measured reflection–absorbance band intensities increased linearly with increasing density of fumaric acid until its density reached $800 \, mg \, m^{-2}$. The observation that this linearity of the reflection–absorbance against density did not hold at $1000 \, mg \, m^{-2}$ may be explained by the following two reasons: (i) due to the increased thickness of the layer of fumaric acid, the intensity of incident radiation reaching the metal substrate decreases, and the amplitude of the combined electric-fields at the metal surface accordingly decreases and (ii) the effect of the enhanced electric-field does not extend to the surface of the fumaric-acid layer.

The reflection–absorption measurement can be applied to the determination of the thickness of a fluorocarbon lubricant applied to the surface of a magnetic disk. The lubricant of a few nanometers to a few tens of nanometers in thickness is applied to the surface of a magnetic disk in order to reduce damage to the magnetic disk when it is touched by a magnetic head reader. (Today it is common to use carbon instead of fluorocarbon as a lubricant.)

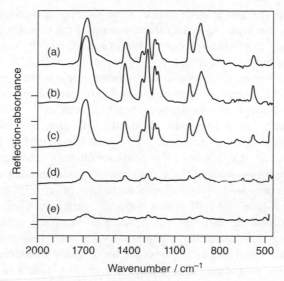

Figure 10.8 *Density dependence of the reflection–absorbance spectrum of fumaric acid on an iron substrate. Density of fumaric acid: (a) 1000 mg m⁻², (b) 800 mg m⁻², (c) 500 mg m⁻², (d) 50 mg m⁻², and (e) 30 mg m⁻². Spectra have been offset for clarity.*

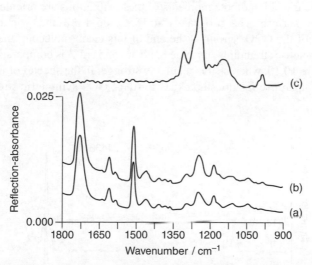

Figure 10.9 *Reflection–absorbance spectrum of fluorocarbon lubricant applied to the surface of a magnetic disk. (a) Reflection–absorbance spectrum of a magnetic disk without the fluorocarbon on it (b) reflection–absorption spectrum of the same disk with the fluorocarbon on it, and (c) a difference generated by subtracting spectrum (a) from spectrum (b). The ordinate scale in (c) has been expanded eight times.*

The three spectra shown in Figure 10.9 are (a) the reflection–absorption spectrum of a magnetic disk without fluorocarbon lubricant applied, (b) that with fluorocarbon lubricant, and (c) a difference spectrum between the spectra in (b) and (a). Spectrum (a) shows absorption bands due to polymer components present in the magnetic material on an aluminum base, and spectrum (b), which looks similar, must however also contain the contribution from the fluorocarbon layer. To extract the spectrum of the fluorocarbon from spectrum (b), it was necessary to subtract spectrum (a) from spectrum (b) by using the technique of difference spectroscopy (see Section 6.2.3). Spectrum (c) obtained in this way represents the spectrum of the fluorocarbon used as the lubricant, and it is possible then to determine the thickness of the fluorocarbon layer by measuring the intensity of one of the bands.

The reflection–absorption measurement approach can also be used to study the molecular orientation in a Langmuir–Blodgett (LB) film. Figure 10.10 shows the reflection–absorption spectra of one-, three-, and five-layer LB films of cadmium arachidate $[CH_3(CH_2)_{18}COO^-Cd^+]$ on a gold-coated glass substrate. For comparison, the absorbance spectrum obtained from a transmission measurement of a three-layer LB film of the same compound deposited onto both sides of an infrared transparent ZnSe plate is shown in Figure 10.11. The most noticeable differences between Figures 10.10 and 10.11 arise from the fact that the reflection–absorption spectrum detects only absorptions arising from molecular vibrations having their transition dipole moments perpendicular to the metal surface, whereas the transmission measurement does not have this restriction. The most apparent difference between the spectra in Figures 10.10 and 10.11 is seen in the CH stretching region; the intensities of both the CH_2 antisymmetric vibration band at $2919\,cm^{-1}$ and the CH_2 symmetric vibration band at $2851\,cm^{-1}$ are comparatively much weaker in the spectra of Figure 10.10, whereas the intensities of these bands are relatively very strong as shown in Figure 10.11. These results undoubtedly indicate that the transition dipole moments of these vibrations are oriented nearly parallel to the substrate surface. The two bands at 1555 and $1432\,cm^{-1}$ yield information on the orientation of the COO^- group at the end of this chain molecule; the intensity of the COO^- antisymmetric stretching vibration band at $1555\,cm^{-1}$ is comparatively weak in the spectra of Figure 10.10, whereas it is relatively stronger in the spectra of Figure 10.11, and an opposite comparative relationship exists for the COO^- symmetric stretching vibration

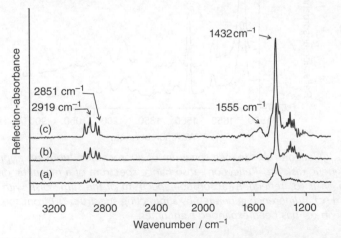

Figure 10.10 *Reflection–absorbance spectra of LB films of cadmium arachidate on a gold–coated glass. (a) Monolayer LB film, (b) bilayer LB film, and (c) three-layer LB film. (Source: Reproduced with permission from Koichi Kobayashi © 2013.)*

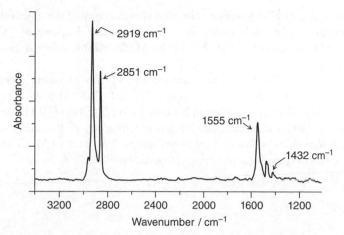

Figure 10.11 *Absorbance spectrum of the three-layer LB films of cadmium arachidate coated onto both sides of a ZnSe plate. (Source: Reproduced with permission from Koichi Kobayashi © 2013.)*

at $1432\,cm^{-1}$. These observations indicate that the COO^- group itself is oriented nearly perpendicular to the substrate surface. Consideration of all these results leads to the conclusion that the arachidate chain as a whole is oriented nearly perpendicular to the substrate with the COO^- group coordinating to Cd^+ being located near the surface of the substrate. A more comprehensive review on studies of the orientation of chain molecules in LB films using RAS may be found in Ref. [10].

In concluding this chapter, it should be emphasized that the reflection–absorption measurement, which is very different from both the conventional transmission and specular-reflection measurements, is the only useful method for measuring high-quality infrared spectra from thin samples ($<100\,nm$ thickness) adsorbed onto metal surfaces.

References

1. Greenler, R.G. (1966) Infrared study of adsorbed molecules on metal surfaces by reflection techniques. *J. Chem. Phys.*, **44**, 310–315.
2. Golden, W.G. (1985) Fourier transform infrared reflection-absorption spectroscopy, In: *Fourier Transform Infrared Spectroscopy*, Vol. **4** (eds J.R. Ferraro and L.J. Basile), Academic Press, New York, pp. 315–345.
3. Tolstoy, V.P., Chernyshova, I.V. and Skryshevsky, V.A. (2003) *Handbook of Infrared Spectroscopy of Ultrathin Films*, John Wiley and Sons, Inc., Hoboken, NJ.
4. Jenkins, F.A. and White, H.A. (1976) *Fundamentals of Optics*, 4th edn, McGraw-Hill, Auckland.
5. Hecht, E. (2002) *Optics*, 4th edn, Addison-Wesley, San Francisco, CA.
6. Born, M. and Wolf, E. (2006) *Principles of Optics*, 7th (expanded) edn, Cambridge University Press, Cambridge.
7. McIntyre, J.D.E. and Aspnes, D.E. (1971) Differential reflection spectroscopy of very thin surface films. *Surf. Sci.*, **24**, 417–434.

8. McIntyre, J.D.E. (1973) Specular reflection spectroscopy of the electrode-interphase, In: *Advances in Electrochemistry and Electrochemical Engineering*, Optical Techniques in Electrochemistry, Vol. **9** (ed R.H. Muller), John Wiley & Sons, Inc., New York, pp. 61–166.

9. Suëtaka, W. (1977) Infrared spectroscopy of investigating metal surfaces. *Bunkou Kenkyu, J. Spectrosc. Soc. Jpn.*, **26**, 251–265. This is a review written in Japanese;Suëtaka never published his thought on Equation (10.1) in English.

10. Umemura, J. (2002) Reflection-absorption spectroscopy of thin films on metal substrates, In: *Handbook of Vibrational Spectroscopy*, Vol. **2** (eds J.M. Chalmers and P.R. Griffiths), John Wiley & Sons, Ltd, Chichester, pp. 982–998.

11

Polarization-Modulation Spectrometry and its Application to Reflection–Absorption Measurements

Koji Masutani
Micro Science, Inc., Japan

11.1   Introduction

In this chapter, a sensitive method for measurement by continuous-scan Fourier-transform infrared (FT-IR) spectrometry called *polarization-modulation spectrometry* is introduced; this is a useful method for measuring, with high signal-to-noise ratio, not only the reflection–absorption spectra of thin films adsorbed onto metal substrates but also other spectra such as vibrational circular dichroism (VCD) spectra. Polarization-modulation spectrometry is a type of "double-modulation FT-IR spectrometry" [1]. In this chapter, descriptions of double-modulation spectrometry are given first, then polarization-modulation spectrometry is discussed, and then its application to the measurement of reflection–absorption spectra of thin films on metal substrates is discussed.

11.2   Principles of Double-Modulation Spectrometry

Double-modulation FT-IR spectrometry may be considered a type of difference-spectrum measurement (see Section 6.2.3), which utilizes the Connes advantage characteristic of FT-IR spectrometry (i.e., measured wavenumbers are accurate and highly reproducible; see Section 4.4.3); the infrared spectra of a sample containing a target material and a reference material are measured independently, and the spectrum of the target material is obtained by calculating the difference between the two spectra. This method is particularly

Introduction to Experimental Infrared Spectroscopy: Fundamentals and Practical Methods,
First Edition. Edited by Mitsuo Tasumi and Akira Sakamoto.
© 2015 John Wiley & Sons, Ltd. Published 2015 by John Wiley & Sons, Ltd.

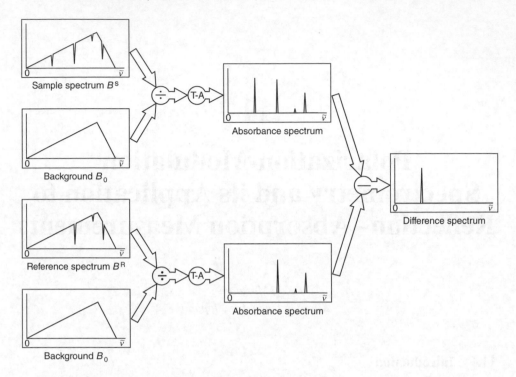

Figure 11.1   *Schematic illustration of the difference-spectrum measurement process.*

effective when the sample spectrum and the reference spectrum have only very small differences between each other. In practice, as illustrated schematically in Figure 11.1, the single-beam sample spectrum B^S and the single-beam reference spectrum B^R are measured independently of each other and then the ratio of each against a common background spectrum B_0 is determined, and their transmittance spectra thus obtained are then converted to their absorbance spectra (which is indicated by T-A in Figure 11.1). The difference spectrum between these two absorbance spectra is then calculated, as shown on the right-hand side of Figure 11.1.

Double-modulation FT-IR spectrometry is effective in cases for which the spectrum of a target sample differs only slightly from that of the reference sample. Unlike the more conventional absorbance difference-spectrum measurement approach, the single-beam sample spectrum B^S and the single-beam reference spectrum B^R are measured as their difference and sum, and their ratio $(B^R - B^S)/(B^R + B^S)$ is calculated, as schematically shown in Figure 11.2. As illustrated, the intensity of the difference spectrum is zero except for regions where absorptions, for example, of the target material, occur, while the intensity of the sum spectrum corresponds to an average of those of the spectra of both the target and reference samples. The final spectrum generated, $(B^R - B^S)/(B^R + B^S)$, shown at the right end in Figure 11.2, is similar to the difference spectrum shown in Figure 11.1, as discussed later.

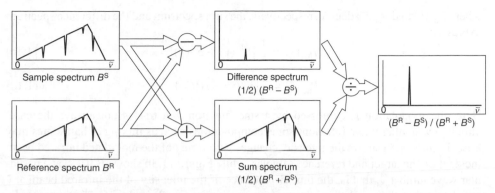

Figure 11.2 *Schematic illustration of the spectrum-measurement process involved in double-modulation spectrometry.*

In this method, an operation is required for alternating between measuring the target sample and the reference sample at a frequency sufficiently higher than the modulation frequency inherent in an interferogram measured by an FT-IR spectrometer. By this operation, the difference spectrum and the sum spectrum can be obtained, respectively, as the alternating-current component and the direct-current component of the modulated signal. These two components can be separated by an electronic filter, and their ratio can then be calculated.

As is already evident, the name *double-modulation spectrometry* comes from the feature of this method, that an external modulation is required to measure signals in addition to the original modulation by an interferometer. An example of an external modulation (perturbation) can be found in applying an alternating-current voltage to a material to examine the function of the material under the effect of an electric current; in this case, the material subjected to the alternating-current voltage is the target sample and the material without the applied voltage is the reference sample. Other examples carried out by this measurement method include studies of materials under external perturbations such as a magnetic field, a laser light, and polarized light (linear and circular). In the rest of the discussion in this chapter, linear polarization of infrared radiation is used for this measurement method in order to enhance the sensitivity of reflection–absorption spectrometry, as described in Chapter 10.

Let us consider a double-modulation FT-IR spectrometry experiment in which a target sample and a reference sample are effectively placed between an infrared source and an interferometer, and the infrared beam from the source irradiates the target and reference samples alternately with a modulation frequency of f_m. The single-beam spectra of the target and reference samples may be denoted, respectively, by $B^S(\tilde{\nu})$ and $B^R(\tilde{\nu})$. As the infrared beam which has passed through the target and reference samples has been modulated with frequency f_m, its spectrum $B_m(\tilde{\nu}, t)$, a function of time t, is expressed in the following form.

$$B_m(\tilde{\nu}, t) = B^S(\tilde{\nu}) \left[\frac{1}{2} \left(1 - \sin 2\pi f_m t \right) \right] + B^R(\tilde{\nu}) \left[\frac{1}{2} \left(1 + \sin 2\pi f_m t \right) \right] \qquad (11.1a)$$

$$B_m(\tilde{\nu}, t) = B_{dc}(\tilde{\nu}) + B_{ac}(\tilde{\nu}) \sin 2\pi f_m t \qquad (11.1b)$$

where $B_{dc}(\tilde{\nu})$ and $B_{ac}(\tilde{\nu})$ denote, respectively, the sum spectrum and the difference spectrum expressed as

$$B_{dc}(\tilde{\nu}) = \frac{1}{2}\left[B^{R}(\tilde{\nu}) + B^{S}(\tilde{\nu})\right] \tag{11.1c}$$

$$B_{ac}(\tilde{\nu}) = \frac{1}{2}[B^{R}(\tilde{\nu}) - B^{S}(\tilde{\nu})] \tag{11.1d}$$

Here, the modulation is expressed by a sine function, but this is not always the case, although with alternative functions some complications such as mixing of harmonics may arise. Figure 11.3a shows the intensity changes with time t of the modulated infrared beam incident on the target and reference samples, while Figure 11.3b shows $B_{m}(\tilde{\nu}, t)$ at a particular wavenumber $\tilde{\nu}$, that is, the time-dependence of the intensity of the infrared beam at $\tilde{\nu}$ which has passed through the target or reference sample. As B^{S} is close to B^{R} in magnitude, B_{dc} is also close to these, and B_{ac} is much smaller than B_{dc}.

The infrared beam modulated before entering the interferometer is again modulated by the interferometer with a modulation frequency proportional to wavenumber $\tilde{\nu}$ (see Section 4.3.1). As a consequence, the interferogram $F(x)$ as a function of the optical path difference (OPD) x is expressed as

$$F(x) = \int \left[B_{dc}(\tilde{\nu}) + B_{ac}(\tilde{\nu})\sin 2\pi f_{m}t\right]\cos 2\pi\tilde{\nu}x\,d\tilde{\nu}$$

$$= \int B_{dc}(\tilde{\nu})\cos 2\pi\tilde{\nu}x d\tilde{\nu} + \sin 2\pi f_{m}t \times \int B_{ac}(\tilde{\nu})\cos 2\pi\tilde{\nu}x\,d\tilde{\nu} \tag{11.2}$$

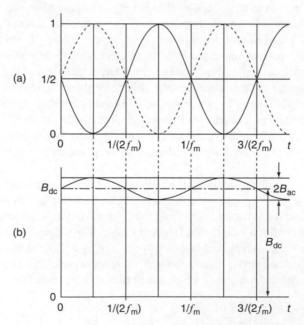

Figure 11.3 *(a) Intensities of the modulated infrared beam incident on the target sample (solid curve) and the reference sample (dashed curve) and (b) intensities of the modulated infrared beam at a particular wavenumber emerging from the target and reference samples.*

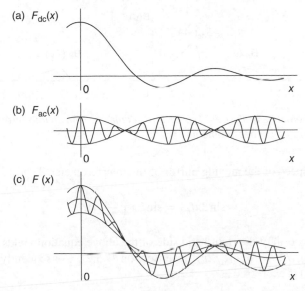

Figure 11.4 *Schematically illustrated interferograms obtained by double-modulation FT-IR spectrometry. (a) $F_{dc}(x)$: unmodulated part of an interferogram, (b) $F_{ac}(x)$: modulated part of an interferogram, and (c) F(x): interferogram of a double-modulation FT-IR measurement.*

In the above equation, $\sin 2\pi f_m t$, which is independent of wavenumber $\tilde{v}$, as shown, can be placed outside of the integral with respect to $\tilde{v}$. The first and second terms on the right-hand side of Equation (11.2) are called $F_{dc}(x)$ and $F_{ac}(x)$, respectively, so that $F(x) = F_{dc}(x) + F_{ac}(x)$. $F_{dc}(x)$ shown in Figure 11.4a is a familiar interferogram shape corresponding to spectrum $B_{dc}(\tilde{v})$. $F_{ac}(x)$ in Figure 11.4b shows that an interferogram corresponding to spectrum $B_{ac}(\tilde{v})$ forms an envelope, which is modulated by a higher-frequency f_m. The modulated infrared beam is then converted to electric signals by a detector, which is required to respond satisfactorily to the frequency range modulated by f_m.

As described above, the ratio $(B^R - B^S)/(B^R + B^S) = B_{ac}(\tilde{v})/B_{dc}(\tilde{v})$ needs to be obtained. For this purpose, $B_{dc}(\tilde{v})$ and $B_{ac}(\tilde{v})$ contained in $F(x)$ must be separately extracted by performing the Fourier-transform of $F(x)$. From the Fourier-transform of the first term of the right-hand side of Equation (11.2), $[B_{dc}(\tilde{v}) + B_{dc}(-\tilde{v})]$ is simply obtained; that is, $B_{dc}(-\tilde{v})$ in the negative wavenumber region is also obtained in addition to the required $B_{dc}(\tilde{v})$, because $B_{dc}(-\tilde{v})$ is symmetric to $B_{dc}(\tilde{v})$ about $\tilde{v} = 0$ (i.e., they are an even function), so that $B_{dc}(-\tilde{v})$ is not distinguished from $B_{dc}(\tilde{v})$ by the Fourier transform. The Fourier-transform of the second term of the right-hand side of Equation (11.2) with respect to x becomes a convolution of the Fourier-transform of $\sin 2\pi f_m t$ and that of $[B_{ac}(\tilde{v}) + B_{ac}(-\tilde{v})]$ (for a discussion of convolution, see Sections 4.4.1.1 and D.2). In practice, because, before performing the Fourier-transform of $F(x)$, $F(x)$ has been converted to a function of x by an analog-to-digital (AD) converter, $\sin 2\pi f_m t$ should be expressed as a function of x by using the relationship between t and x as

$$x = 2vt \qquad\qquad (11.3)$$

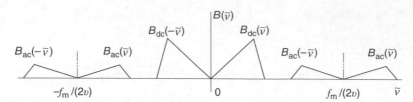

Figure 11.5 *Schematically illustrated spectrum calculated from an interferogram obtained by double-modulation FT-IR measurement.*

where vv is the speed of the moving mirror of the interferometer. Then,

$$\sin 2\pi f_{m} t = \sin 2\pi \left(\frac{f_{m}}{2v} \right) x \qquad (11.4)$$

The Fourier-transform of the right-hand side of the above equation yields a delta function expressed as $[\delta(f_{m}/2v) + \delta(-f_{m}/2v)]$ (see Section D.3.2). Consequently, spectrum $B(\tilde{v})$ contained in $F(x)$ is expressed as

$$B(\tilde{v}) = B_{dc}(\tilde{v}) + B_{dc}(-\tilde{v}) + [B_{ac}(\tilde{v}) + B_{ac}(-\tilde{v})] * \left[\delta \left(\frac{f_{m}}{2v} \right) + \delta \left(\frac{-f_{m}}{2v} \right) \right] \qquad (11.5)$$

where the symbol $*$ indicates a convolution.

The result expressed in Equation (11.5) is schematically illustrated in Figure 11.5, where the spectrum in the negative wavenumber region, which has only mathematical signifi-cance, does not actually exist as electric signals. As is clear in Figure 11.5, $B_{dc}(\tilde{v})$ is a spectrum over the same wavenumber region as measured by an FT-IR spectrometer, and $B_{ac}(\tilde{v})$ appears as a side band centering at a modulation wavenumber of $f_{m}/2v$. Therefore, in order to avoid overlapping of $B_{dc}(\tilde{v})$ with $B_{ac}(\tilde{v})$, modulation frequency f_{m} must be higher than twice the maximum frequency measured by the FT-IR spectrometer, which may be expressed as $2v\tilde{v}_{max}$ ($\tilde{v}_{max}$ is the maximum wavenumber). In practice, in consideration of the facts that both $B_{dc}(\tilde{v})$ and $B_{ac}(\tilde{v})$ cover wide wavenumber regions, and that an elec-tronic filter used for separating these two groups of spectra does not have as sharp a cutoff frequency as required, it is desirable to set f_{m} at a frequency sufficiently higher than the theoretically required value described above. (This situation may change if an optical filter having a sharp cutoff frequency could be used.)

As a result of the above arguments, it becomes necessary to follow the following proce-dures in order to obtain $B_{dc}(\tilde{v})$ and $B_{ac}(\tilde{v})$ separately.

1. To obtain $B_{dc}(\tilde{v})$, the higher-frequency component $B_{ac}(\tilde{v})$ of interferogram $F(x)$ is first removed by passing the signal $F(x)$ through a low-pass electronic filter, and then the Fourier-transform of the resultant lower-frequency component is performed.
2. To obtain $B_{ac}(\tilde{v})$, the signal $F(x)$, after having been passed through a high-pass electronic filter to remove the lower-frequency component $B_{dc}(\tilde{v})$, is then sent to a lock-in amplifier synchronized with a frequency of f_{m}, and then the Fourier-transform of the resultant component is performed. The lock-in amplifier is an instrument used to demodulate the modulated signals. It is desirable to use an appropriate band-pass filter for the frequency range centering at f_{m}, instead of a high-pass filter, by assuming that harmonics of f_{m} are

contained in $F(x)$. In polarization-modulation spectrometry for reflection–absorption measurements, a band-pass filter is used, as higher-frequency harmonics are always contained in $F(x)$. Another point to note is that the cutoff frequency of the internal low-pass filter in the lock-in amplifier to be used for the above-mentioned purpose must be higher than the entire frequency range of $B_{ac}(\tilde{v})$.

The obtained signals $B_{dc}(\tilde{v})$ and $B_{ac}(\tilde{v})$ are used to derive the difference between the absorbances of the target sample and the reference sample, which are denoted, respectively, by A^S and A^R. The following relations hold by definition of absorbance based on Beer's Law (see Sections 1.2.3 and 3.3)

$$B^S(\tilde{v}) = B_0(\tilde{v})10^{-A^S(\tilde{v})} \tag{11.6a}$$

$$B^R(\tilde{v}) = B_0(\tilde{v})10^{-A^R(\tilde{v})} \tag{11.6b}$$

where $B_0(\tilde{v})$ denotes the intensity spectrum of the infrared source. The following equation is obtained by combining Equations (11.1a)–(11.1d) and (11.6a)–(11.6b).

$$\frac{B_{ac}(\tilde{v})}{B_{dc}(\tilde{v})} = \frac{10^{-A^R(\tilde{v})} - 10^{-A^S(\tilde{v})}}{10^{-A^R(\tilde{v})} + 10^{-A^S(\tilde{v})}} \tag{11.7}$$

$$\frac{B_{ac}(\tilde{v})}{B_{dc}(\tilde{v})} = \tanh\left[\left(\frac{1}{2}\ln 10\right)\{\Delta A^{SR}(\tilde{v})\}\right] \tag{11.8}$$

where $\Delta A^{SR}(\tilde{v})$ is what is to be determined finally, which is defined as

$$\Delta A^{SR}(\tilde{v}) = A^S(\tilde{v}) - A^R(\tilde{v}) \tag{11.9}$$

If the value of $\Delta A^{SR}(\tilde{v})$ is sufficiently small, Equation (11.8) may be approximated by the following equation:

$$\frac{B_{ac}(\tilde{v})}{B_{dc}(\tilde{v})} \approx 1.15\Delta A^{SR}(\tilde{v}) \tag{11.10}$$

The error in using this approximate equation is about 1%, when $\Delta A^{SR}(\tilde{v}) = 0.15$.

The procedure for determining the difference between the absorption intensities of the target and reference samples is summarized as follows: (i) the target and reference samples are first modulated by appropriate means, (ii) then the modulated signals from the samples are separated into the alternating-current signals $B_{ac}(\tilde{v})$ and the direct-current signals $B_{dc}(\tilde{v})$ components, and (iii) finally the difference between the absorption spectra of the target and reference samples is obtained from the ratio of $B_{ac}(\tilde{v})/B_{dc}(\tilde{v})$.

11.3 Polarization-Modulation Spectrometry

In this section, polarization-modulation reflection–absorption spectrometry, a representative application of double-modulation spectrometry, is described. When polarization-modulation spectrometry is applied to reflection–absorption measurements for thin films on metal substrates, the polarization of the infrared beam incident on a thin film is alternately switched between being parallel and perpendicular to the plane of incidence, and differences in absorptions for parallel and perpendicular polarizations are measured. The advantages characteristic of this method may be summarized as follows:

1. As the target sample itself also acts as the reference sample, reference signals can be obtained in an ideal form without being affected by instrumental and environmental changes which may occur if the target sample had to be exchanged with a reference sample. As a result, in the case of measuring reflection–absorption spectra of thin films adsorbed onto metal substrates, it becomes possible to detect very small absorption differences associated with the orientation of molecules in adsorbed thin films.
2. As the spectrum to be measured is obtained as a difference between the spectrum of the target sample and that of the reference sample, absorptions due to water vapor and carbon dioxide present in the air are automatically eliminated from the measured spectrum. Even if the AD converter used in the process is a type having a low number of bits, it is unlikely that any significant reduction in the signal-to-noise ratio will occur as a consequence of an insufficient capability of the AD converter.

In this method, a polarizer and a photoelastic modulator (PEM) are used to switch alternately between two linearly polarized infrared beams, the planes of polarization of which are orthogonal to each other. A PEM used in the mid-infrared region consists of an infrared-transparent isotropic crystal (usually ZnSe) and a quartz oscillator. The latter is used to make the ZnSe crystal anisotropic by applying pressure to the ZnSe crystal in one direction (called here the *u axis*); the anisotropic crystal shows different refractive indices in different planes of polarization.

When performing polarization-modulation spectrometry, a beam of infrared radiation linearly polarized in the *y* axis at an angle of 45° to the *u* axis is incident on the PEM, as shown in Figure 11.6. This *y*-polarized radiation can be decomposed into two polarized

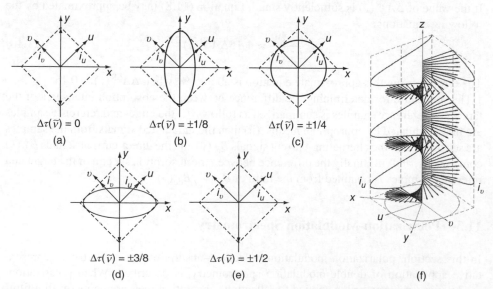

Figure 11.6 *Changes of the polarization of the wave synthesized by linearly polarized i_u and i_v when the phase shift $2\pi\Delta\tau(\tilde{v})$ between them is changed. (a) y-polarized linear polarization, (b) elliptical polarization, (c) circular polarization, (d) elliptical polarization, (e) x-polarized linear polarization, and (f) three-dimensionally depicted circular polarization corresponding to (c).*

radiation beams i_u and i_v which are in phase; the v axis is orthogonal to the u axis (see Figure 11.6). When pressure is applied to the PEM along the u axis, a phase shift of $2\pi\Delta\tau(\tilde{v})$ occurs between i_u and i_v [$\Delta\tau(\tilde{v})$ indicates that the phase shift depends on $\tilde{v}$]. As a result, the polarization of the beam of infrared radiation emerging from the PEM changes as shown in Figure 11.6, in which five cases (a–e) are shown; (a) shows the linearly y-polarized radiation without a phase shift between i_u and i_v, and (e) shows the linearly x-polarized radiation corresponding to a phase shift of π. The case (c) shows that a phase shift of $\pi/2$ gives rise to a circularly polarized radiation in which the amplitude of the combined wave of i_u and i_v rotates about a circle. The illustration in (f) shows three-dimensionally, a circularly polarized radiation in which the amplitude of the resultant wave is depicted by an arrow. A plane of polarization made of i_u and i_v is projected as a broken-line square-like shape at the bottom of (f) on which a phase shift of $\pi/2$ exists between i_u and i_v. The cases (b) and (d) show elliptically polarized radiation corresponding to the phase differences of $\pi/4$ and $3\pi/4$, respectively.

The intensities of the emerging beams of radiation of x- and y polarizations B_x and B_y are given as

$$B_x = B_0 \left[\frac{1}{2} \left\{ 1 - \cos 2\pi\Delta\tau\,(\tilde{v}) \right\} \right] \tag{11.11a}$$

$$B_y = B_0 \left[\frac{1}{2} \left\{ 1 + \cos 2\pi\Delta\tau\,(\tilde{v}) \right\} \right] \tag{11.11b}$$

where B_0 is the intensity of the incident infrared beam. $\Delta\tau(\tilde{v})$ is a function of the modulation frequency f_m expressed as

$$\Delta\tau(\tilde{v}) = \Delta\tau_0(\tilde{v}) \sin 2\pi f_m t \tag{11.12}$$

where $\Delta\tau_0(\tilde{v})$ is the maximum phase shift occurring when the highest possible pressure is applied to the PEM. The intensities B_x and B_y are modulated with the modulation of the phase shift expressed by Equation (11.12) as shown in Figure 11.7. The dashed lines (and curves) linking the figures in (a–c) in this figure are drawn to indicate the conversion of B_x and B_y from the functions of $\Delta\tau(\tilde{v})$ to those of time t. For example, t ① corresponds to t ②; the corresponding $\Delta\tau(\tilde{v})$ is at ③, and the corresponding B_x as a function of $\Delta\tau(\tilde{v})$ at ④ is converted to B_x at ⑤ as a function of t (at t ②). Thus, the curves of B_x and B_y as functions of time t in (c) have shapes that deviate from sinusoidal waves, and their frequencies are twice the applied modulation frequency f_m. To express B_x and B_y as functions of t, $\Delta\tau(\tilde{v})$ in Equation (11.12) is substituted in Equations (11.11a) and (11.11b). To make the results of this substitution more understandable, the following formula is used.

$$\cos(z \sin \phi) = J_0(z) + 2 \sum_{k=1} J_{2k}(z) \cos 2k\phi \tag{11.13}$$

Then, B_x and B_y are expressed as

$$B_x = \frac{1}{2} B_0 \left[1 - J_0 \left\{ 2\pi\Delta\tau_0\,(\tilde{v}) \right\} - 2 \sum_{k=1} J_{2k} \{2\pi\Delta\tau_0(\tilde{v})\} \times \cos 4\pi k f_m t \right] \tag{11.14a}$$

$$B_y = \frac{1}{2} B_0 \left[1 + J_0 \left\{ 2\pi\Delta\tau_0\,(\tilde{v}) \right\} + 2 \sum_{k=1} J_{2k} \{2\pi\Delta\tau_0(\tilde{v})\} \times \cos 4\pi k f_m t \right] \tag{11.14b}$$

where J_0 and J_{2k} are the zeroth and even-numbered Bessel functions.

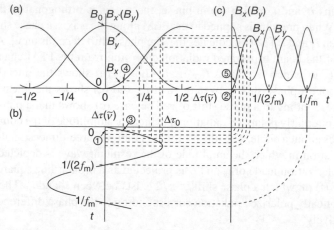

Figure 11.7 *Modulation of linearly polarized radiation emerging from the photoelastic modulator (PEM). (a) Changes of B_x and B_y with the phase shift between i_u and i_v, (b) sinusoidal time dependence of $\Delta\tau(\tilde{\nu})$, and (c) time dependence of B_x and B_y.*

In a similar manner to that developed with Equations (11.1a) and (11.1b) for the intensity of the beam of infrared radiation that has passed through the target and reference samples in double-modulation spectrometry, when polarization-modulation spectrometry is applied to a reflection–absorption measurement of a thin film adsorbed onto a metal substrate, the intensity of reflection $B_m(\tilde{\nu}, t)$ is expressed in the following form as a function of wavenumber $\tilde{\nu}$ and time t:

$$B_m(\tilde{\nu}, t) = B_{dc}(\tilde{\nu}) + B_{ac}(\tilde{\nu}) \cos 4\pi f_m t \tag{11.15a}$$

where $B_{dc}(\tilde{\nu})$ and $B_{ac}(\tilde{\nu})$ are given as

$$B_{dc}(\tilde{\nu}) = \frac{1}{2} B_0(\tilde{\nu})[10^{-A_\perp(\tilde{\nu})} + 10^{-A_{/\!/}(\tilde{\nu})}]$$

$$+ \frac{1}{2} B_0(\tilde{\nu}) J_0\{2\pi\Delta\tau_0(\tilde{\nu})\}[10^{-A_\perp(\tilde{\nu})} - 10^{-A_{/\!/}(\tilde{\nu})}] \tag{11.15b}$$

$$B_{ac}(\tilde{\nu}) = \frac{1}{2} B_0(\tilde{\nu})[2J_2\{2\pi\Delta\tau_0(\tilde{\nu})\}[10^{-A_\perp(\tilde{\nu})} - 10^{-A_{/\!/}(\tilde{\nu})}]] \tag{11.15c}$$

where $A_\perp(\tilde{\nu})$ and $A_{/\!/}(\tilde{\nu})$ are the absorbances of the thin film, respectively, for the incident beams of infrared radiation polarized perpendicular and parallel to the plane of incidence. In deriving the above equations, the fourth term of the Bessel function and higher are removed in consideration of the fact that the corresponding signals must be eliminated by the band-pass filter and the lock-in amplifier.

$B_{dc}(\tilde{\nu})$ and $B_{ac}(\tilde{\nu})$ can be determined by the same procedures as described for double-modulation spectrometry. However, as is clear from a comparison of Equations (11.1b) and (11.15a), polarization-modulation spectroscopy must use $2f_m$ as the synchronizing signal for the lock-in amplifier. By using $B_{dc}(\tilde{\nu})$ and $B_{ac}(\tilde{\nu})$ obtained as described above, the difference between $A_{/\!/}(\tilde{\nu})$ and $A_\perp(\tilde{\nu})$ is determined. The second term of $B_{dc}(\tilde{\nu})$ is negligibly small in comparison with the first term because the difference between $A_{//}(\tilde{v})$ and $A_{\perp}(\tilde{v})$ is very small. Then, the ratio $B_{ac}(\tilde{v})/B_{dc}(\tilde{v})$ can be simplified in the following way:

$$\frac{B_{ac}(\tilde{v})}{B_{dc}(\tilde{v})} = 2J_2\{2\pi\Delta\tau_0(\tilde{v})\}\tanh\left[\left(\frac{1}{2}\ln 10\right)\Delta A_d(\tilde{v})\right] \qquad (11.16)$$

$$\frac{B_{ac}(\tilde{v})}{B_{dc}(\tilde{v})} \approx 2J_2\{2\pi\Delta\tau_0(\tilde{v})\}(1.15)\Delta A_d(\tilde{v}) \qquad (11.17)$$

where $\Delta A_d(\tilde{v})$ is the difference spectrum to be determined, which is defined as

$$\Delta A_d(\tilde{v}) = A_{//}(\tilde{v}) - A_{\perp}(\tilde{v}) \qquad (11.18)$$

As described above, also in polarization-modulation spectrometry, $B_{dc}(\tilde{v})$ and $B_{ac}(\tilde{v})$ are first measured, and then their ratio $B_{ac}(\tilde{v})/B_{dc}(\tilde{v})$ is used to determine $\Delta A_d(\tilde{v})$. These procedures are schematically illustrated in Figure 11.8, in which all the data are depicted in spectral forms.

The coefficient $2J_2\{2\pi\Delta\tau_0(\tilde{v})\}$ on the right-hand side of Equation (11.17) indicates the efficiency of modulation by the PEM. As it has a maximum at $\Delta\tau_0(\tilde{v}) = 1/2$, the intensity of the difference spectrum also becomes a maximum in a wavenumber region in the vicinity of $\tilde{v}$, thereby satisfying this condition. As normally $\Delta\tau_0(\tilde{v})$ is proportional to wavenumber, only a single wavenumber shows a maximum efficiency for each pressure setting of the PEM. Accordingly, the pressure applied to the PEM should be varied with the wavenumber of interest. If the result of measurement is to be utilized for quantitative analysis, it is necessary to divide $B_{ac}(\tilde{v})/B_{dc}(\tilde{v})$ by this coefficient $2J_2\{2\pi\Delta\tau_0(\tilde{v})\}$ in order to create a normalized spectrum.

11.4   Measurement of the Reflection–Absorption Spectra of Thin Films on Metal Surfaces by Polarization-Modulation Spectrometry

In Figure 11.9, a schematic of the procedures for measuring a reflection–absorption spectrum from a thin film on a metal substrate by polarization-modulation spectrometry is illustrated as a block diagram. In this figure, measured signals are indicated as if all of them were spectra (i.e., represented by B), but, of course, each signal before performing a Fourier-transform is actually an interferogram.

The infrared beam emerging from the interferometer of an FT-IR spectrometer enters a polarizer, a PEM, and then an accessory for reflection–absorption measurement. The plane of polarization of the infrared beam after passing through the polarizer may be either parallel or perpendicular to the sample stage of the accessory for reflection–absorption measurement, and the pressure axis of the PEM is oriented at 45° to the plane of polarization of the entering infrared beam. When pressure is applied to the PEM at frequency f_m, the polarized infrared beam is modulated by the PEM, and its polarization is switched at frequency $2f_m$ between that parallel to the plane of incidence on the surface of the sample and that perpendicular to the same plane. As noted previously, it is necessary to set f_m greater than the maximum value of the modulation frequency $2v\tilde{v}$ produced by the interferometer. When a beam of polarized infrared radiation irradiates a thin film on a metal substrate

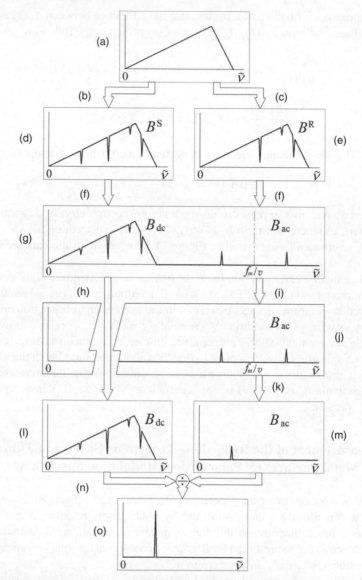

Figure 11.8 *Schematically depicted procedures of reflection–absorption measurement for a thin film on a metal substrate by polarization-modulation spectrometry. (a) Emission spectrum from an infrared source, (b) transmission of infrared beam of radiation polarized parallel to the plane of incidence, (c) transmission of infrared beam of radiation polarized perpendicular to the plane of incidence, (d) spectrum of the sample for infrared radiation of parallel polarization, (e) spectrum of the sample for infrared radiation of perpendicular polarization, (f) switching polarization by using a photoelastic modulator (PEM), (g) spectrum of the sample for polarization-modulated incident infrared beam, (h) signal from (g) after passing through a low-pass filter, (i) signal from (g) after passing through a high-pass filter, (j) output signal from the high-pass filter, (k) passing through a lock-in amplifier, (l) output signal from the low-pass filter, (m) output signal from the lock-in amplifier, (n) processing a division, and (o) absorbance spectrum of the sample finally obtained.*

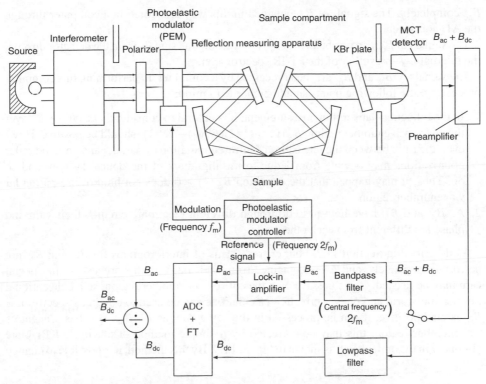

Figure 11.9 *Block diagram of a polarization-modulation FT-IR spectrometer set-up for measuring reflection–absorption spectra from thin films on metal substrates.*

positioned in the accessory in Figure 11.9, infrared radiation polarized parallel to the plane of incidence may be absorbed by the sample, whereas that polarized perpendicularly is not absorbed, as described in Section 11.3. As a result, spectrum B_{ac}, proportional to the difference in absorptions between parallel and perpendicular polarizations, is obtained as the modulated signal, and spectrum B_{dc} corresponding to the direct-current component represents an average of the reflection–absorption spectra for both parallel and perpendicular polarizations.

The modulated signals are converted to electric signals by the detector, which must be able to respond to the frequency range modulated by the frequency $2f_m$. For an electronic system to extract B_{ac}, a band-pass filter and a lock-in amplifier are used. It is necessary to use a band-pass filter which has the band-center frequency at $2f_m$ and has a band width greater than $4v\tilde{v}_{max}$ ($\tilde{v}_{max}$ is the maximum wavenumber to be measured) to cover the entire frequency range of B_{ac} without distorting signals. Only the signals of B_{ac} modulated with the frequency $2f_m$ emerge from this band-pass filter. The modulated signals of B_{ac} are then sent to the lock-in amplifier as the signal wave, and the synchronizing signal of the frequency $2f_m$ obtained from the controller of the PEM is used as the reference signal. The demodulated signals of B_{ac} are obtained from the lock-in amplifier. The lock-in amplifier used in this measurement should respond to a bandwidth covering the frequency $2f_m$, and the internal low-pass filter should have a bandwidth which can pass the output signals of

B_{ac} completely. The signals of B_{ac} obtained in this way are stored in a computer through the AD converter.

The signals of B_{dc} can be obtained simply by passing the signals from the detector through the internal low-pass filter of the FT-IR spectrometer.

The signals of B_{ac} and B_{dc} are then processed by a computer. In performing this computer processing, the following points should be borne in mind:

1. In the reflection–absorption measurement, it is natural to consider that as $B^R \geq B^S$ over the entire wavenumber region, $B_{ac}(\tilde{v}) = (1/2)[B^R(\tilde{v}) - B^S(\tilde{v})]$ should be positive. However, the ratio between the intensities of infrared radiation of parallel and perpendicular polarizations may deviate from 1 due to misalignment of the optical system, and so on. Thus, it may happen that the condition $B_{ac}(\tilde{v}) \geq 0$ does not hold over a particular wavenumber region.

2. As B_{ac} and B_{dc} have both passed through different electronic circuits, their gain and phase are different from each other.

As described in Section 11.2, Fourier-transforms of interferograms for B_{ac} and B_{dc} are performed separately, and ΔA_d is obtained from their ratio. In this process, some special care may be needed. B_{dc} is positive over the entire wavenumber region, so a conventional Fourier-transform and phase-correction calculations can be used to obtain its spectrum. If B_{ac} is also positive, it can be processed in the same way as for B_{dc}. If it has a negative wavenumber region, this may be corrected by a simple method of placing a KBr plate obliquely in the optical path as shown in Figure 11.9. By this method, it is possible to change

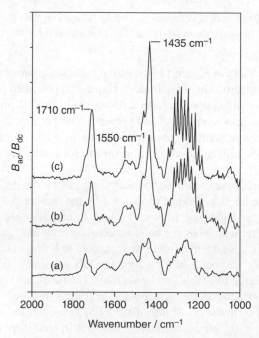

Figure 11.10   *Reflection–absorption spectra of three LB films of cadmium arachidate on silver-coated silicon disks. (a) Monolayer LB film, (b) five-layer LB film, and (c) nine-layer LB film. Maximum-efficiency wavenumber, 1250 cm⁻¹; the angle of incidence, 80°. Spectra have been offset for clarity. (Source: Reproduced with permission of T. Onishi © 2013.)*

the ratio between the intensities of the two differently-polarized beams passing through the KBr plate, so that B_{ac} can be either positive or negative over the entire wavenumber region of interest.

11.5   Application to a Reflection–Absorption Measurement

As an example, the samples used are the Langmuir–Blodgett (LB) films of cadmium arachidate adsorbed onto silver-coated silicon disks with a diameter of 20 mm. For the measurements, a PEM with a modulation frequency $2f_m = 148$ kHz was used; the PEM was set to give maximum efficiency at 1250 cm^{-1}; and the angle of incidence was $80°$.

In Figure 11.10, the reflection–absorption spectra (B_{ac}/B_{dc}) of three LB films of different number layers (one, five, and nine) are shown. The band observed at 1550 cm^{-1} is assigned to the antisymmetric stretching vibration of the COO$^-$ group coordinated to Cd$^+$, the band at 1435 cm^{-1} is assigned to the symmetric stretching vibration of the same group, and the series of bands (called *band progression*) in the 1300–1200 cm^{-1} are due to the CH$_2$ wagging vibrations. The band at 1710 cm^{-1} is considered to be due to the COOH group of residual arachidic acid not involved in the LB film forming process. It is noted that the spectrum (a) of the monolayer film seems to be considerably different from those of the five-layer film in (b) and nine-layer film in (c).

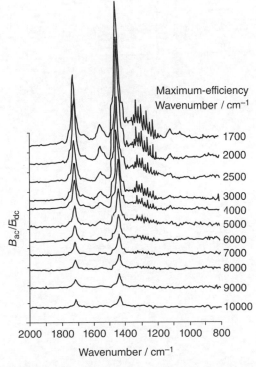

Figure 11.11 *Dependence of the spectral intensity on the maximum-efficiency wavenumber that shifts with the change of pressure applied to the PEM. Sample, nine-layer LB film on a silver-coated silicon disk. Spectra have been offset for clarity. (Source: Reproduced with permission of T. Onishi © 2013.)*

In Figure 11.11, changes of spectral intensities, occurring when the wavenumber at which maximum efficiency occurs is shifted by altering the pressure applied to the PEM, are shown. From these spectral data, it is clear that the PEM pressure setting has a great effect on measured spectra, and the wavenumber region in which bands are definitely observed is limited to the range from 1/2 to 2 times of the maximum-efficiency wavenumber; that is, when the maximum-efficiency wavenumber is $1700\,\mathrm{cm}^{-1}$, the spectrum in the wavenumber region of $3000\text{--}1000\,\mathrm{cm}^{-1}$ would be close to the true one. Thus, it is recommended to control the pressure applied to the PEM by taking the most interesting wavenumber region into consideration.

In concluding this chapter, it should be pointed out that the metal substrates used in reflection–absorption measurements must have optically flat surfaces.

Reference

1. Nafie, L.A. and Vidrine, D.W. (1982) Double modulation Fourier transform spectroscopy, In: *Fourier Transform Infrared Spectroscopy*, Vol. **3** (eds J.R. Ferraro and L.J. Basile), Academic Press, New York, pp. 83–123.

12

Diffuse-Reflection Measurements

Shukichi Ochiai
S. T. Japan, Inc., Japan

12.1   Introduction

A diffuse-reflection measurement is sometimes also called a *powder-reflection sampling method*. It is used for measuring infrared absorption spectra from powder or powdered samples. Sampling for this method is practically easier than making KBr disks or Nujol mull pastes necessary for transmission measurements as described in Section 2.4.3. This method is useful for not only powders but may also be used for materials with coarse (scattering) surfaces such as coatings on paper and glass fibers. The measurement technique may also be used to obtain important information on adsorbates on powders. As an accessory for use with an Fourier transform-infrared (FT-IR) spectrometer, cells for measuring diffuse reflection in vacuum at elevated temperatures are commercially available, and they are utilized as a tool that can perform *in situ* measurements for studying mechanisms of adsorption, desorption, and catalytic reactions in general.

12.2   Diffuse-Reflection Measurements

12.2.1   Principles of Diffuse-Reflection Measurements

When a powder sample is irradiated by an infrared beam, the reflection of radiation occurs over a wide solid angle. What happens is illustrated schematically in Figure 12.1. A proportion of the incident radiation I undergoes specular reflection on the surfaces of the powder particles. As it is impossible to define any specific incidence angle for the radiation at the

Introduction to Experimental Infrared Spectroscopy: Fundamentals and Practical Methods,
First Edition. Edited by Mitsuo Tasumi and Akira Sakamoto.
© 2015 John Wiley & Sons, Ltd. Published 2015 by John Wiley & Sons, Ltd.

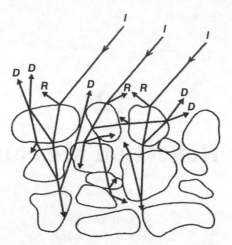

Figure 12.1 *Schematic illustration of diffuse reflection in a powder layer. **I**, incident radiation; **R**, specular reflection; and **D**, diffuse reflection.*

powder surface, the specularly reflected radiation **R** will also be directed over a wide range of directions. The remaining radiation enters by refraction into the insides of powders, and a part of it is transmitted through them after some of it has been absorbed and some has also been internally reflected at the powder surfaces. Such processes are considered to occur randomly within and between powder particles, and, as a whole, diffuse reflection **D** from the powder sample is considered to occur.

As a portion of the diffusely reflected radiation must have been transmitted through the inside of a powder particle at least once, infrared radiation at the wavenumber positions characteristic of the molecular vibrations of the sample has been absorbed; that is, a diffuse reflection spectrum of a sample contains similar information to the infrared absorption spectrum of the sample. Furthermore, as some of the diffuse-reflection radiation must have passed through powder particles more than once, at a wavenumber position where a weak absorption of the sample occurs, the infrared radiation is absorbed multiple times, thereby giving rise to an absorption band seemingly stronger than its intrinsic intensity, as would have been observed with an absorption spectrum recorded in a transmission measurement. Consequently, there are cases where the intensity of a diffuse-reflection band is not proportional to the concentration of the sample giving rise to the band.

To express a diffuse-reflection spectrum, a quantity called the *Kubelka–Munk function* $(K-M)$ $f(R_\infty)$ defined in the following way was introduced in a paper published in 1931 [1–3].

$$f(R_\infty) = \frac{(1 - R_\infty)^2}{2R_\infty} = \frac{K}{S} \qquad (12.1)$$

R_∞ is the absolute reflectance from the incident infrared beam from a sample thick enough such that there is no further change in the measured diffuse reflection with increasing sample depth; K is equivalent to the absorption coefficient α in Equation (1.3) and S is the scattering coefficient relating to the magnitude and shape of the powder particle and the density of packed powder.

In practical measurements, however, it is not possible to measure R_∞. Instead of R_∞, the relative diffuse-reflection coefficient r_∞ is defined as a ratio between the diffuse reflectance from the sample r_∞(sample) and that from a reference r_∞(reference) as given in Equation (12.2). In the mid-infrared region, a fine, dried powder of KCl or KBr is usually used as the reference for which K may be regarded as essentially zero (i.e., $R_\infty \approx 1$).

$$r_\infty = \frac{r_\infty(\text{sample})}{r_\infty(\text{reference})} \qquad (12.2)$$

By substituting R_∞ in Equation (12.1) with r_∞ in Equation (12.2), the K–M function $f(r_\infty)$ actually used for the ordinate of a diffuse-reflection spectrum is given as

$$f(r_\infty) = \frac{(1 - r_\infty)^2}{2r_\infty} \qquad (12.3)$$

For the reference (KCl or KBr), $f(r_\infty)$ is regarded as zero. This function is sometimes expressed as "K–M." K–M is a quantity corresponding to the absorbance A determined from a transmission measurement.

In Figure 12.2, a diffuse-reflection spectra recorded from anthracene is shown in two representations; spectrum (a) gives the K–M function, and spectrum (b) shows the "reflection–absorbance" for the relative diffuse reflectance r_∞, that is, $-\log r_\infty$. By comparing the two spectra, it is clear that the intensities of relatively weak absorption bands are enhanced in spectrum (b).

The K–M function contains uncertain factors which strongly depend on the properties of the sample. The primary factor that discriminates the K–M spectrum from its corresponding transmission spectrum is the presence of specular reflection from the front surface of the sample powder. The specular reflection from the surface of a powder particle is expected to be reflected in all directions, so it is difficult to discriminate it from diffuse reflection

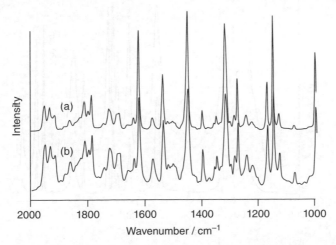

Figure 12.2 Diffuse-reflection spectra of anthracene. (a) K–M spectrum and (b) reflection absorbance spectrum.

which contains the desired information on absorption. The contribution of specular reflection greatly depends on the size of the powder particle and the magnitude of the absorption coefficient. The effect of specular reflection is greater for a powder sample having bands with large absorption coefficients because infrared radiation penetrates only slightly into particles of such a powder and the specular reflection tends to be a large component of the signal detected. If the powder particles are made smaller in size, the ratio of the specular reflection is expected to decrease but will, however, still not be negligible. The other factor that is an important consideration is the scattering coefficient of the sample, which depends on the size of sample powder particles, their shapes, and the condition of packing of the powder.

Diffuse-reflection spectra were measured for fumaric acid powders diluted with KBr at various concentrations, and the concentration dependencies of the intensities of the bands at 720 and 1321 cm^{-1} followed. The diffuse-reflection spectra of fumaric acid of 0.5% and 6.9% by weight in KBr are shown in Figure 12.3. The concentration dependencies of the K–M intensities of the 720 and 1321 cm^{-1} bands for a series of concentrations are plotted in Figure 12.4, which shows the K–M intensities for both these bands are linearly proportional to the concentration up to 10% by weight.

12.2.2   Apparatus for Diffuse-Reflection Measurements

As diffuse reflection spreads out over a large solid angle from the surface of a powder sample, an optical system is required to collect reflection radiation as efficiently as possible over a large solid angle. An example of an accessory for diffuse-reflection measurements is schematically illustrated in Figure 12.5. Two types of cups for diffuse-reflection measurements are commonly commercially available; the diameter of the larger (ordinary) one

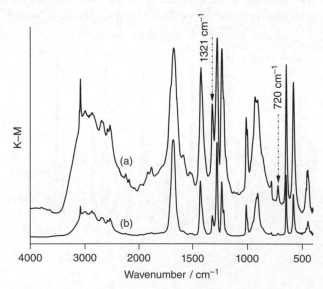

Figure 12.3   *Diffuse-reflection spectra in the K–M format of fumaric acid at two different concentrations. (a) 6.9% and (b) 0.5% by weight.*

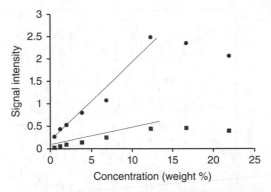

Figure 12.4 *Concentration dependencies of the K–M intensities of the bands at 720 cm^{-1} (•) and 1321 cm^{-1} (■).*

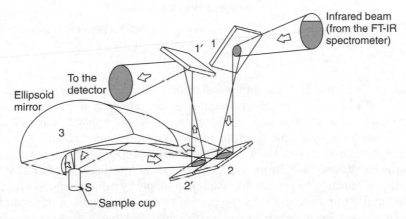

Figure 12.5 *Schematic illustration of an apparatus for measuring diffuse-reflection. S, sample to be measured; 1, 1', 2, 2', plane mirrors; 3, ellipsoid mirror. (Source: adapted with permission from the catalog of Pike Technologies © 2013.)*

is about 10 mm and that of the smaller (micro) one is several millimeters, and these cups can contain powder samples up to a few millimeters depth. For low-concentration samples, it is advisable to use a thick sample layer, as described later.

Diffuse-reflection measurements are often used to study *in situ* reactions on catalyst surfaces. Apparatus for such studies are also commercially available, which make it possible to elevate the sample temperature in vacuum or in a stream of an appropriate gas.

12.2.3 Methods of Measurements

12.2.3.1 Methods of Sample Preparation

Organic compounds for diffuse-reflection measurements are usually mixed with the powders of either KCl or KBr. Measurements of diffuse reflection from powders of neat compounds alone are strongly influenced by specular reflection, so high-quality spectra cannot

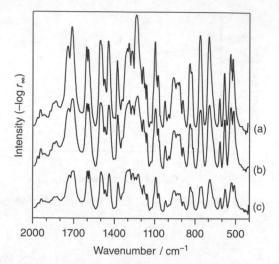

Figure 12.6   *Diffuse-reflection spectra of phenoxyacetic acid at different concentrations. (a) 5%, (b) 45%, and (c) 100% by weight. Spectra have been offset for clarity.*

be obtained. In Figure 12.6, the diffuse-reflection spectra of phenoxyacetic acid at 5%, 45%, and 100% by weight are shown. Spectral contrast is highest in spectrum (a) obtained from the lowest concentration sample containing 5%. As this example shows, it is advisable to prepare ordinary samples for diffuse-reflection measurements at dilutions of 1–5% by weight. A mixture of powders of a sample and a diluent (KCl or KBr) may be prepared in an agate mortar in essentially the same way as used for making a KBr disk (see Section 2.4.3). The mixture may be put into the sample cup simply by using a spatula. It is advisable to make the size of powder particles comparable to the wavelengths of infrared radiation used for the measurement. When the spectra of similar compounds are to be measured and compared, it becomes particularly necessary to minimize variations in the size of sample particles by using sieved samples and to put nearly equal amounts of the sieved samples into the sample cup. If the sample compound has a possibility of deteriorating during grinding, it is better to prepare powders by chopping the compound and mixing it with the diluent powders. In this case, it is better to make the diluent powders as fine as possible, so that they fill up the space between powder particles of the target compound.

Although alkali halides are generally used as diluents, the powders of diamond and silicon are also used when the target compound has a strong interaction with alkali halides, particularly when the purpose of measurement is to study the behavior of the OH group. However, measurements over the wavenumber regions in which the powders of diamond and silicon have absorptions should be avoided or made with due care.

12.2.3.2   *Selection of a Sample Cup and the Sample-Layer Thickness*

For measurements of samples with 1–5% by weight, an ordinary cup with a sample layer of about 1 mm will suffice, but for measuring more dilute samples, the use of a 2–3 mm-thick sample layer will be more appropriate. In Figure 12.7, the diffuse-reflection spectra of a powder sample of β-naphthylmethylether (0.047% by weight diluted in KBr) measured

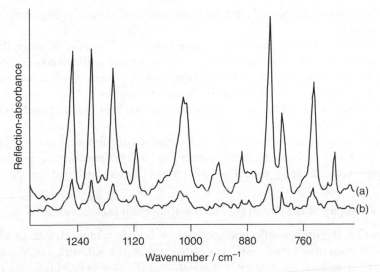

Figure 12.7 *Diffuse-reflection spectra of β-naphthylmethylether measured by using two different sample layers (and cups). (a) 2 mm-thick sample layer in a micro cup and (b) 1 mm-thick sample layer in an ordinary cup. Spectra have been offset for clarity.*

with a 2 mm-thick sample layer in a micro cup and a 1 mm-thick sample layer in an ordinary cup are shown. This example indicates that, when the concentration of a sample is low, the diffusion of infrared radiation extends further and deeper into the sample. If the quantity of a sample is small, the use of a micro cup usually gives a spectrum of a higher quality, particularly by increasing the number of spectra accumulations in order to obtain a higher S/N ratio.

12.2.3.3 Procedures of Spectral Measurements

In a diffuse-reflection measurement, the ratio r_∞ between the reflection intensity of the sample and that of a neat reference material (usually KCl, KBr, etc.), should be determined. This is performed by the following procedure:

1. The diffuse-reflection spectrum of a reference material (r_∞ reference) is measured.
2. The diffuse-reflection spectrum of the diluted sample (r_∞ sample) is measured.
3. The ratio $r_\infty = (r_\infty$ sample)/(r_∞ reference) is determined, and the K–M function is calculated from this value. For the sake of convenience, r_∞ itself or the reflection–absorbance ($-\log r_\infty$) is often used for the ordinate of a diffuse-reflection spectrum. Diffuse-reflection spectra in the mid-IR region are mostly plotted with the K–M function, whereas near-IR spectra often use the ($-\log r_\infty$).

12.2.3.4 Points to Note in Measuring a Diffuse-Reflection Spectrum

1. As the diffuse-reflection measurement is a sensitive method, care should be taken over impurities which may exist in a diluent or a reference material, particularly when the quantity of the target compound is small. In a series of similar measurements, a diluent

or a reference material should have the same origin and should be prepared as a powder at the same time.

2. As alkali halides like KCl and KBr often contain a small amount of water, they should be dried and kept in a desiccator before use; the powder of diamond or silicon should be used as a diluent if the purpose of diffuse-reflection measurement is to specifically obtain information on the OH group in the target compound.

3. Sample powders should be carefully prepared to make the size of powder particles as uniform as possible.

4. Sample powders should always be put in a sample cup up to the same level. The surface of sample powders should be lightly compressed and flattened with a spatula.

5. The water vapor in the air should be fully purged from the inside of the spectrometer.

12.2.4   Examples of the Results of Diffuse-Reflection Measurement

In Figure 12.8, the diffuse-reflection spectrum of coal is compared with its absorbance spectrum obtained from a transmission measurement using a KBr disk. For black materials like coal and toner used for printing, it is generally difficult to obtain high-quality infrared spectra by the transmission method, but sampling for the diffuse-reflection method is easier and spectra of higher quality can often be measured by this method.

For qualitative analysis of a compound separated by TLC (thin-layer chromatography), it is a usual practice to extract the compound with a solvent from a spot in the chromato-graph, and then examine it to identify its chemical structure by an appropriate analytical instrument such as a mass spectrometer, an FT-IR spectrometer, or an NMR spectrometer. In favorable circumstances, the diffuse-reflection measurement may provide an FT-IR spectrum of the compound without the troublesome process of solvent extraction. In Figure 12.9, the infrared spectrum of cholesterol acetate measured by the diffuse-reflection

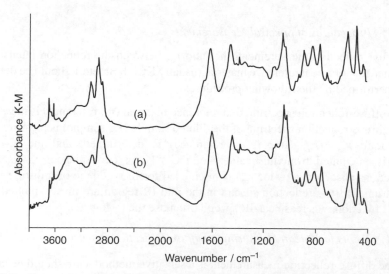

Figure 12.8 *Comparison of the infrared spectra of coal measured by two different methods. (a) Diffuse-reflection spectrum and (b) transmission spectrum from a KBr disk.*

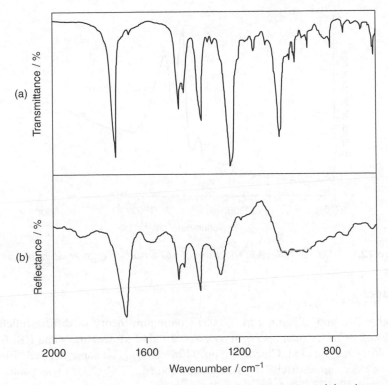

Figure 12.9 *(a) Infrared spectrum of cholesterol acetate measured by the transmission method for a KBr disk and (b) spectrum obtained from a spot containing cholesterol acetate in a thin-layer chromatograph by the method of difference spectrum.*

method is compared with that measured by the transmission method for a KBr disk of this sample. The diffuse-reflection spectrum in Figure 12.9b corresponds to a difference between the diffuse-reflection spectrum measured for powders scraped from a spot in a TLC and that measured for powders scraped from a part without any spot in the same TLC. The powders from the part without any spot, which consist of SiO_2, have strong absorptions in the region of $1300–800\ cm^{-1}$, so this region in Figure 12.9b is not reliable. However, the diffuse-reflection spectrum in Figure 12.9b and the transmission spectrum in Figure 12.9a are essentially the same in the $2000–1300\ cm^{-1}$ region, indicating that the diffuse-reflection measurement is a method which is sometimes convenient for identifying the compound in a spot separated by TLC.

Diffuse-reflection measurements are often utilized for studying catalyst reactions such as decomposition, adsorption, and desorption by using variable-temperature cells. By combining a visible-light irradiation unit to a variable-temperature cell, the photodegradation process of toluene over $TiO_{2-x}N_x$ was followed [4]. In this study, it was confirmed that toluene was finally decomposed into H_2O and CO_2. As this example showed, it is necessary to modify or improve a commercial apparatus particularly for such *in situ* measurements. In Figure 12.10, the diffuse-reflection spectrum of CO adsorbed on an alumina catalyst is shown. In a variable-temperature cell which can be evacuated, the catalyst was first reduced with hydrogen at $300\ °C$, and then CO was introduced at $50\ °C$.

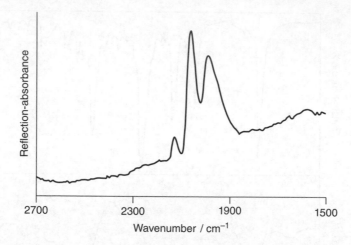

Figure 12.10   *Diffuse-reflection spectrum of CO adsorbed onto an alumina catalyst.*

References

1. Griffiths, P.R. and Olinger, J.M. (2002) Continuum theory of diffuse reflection, In: *Handbook of Vibrational Spectroscopy*, Vol. **2** (eds J.M. Chalmers and P.R. Griffiths), John Wiley & Sons, Ltd, Chichester, pp. 1125–1139. [The paper entitled "Ein Beitrag zür Optik der Farbanstriche" by Kubelka, P. and Munk, F. (1931) was published in Z. *Tech. Phys.* **12**, 593–601].
2. Dahm, D.J. and Dahm, K.D. (2002) Discontinuum theory of diffuse reflection, In: *Handbook of Vibrational Spectroscopy*, Vol. **2** (eds J.M. Chalmers and P.R. Griffiths), John Wiley & Sons, Ltd, Chichester, pp. 1140–153.
3. Yang, L. and Miklavcic, S.J. (2005) Revised Kubelka–Munk theory. III. A general theory of light propagation in scattering and absorptive media. *J. Opt. Soc. Am. A*, **22**, 1868–1873.
4. Irokawa, Y., Morikawa, T., Aoki, K., Kosaka, S., Ohwaki, T. and Taga, Y. (2006) Photodegradation of toluene over $TiO_{2-x}N_x$ under visible light irradiation. *Phys. Chem. Chem. Phys.*, **8**, 1116–1121.

13

Attenuated Total Reflection Measurements

Shukichi Ochiai
S. T. Japan, Inc., Japan

13.1 Introduction

The attenuated total reflection measurement method is often abbreviated and commonly referred to as the *ATR method*. This method was first proposed in the 1960s by Harrick [1] and Fahrenfort [2]. It is a type of reflection spectrometry that is particularly useful for measuring mid-infrared spectra from thick polymer films, paints, paper, fibers, and so on, for which the transmission measurement method is not easily applied. If an infrared spectrum of a thick polymer film or a paint on a metal plate is to be measured by the transmission measurement method, if soluble, a part of the target needs to be scraped off and dissolved in a suitable solvent, and made into a thin cast film (10–30 μm thick) by then evaporating off the solvent, or a KBr disk could be formed from the scraped powder. Other common practices include compression molding of a part of the thick sample into a thinner film appropriate for a transmission measurement, or, in some circumstances, sectioning a thin layer from the sample with a microtome. Providing a surface layer spectrum can be considered as representative of the bulk specimen, then by the ATR method, a required spectrum may be obtained in a much simpler way; that is, the target, which should be cut to an appropriate size, needs only to be brought into close contact with an internal-reflection element (IRE), which is described later. Thus, no time-consuming pre-treatments of the target are required for measuring its mid-infrared spectrum, the quality of which is usually high or at least adequate for a simple identification purpose. This is a definite advantage of the ATR method over the optimum sample requirements for a measurement by the transmission sampling technique.

Today, ATR-based methods are very commonly used procedures (particularly within industrial laboratories) for measuring mid-infrared spectra from a wide variety and diverse

Introduction to Experimental Infrared Spectroscopy: Fundamentals and Practical Methods,
First Edition. Edited by Mitsuo Tasumi and Akira Sakamoto.
© 2015 John Wiley & Sons, Ltd. Published 2015 by John Wiley & Sons, Ltd.

range of samples in various physical forms. ATR methods are utilized not only for bulk analysis but also continue to be used for more specialized purposes such as surface analysis, including depth profiling. It is possible to obtain information on a surface layer of the order of a few micrometers thickness with the help of the high accuracy and sensitivity of Fourier-transform infrared (FT-IR) spectrometry. Over the last few decades, the usefulness of ATR methods has been greatly enhanced by using synthetic diamond as the IRE; this has been accompanied by the development of various types of convenient-to-use accessories. Both small and portable FT-IR spectrometers have been designed specifically to make ATR measurements and have been developed commercially for use in process measurements at production sites within industry and for many in-the-field applications.

13.2   Methods of ATR Measurements

13.2.1   Principles of ATR Measurements

Under normal conditions, when a beam of radiation passes through air (with index of refraction n_1) and irradiates a medium with an index of refraction $n_2 (n_2 > n_1)$, a part of the radiation enters the medium and is refracted, and the remainder is reflected, as shown in Figure 13.1a. The relationship between the angle of incidence θ_i and the angle of refraction θ_t is expressed by Snell's law as

$$n_1 \sin \theta_i = n_2 \sin \theta_t \tag{13.1}$$

If the incident beam irradiates the surface through a medium of a higher refractive index $(n_1 > n_2)$, what happens is illustrated in Figure 13.1b.

Equation (13.1) may be rewritten in the following form.

$$\sin \theta_t = \left(\frac{n_1}{n_2}\right) \sin \theta_i \tag{13.2}$$

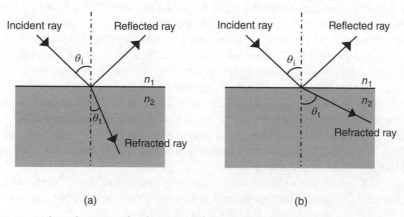

(a)   (b)

Figure 13.1   *(a,b) Schematic of radiation incident on the surface between medium 1 (upper) and 2 (lower) and refracted radiation. θ_i, angle of incidence; θ_t, angle of refraction; n_1, refractive index of medium 1; and n_2, refractive index of medium 2.*

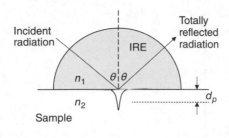

Figure 13.2 *Schematic picture of attenuated total reflection (ATR) occurring with a hemisphere internal-reflection element (IRE). d_p, depth of penetration. The downward projection in the center schematically represents the evanescent wave.*

If the angle of incidence θ_i is equal to θ_c, which is given as

$$\theta_c = \sin^{-1}\left(\frac{n_2}{n_1}\right) \tag{13.3}$$

the angle of refraction θ_t becomes $90°$; θ_c is called the *critical angle*. If $\theta_i > \theta_c$, no refraction occurs, and the incident beam will be totally reflected. This is called *total reflection*.

In the absence of absorption, the refractive index of most organic compounds is close to 1.5 and their absorption indices (k) are <0.2. For such samples, it is often difficult to observe infrared specular-reflection spectra with clear contrast from which useful information can be obtained (see Chapter 8).

In infrared ATR measurements, an optical medium which is transparent in the infrared region and has a high refraction index is used instead of the air to provide a surface in contact with the target sample, and the incident beam irradiates the surface through the optical medium at an angle of incidence larger than θ_c, so that total reflection occurs at the surface, as shown in Figure 13.2.

When the phenomenon of total reflection is examined in detail, it becomes clear that the "evanescent radiation" plays a central role in ATR spectrometry [3, 4]. The electric field of this evanescent wave penetrates the sample and decays exponentially with increasing depth of penetration. If no absorption of the incident radiation occurs, the radiation is totally reflected, but if the energy of radiation is transferred to the sample at a wavenumber at which an absorption by the sample occurs, the reflectance at this wavenumber is reduced by the amount of the absorbed energy. Accordingly, if the spectrum of the total reflection is measured, a spectrum similar to a transmission spectrum is obtained.

If the infrared beam irradiates an ordinary organic compound ($n_2 \approx 1.5$) after passing through an IRE material like Ge ($n_1 = 4.0$), the critical angle θ_c is about $22°$; that is, total reflection occurs even at a small angle of incidence.

How deep does the incident infrared beam penetrate the sample below the surface? The depth of penetration is defined as the distance from the surface to a place where the intensity of radiation becomes $1/e$ of that at the surface. If no absorption exists in the sample, it is given as [3, 4]

$$d_p = \frac{\lambda_1}{2\pi(\sin^2\theta_i - n_{21}^2)^{\frac{1}{2}}} \tag{13.4}$$

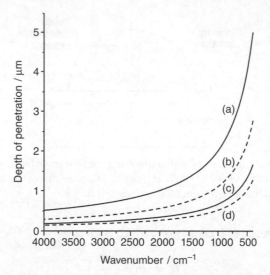

Figure 13.3   *Wavenumber dependence of the depth of penetration. (a) IRE, synthetic diamond; angle of incidence, 45°; (b) IRE, synthetic diamond; angle of incidence, 60°; (c) IRE, Ge; angle of incidence, 45°; and (d) IRE, Ge; angle of incidence, 60°. The refractive index of the sample is assumed to be 1.5. The Ge IRE is practically usable over the region of 4000–600 cm^{-1}.*

where $n_{21} = n_2/n_1$ and $\lambda_1 = \lambda/n_1$, that is, the wavelength of the infrared radiation in the IRE in use (Ge, etc.).

Equation (13.4) shows that, with increasing angle of incidence θ_i and refractive index of the IRE n_1, d_p becomes smaller, and the measured spectrum will contain information on a narrower surface layer. In addition, as is evident from Equation (13.4), the absorption intensity in an ATR spectrum depends on the wavelength; thus, the relative intensities of bands toward longer wavelength (lower wavenumber) tend to be stronger when compared to those in a transmission spectrum. In Figure 13.3, the wavenumber dependence of d_p is shown for Ge ($n_1 = 4.0$) and diamond ($n_1 = 2.4$) at two angles of incidence for each IRE. In this figure, the refractive index of the sample n_2 is assumed to be 1.5. The results in this figure indicate that depth profiling becomes possible by selecting an angle of incidence and an IRE. Table 13.1 lists materials which may be used as the IRE due to their high refractive index and transparency to mid-infrared radiation. It may be worth mentioning here that of late, the use of KRS-5 as an IRE material is decreasing, because it is slightly poisonous.

Recently, synthetic diamond has become commonly used as the IRE for measuring the ATR spectra of organic materials in general, although formerly both KRS-5 and ZnSe were frequently used. Such a change is attributed to the progress in manufacturing synthetic diamond. The refractive index of synthetic diamond is nearly equal to those of KRS-5 and ZnSe, but synthetic diamond is more durable.

For samples with high refractive indices such as a rubber sheet and inorganic materials, Ge (or Si) which has a higher refractive index is used as the IRE. To measure the ATR spectra of aqueous solutions, it is preferable to use synthetic diamond, ZnSe, AS$_2$Se$_3$, or Ge as the IRE.

Table 13.1   *High refractive-index materials used for the internal reflection element (IRE)*

Material	Refractive index[a]	Wavenumber region/cm^{-1}	Comments
Ge	4.0	5000–600	Insoluble in water Brittle
Si	3.4	8000–650	Insoluble in water Brittle
As_2Se_3	2.8	12 500–800	Insoluble in water Slightly brittle
Synthetic diamond	2.4	20 000–200	Resistant to acids and alkalis Hard and durable
KRS-5	2.4	16 000–300	Mixture of TlBr and TlI Slightly soluble in water Poisonous
ZnSe	2.4	20 000–500	Insoluble in water
AgBr	2.2	4500–300	Sensitive to ultraviolet radiation
AgCl	1.9	5000–450	Sensitive to ultraviolet radiation
Sapphire (Al_2O_3)	1.8	5500–1550	Insensitive to ultraviolet radiation

[a]Most values refer to those at 10 μm.

13.2.2 Apparatus for ATR Measurements

Figure 13.4a–c shows schematically, apparatus using IREs of trapezoidal shape, in which internal reflection occurs multiple times; these have been widely used for many years in combination with mid-infrared spectrometers. The advantage of using an IRE with multiple internal reflection lies in the wide diversity and variability of its use; as shown in Figure 13.4b, the sample can be in close contact with both the upper and lower surfaces of the IRE, and as shown in Figure 13.4c, a small-area sample can also be handled. In other

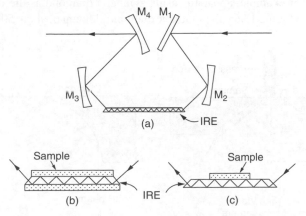

Figure 13.4   *Schematic pictures showing (a) an ATR apparatus with a multiple-reflection IRE, (b) the case where the sample is fully in contact with both surfaces of the IRE, and (c) the case where the sample is in contact with only part of one side of the IRE.*

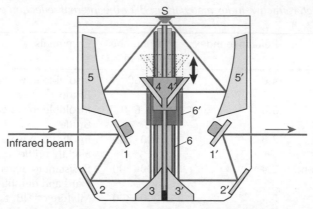

Figure 13.5 *Schematic picture of an optical system of a single-reflection ATR apparatus with a mechanism for varying the angle of incidence. S, sample over the IRE; 1, 1', 2, 2', 4, 4', plane mirrors; 3, 3', 5, 5', parabolic mirrors; 6, 6', slide rail. The angle of incidence to the IRE is varied by moving plane mirrors 4 and 4' on the slide rail. (Source: Adapted with permission from the catalog of Pike Technologies © 2013.)*

words, it is in principle possible to control the absorption intensity by manipulating the amount of the sample in contact with the IRE. However, if the sample cannot be brought into close contact with the IRE, an increase of the absorption intensity expected for a large sample cannot be obtained.

In Figure 13.5, the optical system of an accessory which can vary the angle of incidence for a single reflection is shown schematically. In this apparatus, the infrared beam irradiates a hemispherical or semicylindrical IRE and the direction of the beam to the IRE, that is, the angle of incidence, can be varied continuously. Trapezoid-shaped IREs, for which the angle of incidence is fixed at either 30°, 45°, or 60°, are also commonly used. By varying the angle of incidence, spectra corresponding to different penetration depths can be measured.

Recently, many types of accessories for ATR measurements have been developed to meet various needs. For example, apparatus using synthetic diamond as the IRE are shown in Figure 13.6a,b. In Figure 13.6a, the synthetic diamond element of about 500 μm in thickness

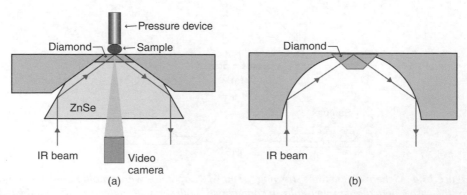

Figure 13.6 *Schematic picture of single-reflection ATR apparatus with a diamond IRE. Diamond IRE (a) supported by a ZnSe crystal and (b) without a ZnSe supporting crystal.*

is supported by a less expensive crystal of ZnSe, the refractive index of which nearly matches that of synthetic diamond, in order to decrease the possibility of mechanical damage to synthetic diamond. (Synthetic diamond is the most robust substance, but its thin element may develop a crack if strong inhomogeneous forces are applied to it.) The ZnSe crystal also plays the role of focusing the infrared beam onto the IRE. A device for applying pressure to the sample is also shown. Applying pressure is necessary to make close contact between the sample and the IRE. As shown in this figure, some accessories have a video camera below the IRE to monitor the state of contact between the sample and the IRE. In Figure 13.6b, an apparatus with only a diamond IRE is shown schematically.

The characteristics of synthetic diamond as the IRE are as follows.

1. Due to the high degree of hardness, it is often possible to measure high-quality ATR spectra from hard materials such as polymer pellets, minerals, and paints on metals by strongly pressing the surfaces of such materials against the diamond IRE.
2. As it is chemically stable, the diamond IRE can be used for measuring spectra from samples dissolved in corrosive solvents as well as from acidic and alkaline solutions.
3. It has a small coefficient of friction, so that it is easy to keep the diamond IRE clean. If something adhering to its surface cannot be dissolved by any solvent, one can scrape it away by using a knife, without causing damage to the diamond IRE surface.
4. Synthetic diamond has infrared absorption bands in the region of $2300-1900\,\text{cm}^{-1}$. If the number of reflections is small (either one or three), the absorption bands due to synthetic diamond can be canceled by dividing the spectrum measured from the sample by a reference spectrum measured without placing any sample in the same apparatus, but if the number of reflections is large (for example, nine), it is difficult to cancel them in this way.

The ATR apparatus with a diamond IRE, which is not supported on ZnSe, is less robust in its mechanical strength, but it has the advantage of offering an extended spectral range from $4000\,\text{cm}^{-1}$ down to about $400\,\text{cm}^{-1}$ when used in combination with a triglycine sulfate (TGS) detector. A similar accessory using Ge as the IRE is also commercially available. ATR apparatus designed specifically for use with FT-IR spectrometer microscope systems have also been developed and are commercially available.

A special ATR apparatus for detecting organic species contaminating a silicon wafer and SiH_n on the wafer surface has also been developed. As a silicon wafer has a high refractive index (about 3.4), it is necessary to use Ge as the IRE at an angle of incidence >59°. To perform the most sensitive measurements, an IRE of the multiple-reflection type is used. Close contact between the IRE and the wafer is necessary, and consequently, a special pressurizing mechanism that ensures total and uniform contact has been developed for this purpose.

Many other applications and developments of the ATR method and accessories have been undertaken, including devising special apparatus for specific purposes. Monitoring chemical reactions in solution is an important application of the ATR method, and a variety of studies have been reported. An important example in this field of research is a combination of the ATR method with the "surface-enhanced infrared absorption (SEIRA)" [5] technique; SEIRA is a phenomenon by which the intensities of the infrared absorption bands of a chemical species adsorbed onto the surface of a thin metal layer consisting of very small metal particles are enhanced. When the ATR method is used for measuring SEIRA, either the thin metal layer or the sample can be in contact with the IRE. This makes it possible to use the thin metal layer in contact with the IRE as an electrode in a cell and to utilize it for not only detecting chemical species adsorbed onto the metal layer but also for studying electrochemical reactions occurring within the cell. Historically, the phenomenon, now usually called *SEIRA*, was first reported by Hartstein *et al.* as early as 1980 by utilizing the ATR method [6] and its usefulness was subsequently applied to mainly polymer studies during the latter half of the 1980s [7, 8].

13.2.3 Points to Note in ATR Measurements

It is important to take note of the following points when measuring ATR spectra.

1. *Close contact between the IRE and a sample*. ATR measurements are made possible by penetration of the incident infrared beam into the sample, and the depth of penetration is shorter than the wavelength of the incident radiation. If a small gap exists between the IRE and the sample, an ATR spectrum will not be obtained. Thus, it is often necessary to press the sample to ensure it has good contact with the IRE. For hard samples, an IRE made of synthetic diamond should therefore be used.

2. *Choice of measuring conditions*. Although ATR spectra are similar to their corresponding transmission spectra, the relative intensities of bands in an ATR spectrum depend on the wavenumber; so, toward lower wavenumbers, the intensities of bands in an ATR spectrum appear relatively stronger when compared to those of the corresponding band intensities in the transmission spectrum, because the depth of penetration becomes larger with decreasing wavenumber (increasing wavelength). It should be noted that the depth of penetration is determined not only by the refractive index of both the sample and the IRE but also the angle of incidence.

   In Figure 13.7, internal reflection spectra recorded from a high-density polyethylene sample measured at various angles of incidence by using an ATR apparatus with a ZnSe IRE are shown. When the angle of incidence is >40°, which is close to (but greater than) the critical angle, the ATR conditions are satisfied. With increasing angle of incidence, the shape of the observed spectrum becomes less distorted, and the spectrum measured at 55° has a shape comparable to that observed in a good-quality transmission spectrum. The distortion of the band shape is related to the anomalous dispersion of the refractive index in the region of an absorption band. Absorption and refraction are closely related phenomena, and the complex refractive index described in Section 1.2.4 is used to discuss these phenomena. The complex refractive index $\hat{n}$ is defined as $\hat{n} = n + ik$, but the refractive index n and the absorption index k are not independent of each other (see Section 8.3). Even within the spectrum of one sample, the distortion is more noticeable for bands of strong intensity. If the ATR condition for a sample is fully satisfied, the peak maxima of bands in an ATR spectrum occur very close in wavenumber to those observed in the transmission spectrum. With increasing distortion of the band shape in an ATR spectrum, the peak position seemingly shifts to a lower wavenumber. In practice, the infrared beam actually incident on the ATR apparatus is not parallel; for example, a beam with an angle of incidence of 45° contains rays with angles of incidence both smaller and larger, and these rays may have angles of incidence smaller or larger than the critical angle and hence distort band shapes.

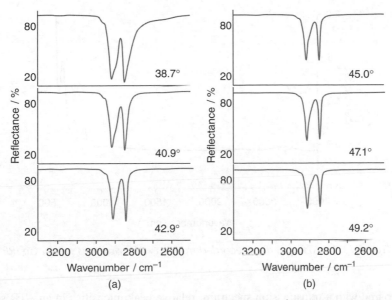

Figure 13.7 *(a,b) Dependence of ATR spectra on the angle of incidence. ZnSe was used as the IRE. The angle of incidence is given for each spectrum.*

3. *The refractive index of the sample relative to that of the IRE.* The intensities of bands in an ATR spectrum increase as the refractive index of the sample becomes closer to that of the IRE. If the intensities of a band measured by using KRS-5, synthetic diamond, and Ge as the IRE at the same angle of incidence are compared, those for KRS-5 and synthetic diamond are greater than that for Ge. This is because the penetration depths for KRS-5 and synthetic diamond are larger than that for Ge, as is expected from Equation (13.4). A material having a higher refractive index such as Ge is more suited for surface analysis, as the ATR spectrum obtained by using such a material as the IRE contains more specific information on a layer depth closer to the surface.

4. *Absorption bands arising from IREs.* KRS-5 has no absorption bands in the region of $4000-400\,cm^{-1}$, but Ge and ZnSe have some absorptions in the low-wavenumber region. Synthetic diamond has absorption bands in the region of $2300-1900\,cm^{-1}$, as mentioned earlier. Figure 13.8 shows the infrared spectrum recorded from an ATR apparatus using Ge as the IRE; the spectrum emitted by the infrared source was used as the background reference spectrum (see Section 3.3). The absorption bands in the low-wavenumber region are due to Ge itself. If an absorption band of a sample overlaps with the bands of Ge, the sample absorption will have a lower signal-to-noise (S/N) ratio. In the region below $600\,cm^{-1}$, where the Ge absorptions are very intense, it is practically impossible to measure reliable absorption spectra from samples over this region.

13.2.4 Points to Note in Quantitative Analysis

Quantitative analysis by the ATR method is possible, but it is necessary to remember and pay attention to the following points.

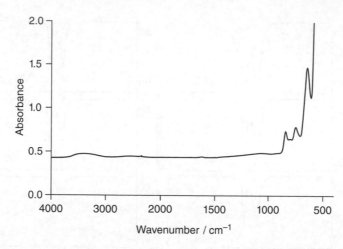

Figure 13.8   *ATR spectrum recorded from an ATR apparatus with a Ge IRE.*

1. Compared with a transmission spectrum, relative peak intensities in an ATR spectrum depend on the wavenumber of the peak.
2. The ATR condition must be fully satisfied, in order not to distort the shape of the band to be used for quantitative analysis.
3. A calibration line should be made by using data obtained exclusively by the ATR method under well-defined experimental conditions. Even if a calibration line made by the transmission method happens to be available, it should not be applied to the results obtained using the ATR method.
4. Reliable results cannot be obtained if a sample is not in close contact with an IRE.
5. In the case of a solution sample, possible dependence of its refractive index on the solute concentration may affect the linearity between solute concentration and measured absorption intensity.

In particular, maintaining a close contact between a sample and an IRE is imperative because without a close contact, ATR spectra are neither reproducible nor confirmable. Accordingly, in any quantitative analysis, it is necessary to have a band that can be used as an internal standard, so that the measured intensity of the analyte band can be normalized. It is desirable to have such an internal-standard band occurring in a spectral position close to the band to be used for the compositional analysis.

In Figure 13.9, the ATR spectra measured from mixtures of cyclohexane and carbon tetrachloride at various cyclohexane concentrations are shown. The peak intensity (in absorbance) of the strongest band of cyclohexane at $1450\,\mathrm{cm}^{-1}$ is plotted against the concentration of cyclohexane in Figure 13.10. In this measurement, the full contact between the sample and the IRE (ZnSe) is established and maintained, as the sample is a solution. Although no particular molecular interaction occurs between cyclohexane and carbon tetrachloride, as can be seen from Figure 13.10, the absorbance of the $1450\,\mathrm{cm}^{-1}$ band is, however, not directly proportional to the concentration of cyclohexane. The origin of this deviation from linearity is probably attributed to the difference in the refractive index between cyclohexane ($n_{\mathrm{d}} = 1.426$) and carbon tetrachloride ($n_{\mathrm{d}} = 1.461$); with

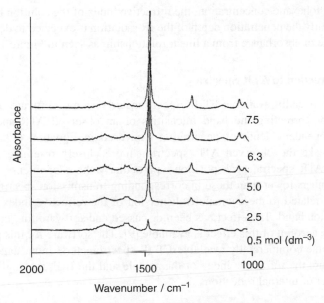

Figure 13.9 *Spectra showing the concentration dependence of the ATR spectrum of cyclohexane dissolved in carbon tetrachloride. The concentration of cyclohexane is given for each spectrum.*

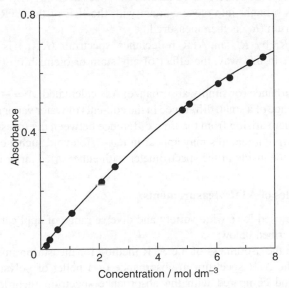

Figure 13.10 *Absorbance of the cyclohexane band at 1450 cm^{-1} plotted against the concentration of cyclohexane.*

increasing cyclohexane concentration, the refractive index of the solution becomes lower, and consequently the penetration depth of the IR radiation is expected to decrease, leading to the decrease of absorbance from a linear relationship as seen in Figure 13.10.

13.2.5   Correction to ATR Spectra

In many commercially available FT-IR spectrometer systems, software (called the *ATR correction*) for correcting the band intensities of an observed ATR spectrum for the wavelength-dependent depth of penetration expressed in Equation (13.4) is installed, in order to make the observed ATR spectrum more closely resemble a transmission spectrum. In ATR spectra, peak positions, particularly those of intense bands, tend to have wavenumbers lower than those in corresponding transmission spectra. As described earlier, this is related to the anomalous dispersion of the refractive index in the vicinity of the absorption band. The effect has been discussed and software for "correcting" ATR spectra to take account of this effect is available [9]. The software for this purpose is also installed in some commercially available FT-IR spectrometers. Input data necessary for this software are the refractive index of the sample and the IRE, the angle of incidence, and the number of internal reflections.

13.2.6   Measurements of ATR Spectra

An ATR spectrum is measured in the following way.

1. The (single-beam) infrared spectrum of the ATR apparatus to be used is measured without a sample, and the result is used as the reference spectrum (R_r).
2. The sample is brought into close contact with the IRE of the ATR apparatus, and its infrared spectrum (R_s) is then measured.
3. By dividing R_s by R_r, an ATR reflectance spectrum (R_{ATR}) is obtained; that is, $R_{ATR} = R_s/R_r$. In this way, the effect of any stain or blemish on the IRE, if present, is removed.
4. The ATR absorbance (or simply absorbance) A is calculated: $A = -\log(R_s/R_r)$.
5. As a consequence of a small difference in the contents of water vapor and carbon dioxide in the atmosphere arising from the time difference between the measurements of R_r and R_s, weak absorption features may appear in R_{ATR}. To avoid such a case, it is important to fully purge the inside of the spectrometer with either dried air or nitrogen.

13.2.7   Examples of ATR Measurements

ATR methods are used for a wide variety and diverse range of applications. Some typical examples are described below.

To demonstrate the usefulness of the ATR method for measuring infrared spectra from hard materials, the ATR spectrum in absorbance of a pellet of polystyrene is shown in Figure 13.11a, and compared with the absorbance spectrum recorded in transmission from a thin polystyrene film shown in Figure 13.11b. The film was formed from the pellet by using a hot compression molding press. Comparison of these two spectra clearly indicates that the relative intensities of bands toward higher wavenumbers in the ATR spectrum are weaker than those in the transmission spectrum, and conversely the relative

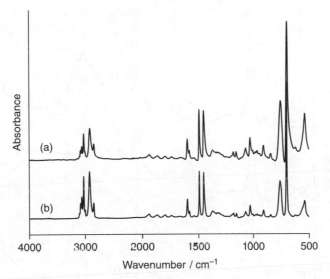

Figure 13.11   *(a) ATR spectrum of a polystyrene pellet and (b) transmission spectrum of a polystyrene film.*

intensities of bands at lower wavenumbers in the ATR spectrum are higher than those in the transmission spectrum.

ATR spectra of a rubber sample containing carbon black measured with a KRS-5 IRE at angles of incidence of 45° and 60° are shown in Figure 13.12a,b, respectively. Spectrum (a) is severely distorted, but the distortion is very much less in spectrum (b). These results indicate that carbon black increases the effective refractive index of rubber, and the angle of incidence of 45° does not satisfy the ATR condition.

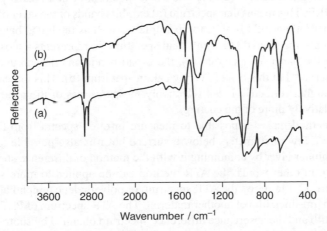

Figure 13.12   *Spectra showing dependence of the ATR spectra of rubber containing carbon black on the angle of incidence; IRE, KRS-5; (a) angle of incidence, 45°, and (b) angle of incidence, 60°.*

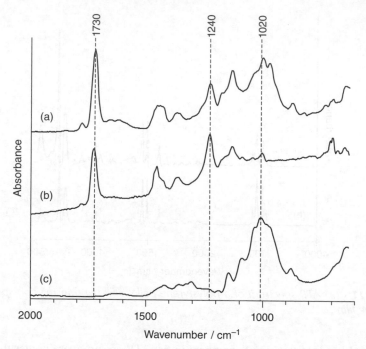

Figure 13.13   *(a,b) ATR spectra of coated paper; (a) IRE, synthetic diamond; angle of incidence, 45°; and (b) IRE, Ge; angle of incidence, 45°. (c) ATR spectrum of uncoated paper;
IRE, Ge; angle of incidence, 45°.*

In Figure 13.13, ATR spectra of coated paper measured by using synthetic diamond and
Ge as the IRE are shown, respectively, in (a) and (b). Spectrum (a) seems to be an addition
of spectrum (b) and spectrum (c); the latter is a spectrum of non-coated paper measured
using the Ge IRE. This means that spectrum (a) exhibits bands of the layer of coating on the
paper (seen at 1730 and 1240 cm^{-1}, for example) as well as the broad band characteristic
of the paper centering around 1020 cm^{-1}, but spectrum (b) represents almost only the layer
of coating on the paper. In spectrum (a), the broad band centering around 1020 cm^{-1} is
definitely observed, but this band is very weak in spectrum (b). This indicates that the Ge
IRE with its higher refractive index selectively measures less of the paper and therefore
apparently relatively more of the coating.

The ATR method makes it possible to measure infrared spectra from the layer of the
order of up to a few micrometers below a surface but it is also possible to extract information on a thinner layer by combining it with the method of difference spectrometry (see
Section 6.2.3); in other words, the ATR method can be applied to more surface specific
analysis. Figure 13.14a shows the ATR spectrum of polyethylene terephthalate (PET) covered with 0.05 μm-thick film of another material. The ATR spectrum of PET itself is shown
in Figure 13.14b and the absorbance difference spectrum obtained by subtracting the spectrum in (b) from that in (a) is shown in Figure 13.14c. This spectrum is mostly due to the
thin film covering the PET, and it is inferred from this spectrum that this surface-layer film
is made of some kind of polyurethane.

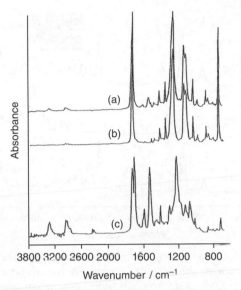

Figure 13.14   *ATR analysis of surface-treated poly(ethylene terephthalate) (PET). ATR spectra of (a) surface-treated PET and (b) untreated PET. (c) Difference between spectra in (a) and (b).*

Figure 13.15a shows the ATR spectrum recorded from a small area (indicated in a circle) printed on a business card and containing some of the paper part surrounding the dark grey point. The ATR spectrum of the paper of the name card is shown in Figure 13.15b. The absorbance difference spectrum between these is shown in Figure 13.15c. This difference spectrum indicates that the main component of the dark gray point is a kind of polyamide, probably a nylon.

In addition to films, fibers may also be analyzed conveniently by the ATR method. Figure 13.16a,b shows, respectively, the ATR spectra of an optical fiber (diameter: 200 μm) and the same fiber coated with a silicone resin. The resin layer seems to be so thick that the spectrum of the fiber is not clearly observed in spectrum (b). When the resin was dissolved by a solvent and removed, a "clean" fiber was obtained; its ATR spectrum is shown in (c). Spectrum (c) is close in appearance to spectrum (a), whereas spectrum (d) which was obtained by subtracting (a) from (c) agreed with the spectrum of the resin, indicating that a small amount of the resin still remained on the fiber, and the coating was not completely removed from the fiber by the solvent treatment.

In another example, useful information was obtained by observing small spots with a video camera and measuring the ATR spectra of those spots. A hair consists of cuticles on its surface, cortex as an interleaving layer, and medullas in its center. It is said that commercially available hair treatments work on the cortices. In order to determine whether a commercially available hair treatment enters the inside of a hair, it was examined by an FT-IR microscopic ATR method. To examine how deep the hair treatment permeates into the hair, the treated hair was sliced with a knife like a flat chisel having a synthetic diamond edge, and microscopic ATR spectra were then measured from an exposed cross section of the hair. The study was also to determine if the hair treatment remained on the cuticles.

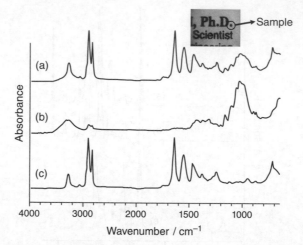

Figure 13.15   *ATR analysis of a spot on a business card. ATR spectra of (a) a spot on a business card and (b) paper surrounding the spot. (c) Difference between spectra in (a) and (b) Include source reference. (Source: Adapted with permission from the catalogue of Pike Technologies © 2013.)*

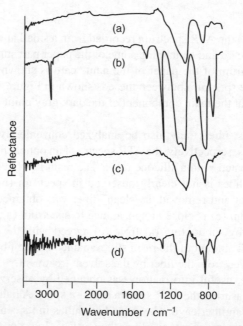

Figure 13.16   *ATR analysis of an optical fiber. ATR spectra of (a) an optical fiber, (b) a resin–coated optical fiber, and (c) a resin-removed optical fiber. (d) Difference between spectra in (c) and (a).*

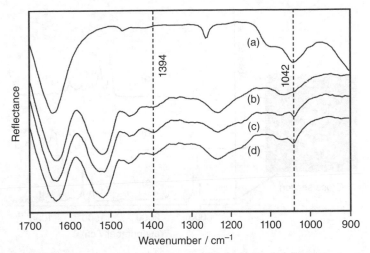

Figure 13.17 *ATR analysis of hair treatment. ATR spectra of (a) a hair treatment, (b) the cortex of an untreated hair, (c) the cortex of a treated hair, and (d) the cuticle of a treated hair.*

As visible light passes through the synthetic diamond used as the IRE, the cross section was observed through the IRE by a microscope with a charge-coupled device (CCD) camera, and the points from which the ATR spectra were to be measured were determined. The minimum size from which an ATR spectrum can be measured is $\sim 12\,\mu m \times 12\,\mu m$, by limiting the aperture size with movable slits. As the diameter of a hair is $80 - 100\,\mu m$, it is possible to measure the ATR spectrum of a particular spot on the cross section of the hair.

In Figure 13.17, the ATR spectra of a hair treatment, the cortex of a hair without a treatment, the cortex of a hair with a treatment, and the cuticle of a hair are shown. The band at $1042\,cm^{-1}$ due to the hair treatment was not found to be present in the spectrum of the cortex of the hair without the hair treatment (spectrum (b)), but this band was observed in both the cortex and cuticle of the hair with the hair treatment (spectra (c) and (d), respectively). The hair used for the spectral measurements was left untouched for 12 h after being exposed to the hair treatment. The results obtained in these measurements made it clear that the hair treatment not only worked on the cuticles but also permeated the cortex.

The band at $1394\,cm^{-1}$ is present in all three spectra of the hair in Figure 13.17b–d, although it is not seen in the spectrum of the hair treatment in (a) of the same figure. This band is probably due to a hair dye, which had been applied several days before. As this band is observed also in the cortex, the hair dye seems to have permeated the inside of the hair. The measurements described above were performed at several points along the hair, and essentially the same results were obtained at each point. These results seem to indicate that the distribution of both the hair treatment and the hair dye was uniform along the hair.

In the example shown in Figure 13.18, the ATR spectra of layers of a laminate film cut obliquely by using a knife with an edge of synthetic diamond are shown. As an FT-IR spectrometer with a microscopic ATR apparatus was used, each of the measured spectra corresponds to an individual component without overlapping of the spectra by other layers. The individual layers were readily identified as a nylon, a polyester, probably PET, and a polypropylene.

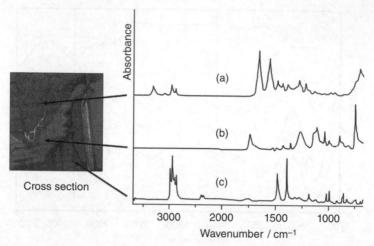

Figure 13.18 *Picture of the cross section of an obliquely cut laminate film (left) and ATR spectra of three exposed layers (right). (a) Nylon, (b) polyethylene terephthalate (PET), and (c) polypropylene (PP).*

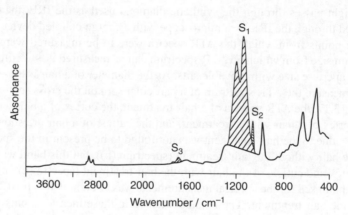

Figure 13.19 *ATR spectrum of a membrane filter and the bands used for quantitative analysis (S_1, S_2, and S_3).*

As has been already mentioned, in some circumstances, the ATR method can also be applied to depth profiling. An example of such an application is shown in Figure 13.19, in which the ATR spectrum recorded from a membrane filter is shown. This filter's base is a type of fluororesin, and is formed by adding two components. When it is formed, one side of the filter is exposed to the air and the other side is in contact with a metal. This causes a difference in the depth distribution of the added components. The two components have different functional groups, both giving rise to strong infrared absorption bands; one has the COOH group and the other SO_3H group. The depth profiling of the two components was performed by measuring the absorbance intensities of characteristic infrared bands of these groups using a microscopic ATR method. By rubbing the membrane filter with sheets

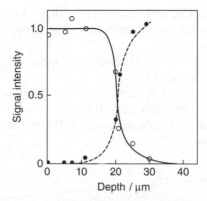

Figure 13.20   *Depth profiling by using the intensities of bands due to the COOH and SO_3H groups in the membrane filter.* ○*, band due to the COOH group; and* ●*, band due to the SO_3H group.*

of fine sandpaper, the thickness of the filter was decreased by $1-3\,\mu m$. This process was repeated, and each time the ATR spectrum of the sample was measured. To decrease the thickness of the sample uniformly by rubbing, the size of the sample was limited ($4\,mm \times 4\,mm$), and a microscopic ATR apparatus was used. By taking the intensities of the two CF stretching bands (S_1 in Figure 13.19) as the intensity reference, the depth profiling was carried out by measuring the intensities of the bands due to the SO_3H group (S_2) and COOH group (S_3). The results are shown in Figure 13.20. The intensities of both the S_2 and S_3 bands changed rapidly at the depth of about $20\,\mu m$ from the surface, but the changes occurred in opposing directions.

Due to marked progress in techniques for measuring ATR spectra, the quality of the spectra measured by this method has greatly improved over the last few decades, and the sample size needed for spectral measurements by this method has decreased considerably. There is no doubt that the ATR method plays a major role in answering various needs within analytical and materials science, and, indeed, in many other areas of application.

References

1. Harrick, N.J. (1960) Surface chemistry from spectral analysis of totally internally reflected radiation. *J. Phys. Chem.*, **64**, 1110–1114.
2. Fahrenfort, J. (1961) Attenuated total reflection: a new principle for the production of useful infra-red reflection spectra of organic compounds. *Spectrochim. Acta*, **17**, 698–709.
3. Harrick, N.J. (1967) *Internal Reflection Spectroscopy*, Interscience Publishers, New York; available from Harrick Scientific Products, Pleasantville, New York.
4. Mirabella, F.M. (2002) Principles, theory and practice of internal reflection spectroscopy, In: *Handbook of Vibrational Spectroscopy*, Vol. 2 (eds J.M. Chalmers and P.R. Griffiths), John Wiley & Sons, Ltd, Chichester, pp. 1092–1102.
5. Osawa, M. (2002) Surface-enhanced infrared absorption spectroscopy, In: *Handbook of Vibrational Spectroscopy*, Vol. 1 (eds J.M. Chalmers and P.R. Griffiths), John Wiley & Sons, Ltd, Chichester, pp. 785–800.

6. Hartstein, A., Kirtley, J.R. and Tsang, J.C. (1980) Enhancement of the infrared absorption from molecular monolayers with thin metal overlayers. *Phys. Rev. Lett.*, **45**, 201–204.
7. Ishino, Y. and Ishida, H. (1988) Grazing angle metal-overlayer infrared ATR spectroscopy. *Appl. Spectrosc.*, **42**, 1296–1302.
8. Ishino, Y. and Ishida, H. (1990) Spectral simulation for infrared surface electromagnetic wave spectroscopy. *Surf. Sci.*, **230**, 299–310.
9. Nunn, S., and Nishikida, K. (2008) Advanced ATR Correction Algorithm. Thermo Scientific Application Note 50581.

14

Photoacoustic Spectrometry Measurements

Shukichi Ochiai
S.T. Japan Inc., Japan

14.1 Introduction

Photoacoustic spectrometry (PAS) was first developed for applications in the ultraviolet–visible region; its practical use for measurements in the mid-infrared region became popular with Fourier transform (FT) spectrometers in the 1980s.

Fourier transform infrared (FT-IR) spectrometry is the most useful method for measuring a high-quality infrared spectrum from a sample over the entire infrared region. PAS in the infrared region utilizes an infrared source and an FT-IR interferometer system; it detects the acoustic signal generated by a sample heated by the modulated infrared radiation from the interferometer. PAS is a useful sampling technique for examining bulk solid samples, particularly for samples which may change their properties when scraped or pulverized, or by other sampling methods that require "damaging" or altering a sample's form. It can also be used for depth profiling of the order of several micrometers or greater.

In this chapter, a brief account of PAS measurements in the mid-infrared region is described. More detailed information is available in Refs [1, 2].

14.2 Photoacoustic Spectrometry

14.2.1 Principles of Photoacoustic Spectrometry

In Figure 14.1, how a photoacoustic (PA) signal is generated within a cell (PA cell) is schematically shown (a more realistic design of a PA cell will be shown later). A sample

Introduction to Experimental Infrared Spectroscopy: Fundamentals and Practical Methods,
First Edition. Edited by Mitsuo Tasumi and Akira Sakamoto.
© 2015 John Wiley & Sons, Ltd. Published 2015 by John Wiley & Sons, Ltd.

is placed on the supporting base (sample holder), which consists of a material with a low thermal conductivity and which should not have any infrared absorption, and then the cell is tightly closed. The incident infrared radiation enters the cell through a polished infrared-transparent window, passes through a non-absorbing coupling gas, and is then incident on the sample. The incident infrared radiation is absorbed by the sample and converted to heat. Subsequently, diffusion of the generated modulated thermal energy occurs, and a part of it is transferred to the gas (usually He) which is in contact with the sample surface.

The process described above may be roughly formulated in the following way. The intensity of the infrared radiation entering the gas is denoted by I_0; reflection at the interface between the gas and the sample is disregarded. If I_0 is considered to represent the energy of radiation (in units of joules) passing through a unit area per unit time, and the absorption coefficient of the sample is denoted by α (in units of per meter), the intensity of radiation is reduced to $I_0 \exp(-\alpha x)$ at a point of distance x (in units of meters) from the surface of the sample (see Section 1.2.3). The energy of absorption by a thin layer of thickness dx may be expressed as $I_0 \exp(-\alpha x) dx$. If this is converted to heat, the temperature T changes proportionally with it. The temperature change ΔT_x in the thin layer dx is expressed as

$$\Delta T_x = k_1 I_0 \exp(-\alpha x) dx \qquad (14.1)$$

where k_1 is a proportionality constant. The incident radiation is modulated (as described later); so is ΔT_x.

The heat generated in the thin layer dx in Figure 14.1 diffuses in a direction opposite to that of the incident radiation until it reaches the sample surface in contact with the coupling

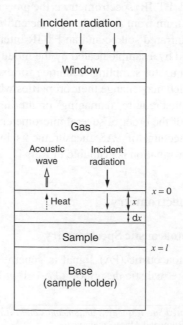

Figure 14.1   *Schematic picture of the PA signal generation process.*

gas. As the thermal energy decays during this process, the temperature change ΔT_s at the surface may be given as

$$\Delta T_s = k_1 k_2 I_0 \exp[-(\alpha + a_s)](-\alpha x)dx \tag{14.2}$$

where k_2 is a proportionality constant and a_s is a sample-dependent constant representing the decay of the diffusing thermal energy. The relation between this constant and thermal diffusion is discussed later.

If the incident radiation is converted to heat in the sample at a rate much faster than the modulation frequency of the incident radiation, the thermal energy will also be transmitted to the gas periodically. As a result, compressional (or acoustic) waves are generated in the gas and detected by a highly sensitive microphone; PAS is based on this process. The infrared radiation from the interferometer of an FT-IR spectrometer has a modulated frequency $f = 2v\tilde{v}$ (see Equation (4.9)). PAS uses this modulated infrared radiation and the acoustic signals are converted to electric signals by the microphone.

The generation of heat and its diffusion within a PA cell was first discussed in detail by Rosencwaig and Gersho [3]. The outline of the theoretical analysis described in Refs [2, 3] may be explained as follows.

To express the distance through which the incident infrared radiation passes through the sample, the inverse of the absorption coefficient $1/\alpha$ is used and denoted by l_α. The thermal diffusion length is defined by $1/a_s$ and denoted by l_d. Coefficient a_s is related to the thermal diffusivity a (in units of $m^2 s^{-1}$) and the modulation frequency f is expressed as [1]

$$a_s = \left(\frac{\pi f}{a}\right)^{\frac{1}{2}} = \left(\frac{2\pi v\tilde{v}}{a}\right)^{\frac{1}{2}} \tag{14.3}$$

The thermal diffusivity a is expressed by the thermal conductivity λ (in units of watts per meter per kelvin), density ρ, and specific heat capacity at constant pressure c_p as

$$a = \frac{\lambda}{\rho c_p} \tag{14.4}$$

Six cases may be considered to occur for PA absorption bands, depending on the interrelationship between l (sample thickness), l_α, and l_d, as shown in Figure 14.2 [3]. Qualitatively, (a)–(c) in this figure correspond to cases for which the sample's absorption may be considered weak, and (d)–(f) to cases where the sample's absorption is strong. PA spectral features expected for these cases are as follows.

(a) $l < l_\alpha < l_d$. The intensities of bands in the PA spectrum are proportional to α (strictly speaking $\alpha(\tilde{v})$) and l, and the relative intensities of the bands observed in the PA spectrum resemble those normally seen in the absorption spectrum of the sample.

(b) $l < l_d < l_\alpha$. In the same way as in (a), the PA spectrum reflects properly the absorption spectrum of the sample.

(c) $l_d < l < l_\alpha$. Bands in the PA spectrum are proportional to α and l_α. The PA spectrum reflects the absorption spectrum of the sample.

(d) $l_\alpha < l < l_d$. The characteristic absorption spectrum of the sample is not measured. The measured PA spectrum corresponds to the spectrum of the emission from the source of radiation. This is explained in the following way. When the sample layer is extremely

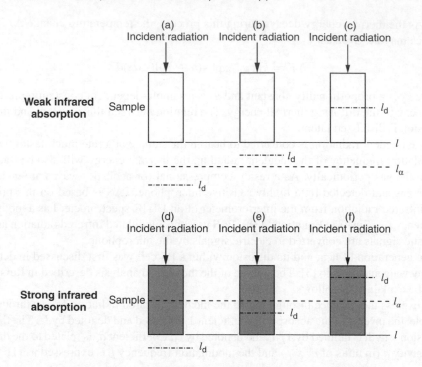

Figure 14.2   *(a–f) Six different cases of PA measurements.*

thin, the normal absorption spectrum of the sample is obtained, although the intensities of bands are generally weak. If the thickness of the sample layer is increased, saturation of the more intense absorption bands will occur; if the thickness is increased still further, a situation will arise at which all the absorption bands over the mid-infrared region effectively become saturated, and hence no absorption spectrum characteristic of the sample will be measured. This situation is similar to the case for which an analytically useful absorption spectrum cannot be recorded by the transmission method because the sample is too thick. As PAS measures the energy of heat converted from the incident radiation, the PA spectrum for (d) becomes equivalent to the emission spectrum of the source of radiation.

(e) $l_\alpha < l_d < l$. In the same way as in (d), only the emission spectrum of the source of radiation is measured.

(f) $l_d < l_\alpha < l$. The intensities of bands in the PA spectrum are proportional to α and l_d. If the condition $l_d < l_\alpha$ holds even for an intense band, the band corresponding to this case can be measured by PAS. Experimentally, however, it is not easy to find a case which satisfies this condition.

14.2.2   FT-IR/PAS Measurements and Points to Note

The PA cell and the detector (microphone) are integrated as a unit and called the *PA detection unit*, which may be considered as a complete PAS cell. An ordinary unit and

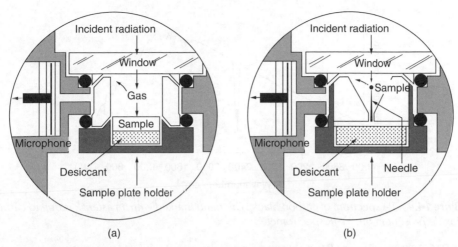

(a)   (b)

Figure 14.3 *Schematic designs of PA cells. (a) Ordinary cell and (b) cell for small samples. (Source: Adapted from the catalogs of MTEC Photoacoustic, Inc. and S. T. Japan, Inc.)*

another for micro sampling are schematically shown in Figure 14.3a,b, respectively. In the micro-sampling cell in (b), the sample is attached to the top of a tungsten needle electrostatically and can be placed together with the needle in the cell. The micro-sampling cell is designed to accommodate the needle easily and decrease the volume of the gas surrounding it. The PA signals (acoustic waves) generated by the sample, after passing through a narrow tube, are detected and converted to electric signals by a microphone, amplified by a preamplifier, and sent to an analog-to-digital converter. After this step, the usual data processing of an FT-IR spectrometer is performed and a PA spectrum is obtained.

As a sample for a reference spectrum, carbon black, glassy carbon, or soot is used. As such black materials come under the case of Figure 14.2d, the spectrum of emission from the source of radiation is obtained as the PA spectrum, which can be used as the (single-beam) reference spectrum. As an example, the PA spectrum of carbon black is shown in Figure 14.4. In this figure, the intense band, marked with an asterisk, is due to carbon dioxide in the optical system of the spectrometer and all the other weaker but sharper bands are absorptions arising from water vapor and carbon dioxide.

The points to note in measuring PA spectra are as follows.

1. As the PA detection unit is very sensitive to small acoustic signals, it is easily affected by environmental noise and vibrations. For example, the noise from the fan of an air conditioner can disturb the PA detection unit. Hence, it is imperative that noise and vibrations from the local surroundings that may adversely affect the measurement are eliminated or kept to a minimum.
2. It is advisable to minimize the volume of the gas in the PA cell, as the magnitude of the PA signal is inversely proportional to the volume of the gas that conveys the acoustic waves. A sample cup should be used and filled for solids or powders.
3. Anything which may move or disturb the PA cell should be removed. For example, electric cables belonging to the PA detection unit should be tightly secured.

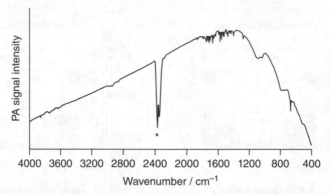

Figure 14.4   *PA spectrum of carbon black. The band marked with an asterisk is due to carbon dioxide present in the FT-IR spectrometer.*

4. Before measuring the PA spectrum of a powder sample, it is necessary to remove water (vapor) by purging the cell with He or dried nitrogen gas for a sufficiently long time. If possible, a small amount of drying agent, for example, magnesium perchlorate, should be put under the sample cup as shown in Figure 14.3 for use as a desiccant.
5. When purging the cell with the gas, the gas should be flowed carefully at an appropriate rate, so as not to blow off or disturb a powder sample.

14.2.3   Procedures of Measurement

1. The PA signal of carbon black (for example) is measured, and it is used as the reference spectrum P_r.
2. If the sample is a powder, its PA spectrum is measured after filling up the cup with the sample as much as possible (to a desirable depth of about 1 mm) and then, as good practice, leveling the top of the powder surface. The measured PA spectrum is denoted by P_s.
3. If the quantity of a sample is small, a metal plate, a component of the PA cell, should be placed under the sample. If the sample is a transparent material, then, in order to avoid reflection from the metal plate under the sample, a black sheet of paper or something similar should be placed beneath the sample.
4. P_s/P_r is calculated, and this is treated as the PA spectrum of the sample. Practically, the PA spectrum P_s/P_r may be treated in the same way as the absorbance spectrum generated from a transmission spectrum. In this chapter, this quantity is called the *PA signal intensity* in the ordinates of the PA spectra shown in Figures 14.4–14.9.

14.2.4   Advantages and Examples of the FT-IR/PAS Measurements

PAS has the following advantages.

1. There is little restriction as to the shape of the sample. The sample can often be handled as received.
2. Time-consuming pretreatment of a sample is scarcely needed.
3. The sample can often be recovered completely.
4. Depth profiling of the order of several micrometers is possible.

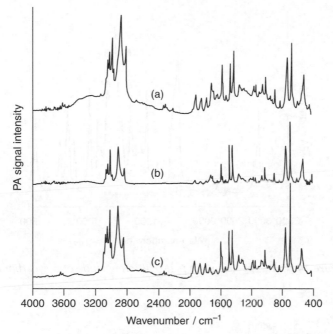

Figure 14.5 *PA spectra of polystyrene in three forms. (a) Foam, (b) thick film, and (c) powder.*

Some examples of the FT-IR/PAS measurements are described.

The PA spectra measured from polystyrene in three different forms are shown in Figure 14.5. The three spectra in this figure, which were obtained without any pretreatments of the samples (other than that they were cut to a size to fit within the PA cell), are not exactly the same as each other, but it is clear that the main component of each sample is polystyrene.

The PA spectrum shown in Figure 14.6 was obtained from the top of a screw made from a polycarbonate material. As this sample was hard and had a complicated structure, it was not possible to measure a high-quality mid-infrared spectrum by other methods. The accumulation time for recording the spectrum shown in Figure 14.6 was 20 s, and it is often possible to measure PA spectra having signal-to-noise (S/N) ratios at this level in such a relatively short period.

In Figure 14.7, the PA spectrum of a coal sample is compared with its diffuse-reflection spectrum. The qualities of the two spectra are similar, but the PA spectrum in (a) was obtained from the coal sample directly, whereas pulverization of the coal to an appropriate particle size followed by dilution in KBr powder was necessary in order to measure the diffuse-reflection spectrum shown in (b). The PA spectrum of the coal sample was measured also by the micro-sampling technique [1], and it was observed that slightly but clearly different spectra were obtained from small pieces taken from various parts, indicating that coal is not a uniform material.

Micro-sampling PAS can also be a useful tool in many forensic studies, as it may more conveniently provide information as to the chemical identity of a chip of paint, a fiber, or

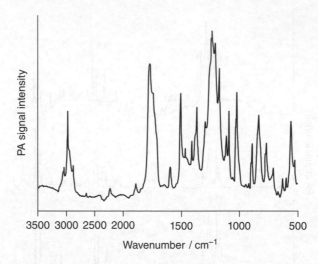

Figure 14.6   PA spectrum of a part of a screw made of polycarbonate.

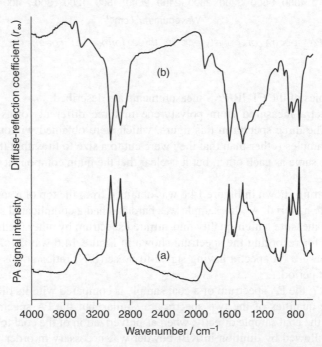

Figure 14.7   Comparison of the PA and diffuse-reflection spectra of coal. (a) PA spectrum and (b) diffuse-reflection spectrum.

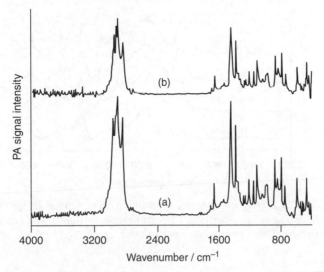

Figure 14.8 *Effects of different acoustic-wave-transmitting coupling gases on PA signals. (a) He and (b) nitrogen gas.*

other small samples from places such as a crime scene. Micro-sampling techniques in PAS are described in detail in Ref. [1].

As the coupling gas for a PAS measurement, nitrogen may also be used; however, as shown in Figure 14.8, the intensities of the PA bands of poly(*trans*-1,4-isoprene) obtained using nitrogen gas are about half of those obtained using He. This indicates that He reflects more sensitively the thermal changes of the sample; the thermal conductivity of He gas is greater than that of N_2.

14.2.5 Depth Profile Analysis

As the thermal diffusion length l_d represents the sampling depth (the distance from which a PA spectrum can be measured) and $l_d = (a/2\pi v \tilde{v})^{1/2}$, it is possible to alter the sampling depth by varying the speed v of the moving mirror of the interferometer. This means that depth profile analysis can be performed with PAS, although the sampling depth depends also on the wavenumber $\tilde{v}$. For materials with a thermal diffusivity of $1 \times 10^{-7} \, m^2 \, s^{-1}$ (a representative value for polymers and organic substances), if v is set at $1 \, cm \, s^{-1}$, l_d is 3–5 μm, and if v is slowed down to $0.025 \, cm \, s^{-1}$, l_d becomes much longer (15–40 μm; l_d becomes increasingly longer toward the lower wavenumber).

In Figure 14.9, the PA spectra of a two-layer film consisting of a 1 mm-thick soft poly(vinyl chloride) basal layer and a 6 μm-thick surface layer are shown. The four spectra were measured by varying the speed of the moving mirror v, which was slower on going from (a) to (d). The band at $1726 \, cm^{-1}$ (due to a plasticizer) is characteristic of the base film. The intensity of this band clearly increased as the speed of the moving mirror was reduced. The results in Figure 14.9 show that information on materials existing at various distances from the surface can be obtained by PAS; in other words, depth profile analysis is possible by this method.

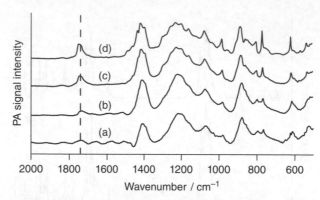

Figure 14.9 *Effects of altering the speed of the moving mirror on the PA spectrum. Sample: a two-layer film consisting of soft poly(vinyl chloride) (base) and poly(vinyl fluoride) (surface). Speed of the moving mirror in centimeter per second: (a) 1.28, (b) 0.64, (c) 0.32, and (d) 0.16.*

PAS is applicable to many kinds of samples, and the fact that almost no pretreatment is needed for samples is a unique practical advantage of this method.

Generally speaking, the sensitivity of PAS in the mid-infrared region has not been particularly high. However, a PA detection unit has been developed in order to realize both a high sensitivity and a large dynamic range [4]. In this apparatus, the expansion and contraction of the gas in the PA cell is transferred to a cantilever, the movement of which is detected by a laser interferometer. With such an improvement in the method of measurement with a conventional rapid-scanning interferometer, PAS may become a more useful tool in infrared spectroscopic analysis.

References

1. McClelland, J.F., Jones, J.W., Luo, S. and Seaverson, L.M. (1993) A practical guide to FT-IR photoacoustic spectroscopy, In: *Proper Sampling Techniques for Infrared Analysis* (ed P.B. Coleman), CRC Press, Boca Raton, FL, pp. 107–144.
2. McClelland, J.F., Jones, J.W. and Bajic, S.J. (2002) Photoacoustic spectroscopy, In: *Handbook of Vibrational Spectroscopy*, Vol. **2** (eds J.M. Chalmers and P.R. Griffiths), John Wiley & Sons, Ltd, Chichester, pp. 1231–1251.
3. Rosencwaig, A. and Gersho, A. (1976) Theory of the photoacoustic effect with solids. *J. Appl. Phys.*, **47**, 64–69.
4. Uotila, J. and Kauppinen, J. (2008) Fourier transform infrared measurement of solid-, liquid-, and gas-phase samples with a single photoacoustic cell. *Appl. Spectrosc.*, **62**, 655–660.

15

Emission Spectroscopic Measurements

Shukichi Ochiai
S. T. Japan Inc., Japan

15.1 Introduction

Emission spectroscopy in the visible region has been utilized in astrophysics for a long time to detect and analyze the radiation emitted by the sun and other stars for the purpose of identifying materials giving rise to the emissions.

When emission spectroscopy is used in materials science and technology, the temperature of the target of analysis is usually lower than 200 °C. In this situation, the radiation emitted by the sample is very weak. By using a Fourier transform infrared (FT-IR) spectrometer, however, it is possible to detect weak infrared radiation emitted by a sample at relatively low temperatures (i.e., in the range from room temperature to 100 °C) with adequate or even high signal-to-noise (S/N) ratios.

Infrared emission from a material contains information characteristic of the material, and the information is obtainable in a form that is basically equivalent to an infrared absorption spectrum. Depending on the purpose of analysis and the shape of the target of analysis, infrared emission spectroscopy measurements can be more informative or convenient than an infrared transmission or reflection spectroscopy measurement, which is described elsewhere in this book. Infrared emission spectroscopy provides a means for the remote detection and analysis of gases, for example, volcanic gas and flue gases [1].

Introduction to Experimental Infrared Spectroscopy: Fundamentals and Practical Methods,
First Edition. Edited by Mitsuo Tasumi and Akira Sakamoto.

15.2   Infrared Emission Measurements

15.2.1   Principles of Infrared Emission Spectrometry

Emission from a body occurs from thermally excited atoms and molecules within the body. The basic principle of thermal emission is described by Kirchhoff's law which states that the ratio between the energy of radiation emitted by a body in a thermal equilibrium and its absorptance is a function of only the temperature of the body and the wavenumber (or wavelength) of the radiation; it does not depend on the material constituting the body. The absorptance mentioned above may be defined as follows. When a body is irradiated, the radiation is partly reflected, partly absorbed, and the remainder passes through the body, if scattering by the body is ignored. If the proportions of the reflection, absorption and transmission are expressed, respectively, by reflectance (r), absorptance (α), and transmittance (τ), the following relation holds: $r + \alpha + \tau = 1$. It is clear that each of the three quantities is a dimensionless constant with a value between 0 and 1. (As each of them is a function of wavenumber $\tilde{\nu}$, they are expressed as $r(\tilde{\nu})$, $\alpha(\tilde{\nu})$, and $\tau(\tilde{\nu})$ when necessary.) Usually, transmittance is denoted by T, but it is not used in this chapter to avoid confusion with temperature T.

Kirchhoff's law is given as

$$\frac{L(\tilde{\nu})}{\alpha(\tilde{\nu})} = \rho(\tilde{\nu}) \tag{15.1}$$

where $L(\tilde{\nu})$ is the radiance in units of watts per steradian per square meter, which means the energy of radiation with wavenumber between $\tilde{\nu}$ and $\tilde{\nu} + d\tilde{\nu}$ emitted from the unit area of the body per unit time, and $\rho(\tilde{\nu})$ is the energy of radiation with wavenumber $\tilde{\nu}$, which does not depend on the material, as explained in the next paragraph.

If a body absorbs all the radiation it receives, it is called a *blackbody*. Molecules in the body are excited by the energy of absorbed radiation to states of higher energy, and return to the original ground state by releasing the energy by the process of emission of radiation. This means that the emission of radiation is the opposite of the absorption of radiation. In fact, a body which shows a strong absorption of radiation gives a strong emission of radiation. The absorptance of an ideal blackbody is unity for radiation over the entire wavenumber region. In this case, Equation (15.1) is simplified as $L(\tilde{\nu}) = \rho(\tilde{\nu})$, where $\rho(\tilde{\nu})$ represents Planck's law of radiation (see below).

Equation (15.1) may be rewritten in the following form:

$$\varepsilon(\tilde{\nu}) = \frac{L(\tilde{\nu})}{\rho(\tilde{\nu})} \tag{15.2}$$

where $\varepsilon(\tilde{\nu})$ is called the *emittance* which is defined as the ratio between the radiation of a body and that of a blackbody. Numerically $\varepsilon(\tilde{\nu})$ is equal to $\alpha(\tilde{\nu})$. As $\varepsilon(\tilde{\nu})$ is a function of temperature, it is expressed as $\varepsilon(\tilde{\nu}, T)$ when necessary.

As a blackbody is the strongest emitter at a temperature, its emission is used as the reference to the emission from another body at the same temperature. In the following subsections, a brief explanation is first given relating to the emission from a blackbody, and next the measurement of emission from a non-blackbody is described.

15.2.1.1 Emission from a Black Body

As described above, a blackbody is defined as a body which absorbs all radiation received by it. A strongly absorbing body is a strongly emitting body. For a blackbody, the following relation holds:

$$\varepsilon_{BB}(\tilde{v}, T) = \alpha_{BB}(\tilde{v}, T) \equiv 1 \tag{15.3}$$

where $\varepsilon_{BB}(\tilde{v}, T)$ and $\alpha_{BB}(\tilde{v}, T)$ are, respectively, the emittance and absorptance of a black-body at wavenumber $\tilde{v}$ and temperature T. Both of these are unity by definition. The emittance of a non-blackbody is smaller than unity.

According to Planck's law of radiation, the energy of radiation from a blackbody $\rho(\tilde{v}, T)$ per unit volume (in units of joule per square meter) is given as

$$\rho(\tilde{v}, T) = 8\pi hc\tilde{v}^3 \left[\exp\left(\frac{hc\tilde{v}}{k_B T}\right) - 1\right]^{-1} \tag{15.4}$$

where c is the speed of radiation, k_B is the Boltzmann constant, and h is the Planck constant. Usually, Planck's law of radiation is expressed as a function of either the wavelength or frequency. Equation (15.4) expresses the law as a function of wavenumber as this usually represents the abscissa unit of a mid-infrared spectrum.

The spectra of blackbody radiation at three different temperatures calculated by Equation (15.4) are shown in Figure 15.1. With lowering temperature, the intensity of emission decreases, and the maximum of the distribution of the emission shifts to lower wavenumber. The radiation emitted from the unit area of such a blackbody in a unit solid

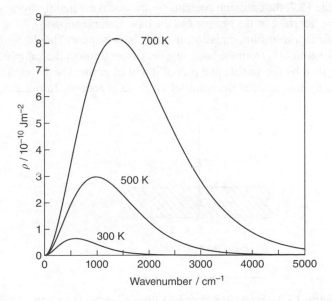

Figure 15.1 *Emission spectra of the blackbody radiation at three different temperatures.*

angle per unit time corresponds to $\rho(\tilde{v})$ in Equation (15.1), which is expressed as $\rho_{BB}(\tilde{v}, T)$ in the later discussion.

Knowing the wavenumber $\tilde{v}_{max}$, which corresponds to the maximum of the distribution of blackbody emission at temperature T, is important for estimating the wavenumber region over which an emission spectrum can be observed. $\tilde{v}_{max}$ is given by the following formula which can be derived using $d\rho(\tilde{v}, T)/d\tilde{v} = 0$.

$$\tilde{v}_{max} = 1.93T \tag{15.5}$$

This relation is called *Wien's displacement law.*

15.2.1.2   *Emission from a Non-Black Body*

The emittance $\varepsilon(\tilde{v}, T)$ of a non-blackbody at temperature T is expressed by rewriting Equation (15.2) as follows:

$$\varepsilon(\tilde{v}, T) = \frac{L(\tilde{v}, T)}{\rho_{BB}(\tilde{v}, T)} \tag{15.6}$$

For a non-blackbody, $\varepsilon(\tilde{v}, T) < 1$, and the reflectance $r(\tilde{v}, T)$ is given as

$$r(\tilde{v}, T) = 1 - \varepsilon(\tilde{v}, T) \tag{15.7}$$

For a blackbody $r_{BB}\varepsilon(\tilde{v}, T) = 0$, and for a non-blackbody $r(\tilde{v}, T) > 0$.

15.2.1.3   *Emission from a Sample*

In the measurement of the emission spectrum from a sample shown schematically by the arrows in Figure 15.2, the emission contains contributions emanating from both the sample surface and the interface of the sample and the base (sample support).

If the sample is a thin film, emission from its back support should also be taken into account. As the emission from the back support passes through the sample layer, a part of it will be absorbed by the sample and part of it will be emitted at the sample surface. The emission actually measured is the result of all these processes. In practice, it is possible

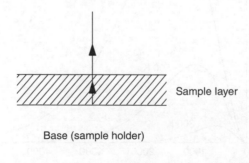

Base (sample holder)

Figure 15.2   *Schematic emissions from a sample layer and its base.*

to reduce the emission from the back support by using a flat plate of a metal with a low emittance.

15.2.2   Emission Spectrometer

In infrared emission spectrometry, very weak infrared radiation of emission from a sample itself is to be measured. Consequently, infrared emission measurements have requirements considerably different from transmission and reflection measurements. It is not desirable to use a complicated optical apparatus for emission measurements.

Because, in emission measurements, the source of infrared radiation and the sample are one and the same, the apparatus for an emission measurement is placed before the interferometer of an FT-IR spectrometer, effectively replacing the conventional source used for a transmission measurement. In order to meet the various requirements of analysis such as measuring the angle dependence of emitted radiation, controlling the temperature of the sample, adsorbing a gas onto the sample, and desorbing a gas from the sample, it is desirable to place an apparatus for emission measurements in a spacious compartment. As an example, an apparatus for emission measurements is schematically shown in Figure 15.3.

15.2.3   Emission Measurements and Points to Note

To measure a high-quality emission spectrum, the sample should be prepared carefully. The following points should be handled in an appropriate manner.

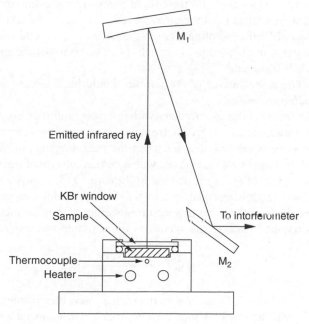

Figure 15.3   *Schematic example of an apparatus for emission measurements. M_1, a spherical mirror; and M_2, a plane mirror.*

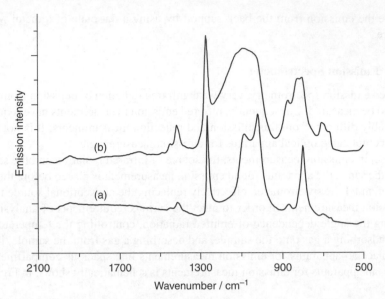

Figure 15.4 *Emission spectra of silicone grease at 140°C. (a) A thin sample layer and (b) a thick sample layer.*

1. If the sample layer is too thick (in the case of a gas, if its pressure or concentration is too high), the wavenumber resolution may be reduced and the band shape may become distorted because of self-absorption. The emission spectra from a thin and a thick sample layer are compared in Figure 15.4, in which it can be seen that the spectrum from the thicker sample is distorted.
2. The temperature at the surface of the sample should be measured accurately with a surface-temperature sensor.
3. Purging of the inside of the spectrometer with dried air or nitrogen is essential in order to remove the unwanted effects arising from the absorptions by water vapor and carbon dioxide. This is particularly important when the measurements of the spectra of the reference and the sample are undertaken with a certain interval of time.
4. Although the emittance $\varepsilon(\tilde{\nu}, T)$ is defined by Equation (15.6), it is a usual practice to determine it by the following equation after measuring the emission spectra of the sample and a blackbody at the same temperature together with other measurements. This procedure is required because the apparatus for emission measurements has its own characteristics.

$$\varepsilon(\tilde{\nu}, T) = \frac{I_S(\tilde{\nu}, T)}{I_{BB}(\tilde{\nu}, T)} \tag{15.8}$$

where $I_S(\tilde{\nu}, T)$ and $I_{BB}(\tilde{\nu}, T)$ are, respectively, the intensities of the emission spectra of the sample and the blackbody at wavenumber $\tilde{\nu}$ and temperature T. $I_S(\tilde{\nu}, T)$ and $I_{BB}(\tilde{\nu}, T)$ include the effects of the instrumental function, but $\varepsilon(\tilde{\nu}, T)$ obtained by taking their ratio becomes free from the instrumental function. However, $I_S(\tilde{\nu}, T)$ and $I_{BB}(\tilde{\nu}, T)$ contain the emission from the background $I_{BG}(\tilde{\nu}, T)$. The simplest way to estimate $I_{BG}(\tilde{\nu}, T)$ is to use the emission spectrum of the metal base in the absence of the sample (see Figure 15.2). Then, $\varepsilon(\tilde{\nu}, T)$ is obtained by the following equation:

$$\varepsilon(\tilde{\nu}, T) = \frac{I_S(\tilde{\nu}, T) - I_{BG}(\tilde{\nu}, T)}{I_{BB}(\tilde{\nu}, T) - I_{BG}(\tilde{\nu}, T)} \tag{15.9}$$

when the temperature of the sample is high enough, a high-quality emission spectrum can be obtained in this way. However, if the temperature of the sample is relatively low, a more accurate method should be adopted to derive $I_{BG}(\tilde{\nu}, T)$ as described later (see Section 15.3.3).

5. The material of the base (sample support) should have a high thermal conductivity, and should not be altered by heating.
6. If a sample, when heated, might give off a gas which may deteriorate or adversely affect the performance of optical elements, the compartment which houses the apparatus for measuring emission should be purged with an appropriate amount of nitrogen.
7. As the detector for emission measurements needs to be sensitive, a highly sensitive mercury cadmium telluride (MCT) detector is usually employed. Its use becomes indispensable if the temperature of the sample is close to room temperature.

15.3 Examples of Emission Spectra and Techniques and Procedures Used in Emission Measurements

15.3.1 Comparison of Emission and Transmission Spectra

In Figure 15.5, the emission spectrum of silicone grease is compared with its transmission spectrum. In this example, the emission spectrum is similar to the transmission spectrum. Generally speaking, however, bands in emission spectra are often distorted due to overlapping of emission from sources other than the sample. Further, if the sample is thick, the wavenumber resolution may become lower due to self-absorption, as mentioned already. Methods for avoiding these drawbacks are described in the following section.

15.3.2 Transient Infrared Emission Spectrometry

A method called *transient infrared emission spectrometry* (*TIRES*) was proposed in order to measure a high-quality emission spectrum from thick samples by reducing the effects of self-absorption [2]. Actually, this method has been separated into two types, namely, the surface-heating "transient infrared emission spectrometry (TIRES)" [3] and the surface-chilling "transient infrared transmission spectrometry (TIRTS)" [4]. In TIRES, a small portion of a thin surface layer of a sample (moving) is transiently heated by a jet of hot nitrogen gas, and the emission from the heated thin surface layer is detected. In TIRTS, by contrast, a small portion of the thin surface layer of a high-temperature sample is transiently chilled by cooled helium gas. In this case, the inside of the high-temperature sample works as an infrared source. As a result, the transmission spectrum of the chilled thin surface layer is observed. In Figure 15.6, the spectra measured by TIRES and TIRTS using a thick polycarbonate sample are compared with those obtained by ordinary emission spectrometry and photoacoustic spectrometry. The results obtained by TIRES and TIRTS are comparable to that obtained by photoacoustic spectrometry.

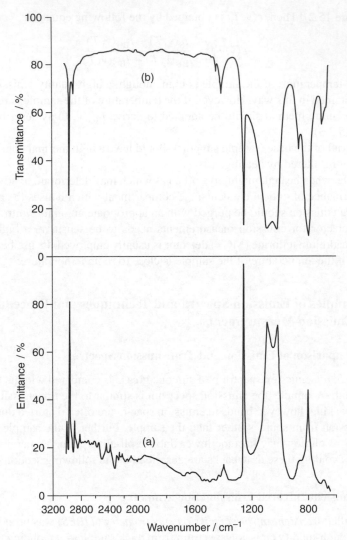

Figure 15.5 *Comparison of (a) the emission spectrum of silicone grease at 180 °C and (b) the corresponding transmission spectrum.*

15.3.3 Emission Measurements by Using Two Blackbodies

Interest in the technology which utilizes infrared radiation emitted by ceramics, and similar, is increasing, and its applications are broadening. For this purpose, it is required to determine accurately the spectroscopic emittance of ceramics and their composites with plastics and fibers. As the temperature at which they are practically used with their properties unchanged from the normal is usually below 100 °C, the emissions from them are very weak, and at a level which is close to that of the emission from their environment (called the *background emission*). When the emission in this temperature range is measured, it is essential to take into account the background emission. To estimate the background

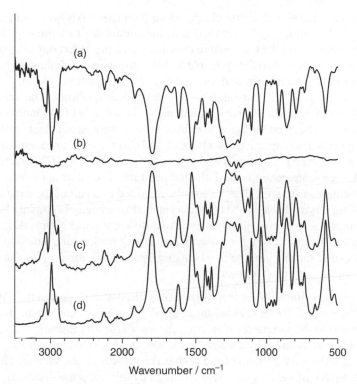

Figure 15.6 *Comparison of the spectra of a 3 mm thick polycarbonate film measured by different methods. (a) Transmission spectrum measured by TIRTS, (b) emission spectrum measured by a conventional method, (c) emission spectrum measured by TIRES, and (d) photoacoustic spectrum. (Source: Reproduced with permission from J. F. McClelland and R. W. Jones.)*

emission contained in the emission measured from a sample, a method developed by the present author (in cooperation with K. Masutani) is described here.

The emission spectrum $I_S(\tilde{\nu}, T)$ measured from a sample at temperature T is expressed as

$$I_S(\tilde{\nu}, T) = [I_S^t(\tilde{\nu}, T) + I_{BG}(\tilde{\nu})]A(\tilde{\nu}) \tag{15.10}$$

In this equation, $I_S^t(\tilde{\nu}, T)$ represents the "true" intensity of emission from the sample, $A(\tilde{\nu})$ is the instrumental function, and $I_{BG}(\nu)$ is the background emission that needs to be determined in this measurement. $I_{BG}(\tilde{\nu})$ is regarded as a spectrum which does not change during the measurement, provided the temperature of the spectrometer is kept constant. If $I_{BG}(\tilde{\nu})$ and $A(\tilde{\nu})$ are known, $I_S^t(\tilde{\nu}, T)$ can also be determined from Equation (15.10). The method developed here to determine $I_{BG}(\tilde{\nu})$ consists of the following two steps: (i) measuring the emissions of two blackbodies set at two different temperatures, $I_{BB}(\tilde{\nu}, T_1)$ and $I_{BB}(\tilde{\nu}, T_2)$; and (ii) solving the following simultaneous equations.

$$I_{BB}(\tilde{\nu}, T_1) = [I_{BB}^t(\tilde{\nu}, T_1) + I_{BG}(\tilde{\nu})]A(\tilde{\nu}) \tag{15.11}$$

$$I_{BB}(\tilde{\nu}, T_2) = [I_{BB}^t(\tilde{\nu}, T_2) + I_{BG}(\tilde{\nu})]A(\tilde{\nu}) \tag{15.12}$$

In the above two equations, at first glance, more than two unknowns seem to exist, but actually $I_{BB}^t(\tilde{v}, T_1)$ and $I_{BB}^t(\tilde{v}, T_2)$ are known, as the emittance of a commercially available blackbody is known, and the temperature dependence of the blackbody emission is given by Equation (15.4). It is therefore possible to solve the above simultaneous equations to derive $I_{BG}(\tilde{v})$ and $A(\tilde{v})$, and then to determine $I_S^t(\tilde{v}, T)$ by Equation (15.10).

In this description, $I_{BG}(\tilde{v})$ is treated as an independent spectrum. The validity of this point should be examined. The background emission consists of: (i) emission from the detector itself, (ii) the fixed background emission from the measurement system (i.e., the wall of the spectrometer, optical elements, etc.), and (iii) the varying background emission which depends on the conditions at the surface of the sample. If an MCT detector operating at liquid-nitrogen temperature is used, the temperature of the detector is constant, so the background emission from the detector may be regarded as a part of the fixed background emission. If the varying background emission can be made negligibly small, the method of using the emissions from two blackbodies at two different temperatures described above would be valid. Then, hardware for reducing the varying background emission is required. Apparatus devised for this purpose is schematically shown in Figure 15.7. The characteristics of this apparatus are as follows.

1. The sample is positioned such that its surface is tilted with respect to the optical axis of the interferometer, in order to avoid any emitted radiation reflected by the interferometer and returning to the sample from entering the interferometer again.
2. The sample is surrounded by a water-cooled blackbody (called the *trapping blackbody*) in order to reduce the emission from within the vicinity of the sample. The reflection from the surface of the sample is also absorbed by this trapping blackbody.
3. Although not necessary for reducing the varying background emission, the emissions from the sample and the two blackbodies at two different temperatures can be conveniently measured at the same conditions by switching mirrors.

The emission from a 30 μm-thick polystyrene film placed on a gold-deposited mirror was measured by using this apparatus, and the measured data were processed by Equations (15.11), (15.12), and (15.10) to obtain $I_S^t(\tilde{v}, T)$, and then the emittance spectrum $\varepsilon(\tilde{v}, T)$ was calculated by Equation (15.9) using $I_{BB}(\tilde{v}, T)$ obtained by Planck's law of radiation. The result is shown in Figure 15.8a. This spectrum is basically similar to the absorbance spectrum recorded from a transmission measurement shown in Figure 15.9, but it has a high background, and the relative band intensities in the spectrum of Figure 15.8a are very different when compared with those of Figure 15.9. These are thought to be due to the emission from the trapping blackbody reflected at the surface of the sample.

Even if the temperature of the trapping blackbody is lowered, the varying background emission cannot be removed when the temperature of the sample is low. In such a case, if the temperature and emittance of the trapping blackbody is known, the varying background emission can be removed by the following calculations.

The intensity of emission from the sample $I_S(\tilde{v}, T)$ obtained by the method described above can be expressed by the following equation, where the reflectance of the sample is denoted by $r_S(\tilde{v}, T)$ and the intensity of emission from the trapping blackbody by $I_{TB}(\tilde{v})$ (the temperature of the trapping blackbody is kept constant in all the measurements conducted

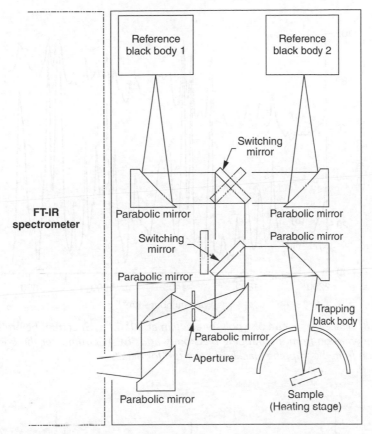

Figure 15.7 *Schematic of an apparatus for emission measurements designed for removing the effects of the background emission by using two blackbodies at two different temperatures. See text for details.*

by using this apparatus).

$$I_S(\tilde{v}, T) = I_S^t(\tilde{v}, T) + r_S(\tilde{v}, T)\, I_{TB}(\tilde{v}) \tag{15.13}$$

By denoting the intensity of emission from the true blackbody at a temperature the same as that of the sample by $I_{BB}^t(\tilde{v}, T)$, and by using Equation (15.7), Equation (15.13) may be rewritten as

$$I_S(\tilde{v}, T) = \varepsilon(\tilde{v}, T)\, I_{BB}^t(\tilde{v}, T) + [1 - \varepsilon(\tilde{v}, T)]\, I_{TB}(\tilde{v}) \tag{15.14}$$

$$\varepsilon(\tilde{v}, T) = \frac{[I_S(\tilde{v}, T) - I_{TB}(\tilde{v})]}{[I_{BB}^t(\tilde{v}, T) - I_{TB}(\tilde{v})]} \tag{15.15}$$

If a blackbody with a known emittance is used, it is possible to calculate $I_{TB}(\tilde{v})$ by this equation and to correct the varying background emission.

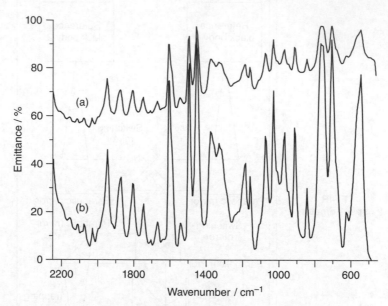

Figure 15.8   *Emission spectra of a polystyrene film at 30°C. (a) Spectrum before the correction for the emission from the trapping blackbody and (b) spectrum after the correction for the emission from the trapping blackbody. See text for details.*

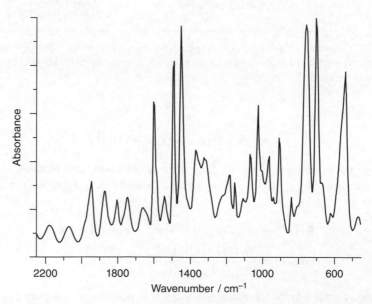

Figure 15.9   *Absorbance spectrum of a thin polystyrene film at 30°C recorded using a transmission measurement.*

As an example similar to that described, the measurement of emission from a 30 μm-thick polystyrene film on a gold-deposited mirror is described. The temperatures of reference blackbodies 1 and 2 were 40.8 and 80 °C, respectively. The emission from the gold-deposited mirror itself is so weak that it may be disregarded in the present measurement. However, the emission from the trapping blackbody, which passes through the sample and is reflected back by the gold-deposited mirror, overlaps the emission from the sample. The trapping blackbody used in this measurement has an emittance of 0.96 in the region of 3–8 μm, and its temperature is set at 9 °C. The emission spectrum shown in Figure 15.8b was obtained by correcting the spectrum in Figure 15.8a by Equation (15.15). The effectiveness of this correction is clearly seen in that the emission spectrum in Figure 15.8b is now very similar to the absorbance spectrum of polystyrene shown in Figure 15.9.

Thus, infrared emission spectrometry is a useful method of analysis over a wide temperature range, if the background emission can be corrected appropriately.

References

1. Mink, J. (2002) Infrared emission spectroscopy, In: *Handbook of Vibrational Spectroscopy*, Vol. **2** (eds J.M. Chalmers and P.R. Griffiths), John Wiley & Sons, Ltd, Chichester, pp. 1193–1214.
2. Jones, R.W. and McClelland, J.F. (1989) Transient infrared emission spectroscopy. *Anal. Chem.*, **62**, 650–656.
3. Jones, R.W. and McClelland, J.F. (1990) Quantitative analysis of solids in motion by transient infrared emission spectroscopy using hot-gas jet excitation. *Anal. Chem.*, **62**, 2074–2079.
4. Jones, R.W. and McClelland, J.F. (1990) Transient infrared transmission spectroscopy. *Anal. Chem.*, **62**, 2247–2251.

16

Infrared Microspectroscopic Measurements

Shukichi Ochiai[1] *and Hirofumi Seki*[2]

[1]*S. T. Japan, Inc., Japan*
[2]*Toray Research Center, Inc., Japan*

16.1 Introduction

In many studies, particularly those related to materials and forensic science, it is frequently necessary to measure a mid-infrared spectrum from a trace amount of a sample or a sample of small size. In some circumstances, this may be accomplished by using a beam-condenser accessory within the conventional sample compartment of a Fourier-transform infrared (FT-IR) spectrometer. Perhaps today though, it is more convenient to use infrared microspectrometry (often commonly referred to as *infrared microspectroscopy* or *even infrared microscopy*). Based on an optical microscope (or infrared microscope) coupled to an FT-IR spectrometer, it is one of the most useful methods for structural analysis of such samples and can often be undertaken in a non-destructive manner [1, 2].

16.2 Mid-Infrared Microspectroscopic Measurements

16.2.1 Infrared Measurements of Trace Amount and Small-Size Samples

No clear definition of a "trace amount" or "small size" seems to exist. Samples of small quantities to be analyzed exist in various shapes and forms. In this chapter, a sample having a diameter of about 10 μm is considered to be typical of the smallest sample size that may

Introduction to Experimental Infrared Spectroscopy: Fundamentals and Practical Methods,
First Edition. Edited by Mitsuo Tasumi and Akira Sakamoto.
© 2015 John Wiley & Sons, Ltd. Published 2015 by John Wiley & Sons, Ltd.

be characterized by mid-infrared microspectrometry; for example, a microcrystal, a foreign body embedded in a plastic material (often called a *fisheye*), or a contaminant on an electric contact, and so on. Where a number of smaller similar samples exist separately, they may be collectively called a *trace amount of a sample*; for example, extremely small particulates which are scattered across a metal surface.

For the analysis of a trace amount of a sample of very small size, it is most important to collect representative specimens in order to increase the total amount to a level that is sufficient to measure its infrared spectrum by either a more conventional method (e.g., placing a beam-condenser accessory into the conventional sample compartment in order to focus the infrared beam onto a small-diameter KBr disk which contains a trace amount of a sample) or by using an infrared microscope. Specific methods of such sampling will be described in Section 16.2.3.1.

Infrared microscopes, which are today frequently used in many laboratories, are optical microscopes combined with FT-IR spectrometers. Nowadays, some FT-IR spectrometers are smaller in size than the infrared microscope units combined with them. This seems to show the importance of infrared microspectroscopy for many and varied application purposes and uses in places other than spectroscopy laboratories.

With an infrared microscope system, the infrared beam is focused onto the sample in order to achieve high lateral spatial resolution; also (as the infrared and visible beams are collinear and parfocal), it is possible to view the sample being examined by using the optical microscope, thereby enabling one to select the part of the sample from which an infrared spectrum should be recorded by masking the unwanted part. In order to minimize diffraction effects, the minimum size of the target sample area is, as mentioned above, about 10 µm in diameter.

A small-size target specimen will often need to be removed from a matrix in which it is buried by using a knife, scalpel, or needle in an appropriate and safe manner, depending on the conditions surrounding it. If the target specimen is covered by a layer of some other material this may need to be scraped off. Infrared microscopic measurements are utilized in many areas of study. In order to satisfy the various needs, in addition to straightforward transmission measurements, various types of reflection measurements such as specular reflection, attenuated total reflection (ATR), and "transmission–reflection" measurements (described later) can be performed with an infrared microscope designed for such purposes.

16.2.2   Infrared Microscopes

In an infrared microscope, the infrared beam from the interferometer of an FT-IR spectrometer is focused onto a sample placed on the sample stage of the microscope and the transmitted radiation (in the case of reflection measurements, the reflected radiation) is then focused onto to the spectrometer detector. An infrared microscope set-up, which is equipped with Cassegrain condensing and objective mirror systems, is schematically shown in Figure 16.1. In transmission measurements, the infrared beam from the interferometer enters the microscope from underneath via a switching mirror, and is focused onto the sample by the Cassegrain condenser. The radiation which has been transmitted through the sample passes through the Cassegrain objective and then to the detector. In reflection measurements, on the other hand, the infrared beam from the interferometer is directed upwards via the switching mirror, and is then focused downwards onto the sample.

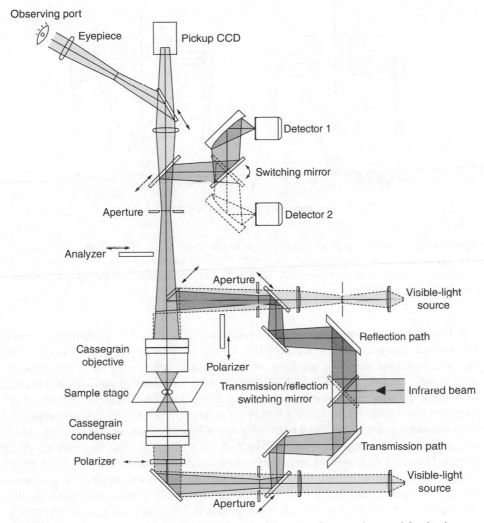

Figure 16.1 *Schematic of an infrared microscope set-up that can be used for both transmission and reflection measurements. (Source: Adapted with permission from the catalog of Bruker Optics, Inc.)*

Figure 16.2 shows schematics of two measurement set-ups using an all-reflection mirror objective design which may be used in an infrared microscope system. In Figure 16.2a, a schematic of the objective operating in ATR mode is shown. The infrared beam, via a mirror situated on the right-hand side of the microscope, enters the right-hand half of the Cassegrain objective, and then the IRE (internal reflection element for ATR) from its right-hand side. The infrared beam, after being attenuated through absorption by the sample, leaves the IRE from its left-hand side, passes to the left-hand half of the Cassegrain objective from which it is reflected, and finally reaches the detector located in the upper part of the microscope. The directions of the infrared beam in the Cassegrain objective are indicated with arrows.

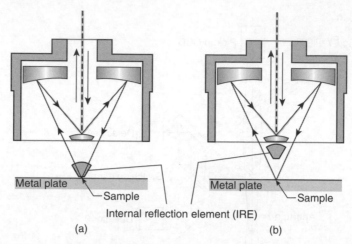

Figure 16.2 *Schematics of an objective for microscopic ATR and transmission–reflection measurements. (a) Arrangement for ATR measurements and (b) arrangement for transmission–reflection measurements. The IRE can be moved up and down. (Source: Adapted with permission from the catalog of Bruker Optics, Inc.)*

In Figure 16.2b, the objective set for a "reflection–absorption" measurement is shown. Dirt on and objects attached to metal plates are frequently targets of infrared microscopic measurements, but generally speaking, as their surfaces are not flat and their refractive indices are not high, no useful high-quality specular-reflection spectra can be recorded from them. In such a case, it is common to try to measure the "transmission–reflection" spectrum (sometimes referred to as a *transflection* measurement) by utilizing the reflection from the metal plate. The infrared beam, which irradiates the sample (dirt or other objects on the metal plate) from the right-hand half of the Cassegrain objective may be partly reflected from the surface of the sample but in many circumstances most of it will transmit through the sample. The transmitted beam will be reflected by the metal plate, pass again through the sample, and then pass through the left-hand side of the Cassegrain objective, and then reach the detector. As a result, it is expected that, by this "transmission–reflection" measurement, a spectrum similar in appearance to a transmission spectrum will be obtained (see also Chapter 10).

An operator may also view an image of the sample through the reflection measurement objective by passing visible light through it. This is useful for selecting specific areas on the sample from which infrared spectra should be measured.

For the detector, a highly sensitive mercury cadmium telluride (MCT) (typically, $250 \, \mu m$ $\times \, 250 \, \mu m$) (labeled position Detector 1 in Figure 16.1) is usually employed. A detector of this type can cover the wavenumber region down to about $700 \, cm^{-1}$. When spectra in the region lower than $700 \, cm^{-1}$ are needed, a broad-band MCT detector (Detector 2 in Figure 16.1), which can extend the measurable region down to about $450 \, cm^{-1}$, is used.

FT-IR spectrometers specifically manufactured for reflection microspectrometry are commercially available. Such spectrometers can be combined with most microscopes having infinity-corrected optics (in which the output from the Cassegrain objective mirror becomes parallel radiation). Since, in the infinity-corrected design, the output is a parallel beam, an image-forming system is needed. The infinity-corrected design has the advantages of: (i) no change in magnification occurring if the distance between the objective and image-forming mirrors is changed, and (ii) no distortion of an image occurring if a parallel plate (for example, a polarizer) is inserted between the objective and image-forming mirrors.

16.2.3   Sample Preparation

16.2.3.1   Trace Amount of a Samples

In this subsection, methods will be considered for preparing trace amounts of samples from which their infrared spectra may be measured, without resorting to using an infrared microscope, by using a beam-condenser accessory positioned in the conventional sample compartment of an FT-IR spectrometer.

If dirt, contaminant spots, and so on, on a metal or plastic plate can be scraped off with a knife or other tool from the plate, and if necessary gathered together, to form a sample of an appropriate small size, it may become a target sample for an infrared microspectroscopic measurement. However, in practice, it is often difficult to carry out such a process. In this case, it may be better to put a small amount of KBr powder onto the dirt or contaminant spots and then rub the KBr powder lightly with a small pestle in order to mix the dirt or contaminant spots with the KBr powder. Then, a KBr disk of a small diameter may be formed from the mixed powder. Presses are available commercially for forming a KBr disk with a diameter as small as 1 mm. When the amount of a sample is small, it is definitely better to use a small diameter disk for infrared measurements. This is demonstrated in Figure 16.3, where the infrared spectra of equal amounts of fumaric acid in two KBr disks

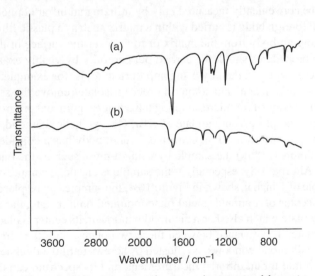

Figure 16.3   *Infrared transmission spectra of an equal amount of fumaric acid in KBr disks with diameters of (a) 5 mm and (b) 13 mm; spectra are shown on the same transmittance scale.*

of different diameters are compared; it is clear that the infrared spectrum of the smaller-diameter KBr disk is of much higher quality.

If the dirt or contaminant spots can be dissolved in a low boiling-point solvent, solvent extraction (removal) may be an effective method for collecting the sample. If the base plate is a plastic material, the solvent must be carefully selected to ensure that none of the polymer support is dissolved. The solution obtained may then be slowly dripped from a syringe onto a thin film, for example, of polyethylene or polytetrafluoroethylene, and after the solvent has evaporated completely, the remaining powder can then be made into a KBr disk. If the solution is dropped directly onto an infrared-transparent window or a metal plate, the spot often tends to spread to form a circle of the sample after evaporation of the solvent, making the sample then not suitable for either the transmission or transmission–reflection measurement method. If the solution is dropped into a small hole (for example, with a diameter of 1 mm and a depth of 3 mm drilled into a metal support) filled with KBr powder, a diffuse-reflection spectrum may be measured from this after complete evaporation of the solvent.

Eluent from a liquid chromatograph can sometimes be handled in the same way as described above, but if the mobile phase is water or contains water, the eluent cannot be dropped onto KBr powder.

Since the size of the sample prepared in the manner described above is not considered very small, an infrared spectrum of a high signal-to-noise (S/N) ratio can be measured if a beam-condenser accessory is used. A beam-condenser is an accessory much simpler than a microscope, and it is used easily to focus the incident infrared beam onto such a small KBr disk.

16.2.3.2   Small Sample Size

This subsection describes methods for preparing small-size samples, the infrared spectra of which may be conveniently measured only by utilizing an infrared microscope.

When a small foreign body is buried within a matrix such as a plastic film, it is necessary to extract the foreign body from the matrix or to uncover the surface of the foreign body. Such sampling procedures should be carried out carefully by visibly examining the sample under a microscope and then by using appropriate tools, for example, a needle, a pair of tweezers, a knife, a plane, and so on, in order to isolate/remove the sample. Recently, tools specifically designed for microsampling have made significant progress. An extracted foreign body may be placed onto a plate of KBr, Ge, synthetic diamond, etc., in order to measure its infrared spectrum. If an extracted foreign body has a considerable thickness, it is better to compress (thin) the sample by using a press such as the one used for making KBr disks. Alternatively, especially if the sample is elastic, a compression cell may be used, an example of which is shown in Figure 16.4; the sample can be placed on a diamond window in the center of a circular metal plate mounted on a rectangular metal plate, and another circular plate which also has a diamond window in its center is placed on top of the other circular plate. The distance between these two plates, with the sample sandwiched between the two diamond windows, is adjusted by screws until the thickness of the sample becomes appropriate for an infrared measurement, and its spectrum can then be recorded with the sample contained within the cell in this compressed state.

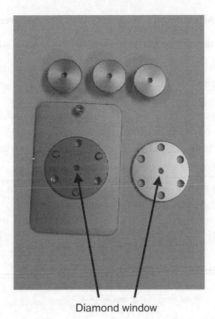

Diamond window

Figure 16.4   *Compression cell with diamond windows. (Source: Reproduced with permission of S. T. Japan, Inc.)*

If the target specimen of microscopic measurements exists on a metal substrate, for example, dirt on electric contacts and magnetic heads, or is a small irregular point on a painted metal plate (sometimes called a *cissing*), the transmission–reflection measurement can sometimes prove to be a useful examination technique. For this purpose, as mentioned earlier, the objective set for reflection measurements as shown in Figure 16.2b is used.

16.2.3.3   Film and Laminate Films

If it is necessary to measure the transmission infrared spectrum from each layer of such as a polymer laminate film, then in some circumstances, a specimen of suitable thickness may be prepared by cross-sectioning a thin layer (typically 10–30 µm) by using a microtome. Layer widths need to be about 10 µm or wider in order to record an infrared transmission spectrum in which artifacts caused by diffraction effects are minimized. If the width of a narrow layer is <10 µm, then this may sometimes be increased by cross-sectioning the laminate film at an oblique angle, thereby creating effectively a wider layer better suited to minimizing diffraction effects in its recorded spectrum. A laminate film may need to be first set in an epoxy resin and then a thin cross section of the laminate film is taken using a microtome, so that the transmission spectrum of each layer can then be measured. An appropriate varying (gradient) thickness film sample for a depth-profiling measurement may be prepared by the method of gradient shaving in order to generate a specimen of varying (sloping) thickness, from which appropriate infrared spectra may be recorded from each layer. For the details of the gradient-shaving method, see Section 16.5.

16.2.4   Mid-Infrared Microscopic Measurement Procedures

Microscopic measurements can be performed in the following way.

1. The target area from which an infrared spectrum needs to be measured is specified, and optical adjustments are performed in order to obtain an image of high contrast.
2. A masking aperture is placed around the selected target area in order to define it, and thereby prevent the spectral characteristics of surrounding material from being measured.
3. An infrared spectrum of the target area is measured.
4. For the reference spectrum, a single-beam background spectrum of the aperture in place, but without the sample, is measured.

It is recommended that a clear visible image of the target area is recorded as it may help in the understanding of the measured infrared spectrum.

Spatial resolution of an infrared microscope is determined by the numerical aperture (NA) of a Cassegrain optic and the wavelength of infrared radiation λ [2, 3].

The minimum distance r between two points which can be distinguished as different images is given by the following equation:

$$r = \frac{0.61\lambda}{NA} \tag{16.1}$$

In this equation, NA is the numerical aperture of the objective (or condenser) mirror defined as

$$NA = n \sin \theta \tag{16.2}$$

where n is the refractive index of the medium through which the sample is irradiated ($n = 1$ in the case of the air) and θ is the angle between the normal to the surface of the sample and the outermost ray going to the objective. The value of NA of a Cassegrain optic is in the range of $0.5-0.7$. In practice, the value of r becomes larger because of aberration in the optical system. If the incident infrared radiation is scattered strongly by the sample, the spatial resolution may be reduced. To minimize this effect, a second aperture is placed after the sample. This so-called "redundant aperture" is installed in many commercially available infrared microscopes.

Since, in microscopic ATR measurements, n is the refractive index of the IRE (which is usually larger than 2), the lateral spatial resolution ought to become higher. However, it should be noted that NA becomes half the normal value, because only half of the Cassegrain optic is actually used in the ATR measurements.

If a masking aperture of a size <20 μm is used in the transmission measurements, spectral information from positions adjacent to the area from which the infrared spectrum should be measured may overlap the spectrum to be measured due to the effects of diffraction. This possibility should be taken into account when analyzing and interpreting the measured spectrum.

16.3   Application Examples

In Figure 16.5, the infrared spectra recorded from different layers in a multi-layer plastic used in manufacturing a bottle are shown. A 30 μm-thick sample for the infrared

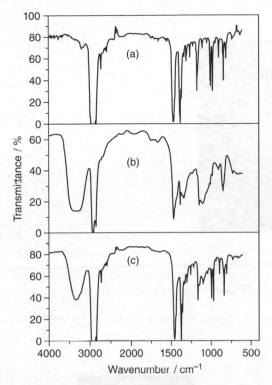

Figure 16.5 *Infrared microscopic spectra recorded from plastic-bottle layers. (a) Innermost layer, (b) second layer from inside, and (c) third layer from inside.*

measurements was obtained by cutting the side of the bottle with a microtome blade. The infrared spectra of the three layers from the inside of the bottle toward the outside were individually successively measured by masking each with a square aperture (50 μm × 50 μm). Since the width of each layer is much wider than 50 μm, the infrared spectrum recorded from each layer is not mixed with that of the adjacent layer(s). Spectrum (a) corresponds essentially to a 300 μm-wide polypropylene layer, which is used for the first (innermost) layer because of its more stable properties. Spectrum (b) shows that the second layer from the inside is a copolymer of ethylene and vinyl alcohol (100 μm wide), which is used as a barrier layer to prevent the ingress of oxygen into the bottle. Spectrum (c), corresponding to the third layer, is a mixture of polypropylene and the copolymer used for the second layer. A dye is contained in this layer for the purpose of making the bottle content invisible from the outside. Although the infrared spectrum of the outermost layer is not shown in Figure 16.5, it is polypropylene-based (500 μm thick).

In Figure 16.6, a microphotograph of a cross section of a drug and the microscopic ATR spectra recorded from its four components, which can be distinguished in the microphotograph, are shown. The areas distinguished in the microphotograph produce different ATR spectra. Although spectrum (a) corresponding to starch is similar to spectrum (c) arising from gum (a mixture of polysaccharides), their microphotographs are definitely different. This example indicates that a clear microphotograph helps complement spectral analysis and understanding.

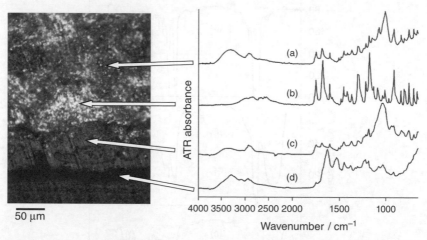

Figure 16.6 *Microphotograph of the cross section of a drug and the infrared microscopic spectra recorded from different areas (indicated by the arrows) of the cross section. (a) Starch, (b) acetaminophen (p-acetylaminophenol), (c) gum, and (d) gelatin (spectra have been offset to aid clarity). (Source: Reproduced with permission from Smiths Detection © 2013.)*

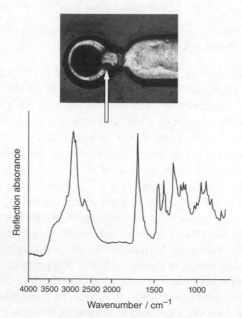

Figure 16.7 *Small foreign matter (indicated by the arrow) on a printed-circuit board and its transmission–reflection spectrum. (Source: Reproduced with permission from Smiths Detection © 2013.)*

In structural analysis of active pharmaceutical ingredients, crystal polymorphism is an important subject of study. Crystal structures are determined by X-ray diffraction studies. Once the correlation between crystal polymorphs and their infrared spectra has been established, it can then become a relatively easy task to distinguish polymorphs by measuring their microphotographs and corresponding microscopic infrared spectra, as infrared microspectrometric measurements can be conveniently performed within a much shorter time than X-diffraction analysis.

In Figure 16.7, the infrared spectrum of a very small amount of a substance that was found as a contaminant on a printed-circuit board is shown. This infrared spectrum was measured by the microscopic transmission–reflection method, utilizing the reflection from the base metal. By a search of spectral data library, it became clear that this substance consisted mainly of abietic acid contained in the solder.

16.4 Mapping Analysis

It is possible to measure infrared spectra from successive points of a sample placed on an x–y motorized stage, movable in a regular manner by computer control. Such a study may be called a *mapping* measurement.

The mapping of a cross section of a rock, a sample form rarely studied by infrared spectroscopy, was undertaken [4]. The sample was a granite consisting of quartz, potash feldspar, and many other minerals. The measurements were performed on a polished piece 30 μm thick (a square-like shape about 3 mm long and wide). A microphotograph of a part of this small piece is shown in Figure 16.8. The target of mapping was the area shown in the nearly square frame (140 μm in the x-direction and 130 μm in the

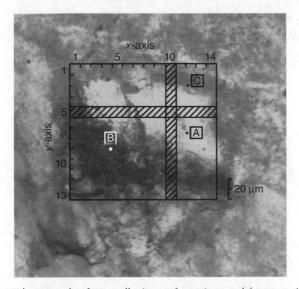

Figure 16.8 *Microphotograph of a small piece of granite used for mapping measurements (see text for details). (Source: Reproduced with permission from Geochemical Journal © 1989.)*

y-direction) in the center of this figure (the frame and its *x*- and *y*-axes were chosen somewhat arbitrarily). The mapping was performed by measuring transmission spectra using a masking aperture of 10 μm × 10 μm and a step size of 10 μm; spectra were recorded at a wavenumber resolution of 8 cm^{-1}. In the framed part, region A was observed to be translucent, region B had a dark color, and region C was less colored. The results of mapping along the *y*-axis of the tenth column in the *x*-axis (indicated with oblique lines) are shown in the 13 infrared spectra in Figure 16.9a, and those along the *x*-axis of the fifth row in *y*-axis (also indicated with oblique lines) are shown in the 14 infrared spectra in Figure 16.9b. Bands in the spectra shown in these figures are due to the OH stretching vibration. The sharp peak at 3480 cm^{-1} may be assigned to the OH stretching band of a component mineral, prehnite, which is generally translucent and has a chemical formula of $Ca(Al,Fe^{3+})_2Si_3O_{10}(OH)_2$. As the 3480 cm^{-1} band is strongly observed in region A and the other translucent region of the microphotograph shown in Figure 16.8, these regions consist predominantly of prehnite. Another broader band at about 3400 cm^{-1} observed particularly in region B is thought to be attributable to a chlorite-group mineral, the formula of which is $(Mg,Al,Fe)_{12}(Si,Al)_8O_{20}(OH)_{16}$.

The infrared spectra recorded from points along either the *x*- or *y*-axis shown in Figure 16.9 may be called the results of a *line analysis*. Such measurements can be

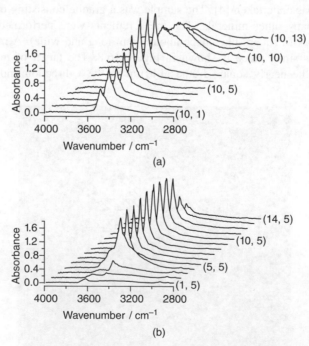

Figure 16.9 *Infrared microscopic spectra of the OH stretching region recorded from areas of the sample as shown in Figure 16.8. (a) Spectra of the 13 points along the y-axis in the tenth column of the x-axis and (b) spectra of the 14 points along the x-axis in the fifth row of the y-axis.*

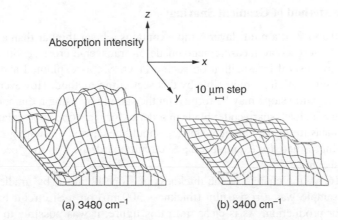

(a) 3480 cm⁻¹   (b) 3400 cm⁻¹

Figure 16.10   *Mapping of the sample area shown in Figure 16.8. (a) Three-dimensional map showing distribution of prehnite obtained by monitoring the 3480 cm⁻¹ band and (b) three--dimensional map showing distribution of a group of chlorites obtained by monitoring the 3400 cm⁻¹ band. (Source: Reproduced with permission from Geochemical Journal © 1989.)*

extended to two dimensions to cover the entire area in the frame. From the results of such measurements, the intensities (denoted here as the z-axis) of the 3480 and 3400 cm⁻¹ bands in the framed area are shown in a three-dimensional presentation in Figure 16.10a,b, respectively. Such a study may be called a *spectral surface mapping* measurement.

   While spectroscopic mapping is utilized more and more in analyzing industrial products, it has a drawback in that its measurements usually take a long time. Infrared spectroscopic imaging, an alternative method of recording spectral information from a small sample area, is introduced in Chapter 17.

16.5   Depth-Profile Analysis by the Method of Gradient Shaving

In the field of science and engineering of industrial materials and semiconductors, it is often important to study the states of compounds near the surface of a material. Such studies include the depth profiling of the composition of chemical components, the examination of effects of surface processing, and so on. The ATR method (see Chapter 13) and photoacoustic spectrometry (see Chapter 14) are useful for such purposes, but it is difficult by these methods to probe thin (less than a few micrometers thick) layers with a sufficiently high depth resolution. Even infrared microscopes have theoretical and practical limits to their spatial resolution and it is difficult to study the states of chemical bonding of a target material of a depth smaller than a few micrometers. To overcome such a limitation, the method of gradient shaving was developed [5]. This method forms a very gentle slope on the target sample by gradient shaving. This pretreatment of the sample may be used to effectively extend the apparent length of a very thin layer to that which is sufficient for probing varying depth-related compositional differences within the thin layer by infrared microscopic measurements.

16.5.1   The Method of Gradient Shaving

As mentioned above, if a multi-layer sample contains a layer thinner than a few micrometers, infrared microscopic measurements made on transverse cross sections of the sample will not give useful information on such a layer. So, as explained above, recording
depth-profile information is impossible by this sectioning method. However, as shown in
Figure 16.11, a gentle slope may be formed on the sample by using a diamond cutter. It is
important to use a clean, sharp cutter blade in order to form a slope appropriate for infrared
microscopic measurements. Gradient shaving should be carried out with great care. If the
sample is a film formed, for example, on a Si circuit board, the film may peel off from the
board while shaving.

In Figure 16.12, the change of the thickness of slope obtained by gradient shaving is
shown. The sample was a resist film (thickness 300 nm) on a Si circuit board used for
semiconductor production. As is clear from this figure, it was possible to form a slope
(film length) of longer than 300 μm. This means that the length of the film now available for

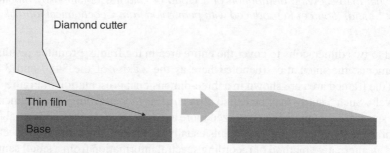

Figure 16.11   *Illustration of how to perform a gradient shaving.*

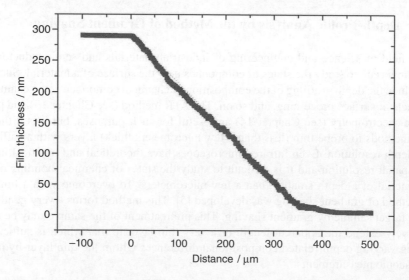

Figure 16.12   *Thickness of the slope prepared by gradient shaving. (See text for details.)*

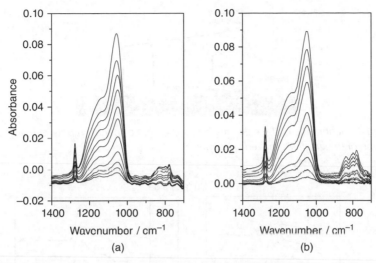

Figure 16.13 *Infrared microscopic transmission spectra of recorded from a film of MSQ on a Si circuit board at different points on the slope specimen prepared by gradient shaving. (a) MSQ film with surface processing and (b) MSQ film without surface processing.*

a depth-profile analysis by infrared microspectroscopy was effectively extended by more than 1000. In this case, the method of gradient shaving caused little change in the properties of the sample. This is considered to be an advantage of this method of sample preparation.

16.5.2 Examples of Measurements

16.5.2.1 Depth-Profile Analysis of an Interlayer-Insulating Film on a Si Circuit Board

In Figure 16.13a,b, the infrared microscopic transmission spectra measured along the direction of the slope formed (by gradient shaving) on an interlayer-insulating film of methylsilsesquioxane (MSQ; thickness 250 nm), with and without surface processing, respectively, are shown. The intensities of bands decreased proportionally as the target of microscopic measurements was shifted along the slope in the direction of decreasing thickness of the slope, indicating that the slope formed was of a good shape. A molecule of MSQ has a cage made of the Si–O linkage with the methyl groups directing outward. The strong band at about $1040 \, cm^{-1}$ is due to the Si–O stretch.

In order to obtain information at a point at a certain distance from the surface, it is necessary to calculate absorbance difference spectra between adjacent points as shown schematically in Figure 16.14. The difference spectra obtained in this way are shown in Figure 16.15. As seen in (b) of this figure, any spectral changes with depth are very small in the film without surface processing. By contrast, a shift of the Si–O stretch toward a higher wavenumber is seen near the surface in (a). This is considered to be a consequence of surface processing.

By analyzing the results obtained in the measurements described above, it was possible to estimate quantitatively the chemical change due to surface processing. In Figure 16.16, the intensity of a band at about $1280 \, cm^{-1}$, which is associated with the Si–CH_3 group, is

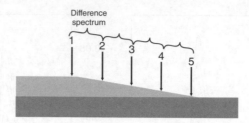

Figure 16.14   *Schematic of the calculation of a difference spectrum between adjacent points on the slope of the gradient-shaved specimen.*

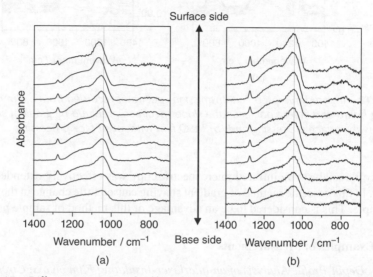

Figure 16.15   *Difference spectra of a film of MSQ obtained at different points on the slope. (a) MSQ film with surface processing and (b) MSQ film without surface processing. See text for details; spectra have been offset for clarity.*

plotted against depth. This result shows that the concentration of Si–CH$_3$ groups greatly decreased near the surface, and this decrease continued for about 100 nm downward from the surface in the surface-processed film, as seen in Figure 16.16a, whereas the number of the Si–CH$_3$ groups did not change in the film the surface of which was not processed, as evident in Figure 16.16b. In this analysis, information on the chemical bonding was obtained at a spatial resolution of 25 nm, which is unusually high for an infrared microscopic analysis.

16.5.2.2   *Analysis of the Degradation of a Polycarbonate Resin*

The result of a light-resistance test of a polycarbonate resin was examined by using the method of gradient shaving. Since the sample was thick, specular reflection from the sample was measured by using a reflection optic. The incident angle was about 30° against the slope prepared by gradient shaving, and the values of absorption index k were calculated from the measured specular-reflection spectrum by using the Kramers–Kronig relations (see Section 6.2.7 and Chapter 8). The absorption index k corresponds to the magnitude of absorption

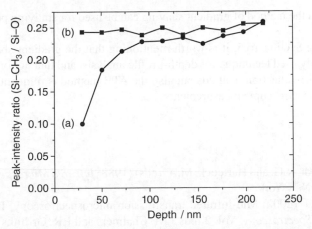

Figure 16.16 *Relative change of the number of the Si–CH₃ groups with depth. (a) MSQ film with surface processing and (b) MSQ film without surface processing. (See text for details).*

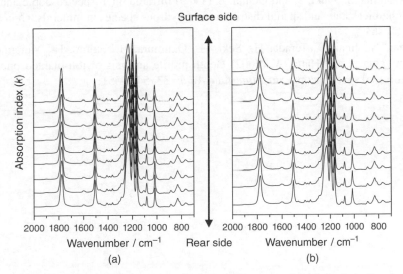

Figure 16.17 *Absorption-index spectra of a polycarbonate resin. (a) Resin before light-resistance test and (b) resin after light-resistance test. (See text for details; spectra have been offset for clarity.)*

(see Section 1.2.4). The absorption-index spectra calculated from the specular-reflection spectra measured from various points along a length of thickness variation of about 5 μm are shown in Figure 16.17. As seen in (a) of this figure, the absorption-index spectrum of the sample before the test of light-resistance is independent of depth. After the test of light-resistance, the spectra shown in (b) obtained from the points near the surface showed changes in the region of $1800-1700 \, \text{cm}^{-1}$, indicating the formation of another C=O group. This indicates that an oxidative change of the resin occurred in the layer very close to the surface and that the method of gradient shaving can be used for studying photodegradation processes.

In concluding Section 16.5, it is worth mentioning that the gradient-shaving method is now a commonly used technique for depth-profile analysis, and that not only the transmission and specular-reflection methods but also the ATR method is frequently used for such infrared microspectroscopic measurements.

References

1. Messershmidt, R.G. and Hartcock, M.A. (eds) (1988) *Infrared Microspectroscopy*, Marcel Dekker, New York.
2. Sommer, A.J. (2002) Mid-infrared transmission microspectroscopy, In: *Handbook of Vibrational Spectroscopy*, Vol. **2** (eds J.M. Chalmers and P.R. Griffiths), John Wiley & Sons, Ltd, Chichester, pp. 1369–1385.
3. Sommer, A.J. and Katon, J.E. (1991) Diffraction-induced stray light in infrared microspectroscopy and its effect on spatial resolution. *Appl. Spectrosc.*, **45**, 1633–1640.
4. Nakashima, S., Ohki, S. and Ochiai, S. (1989) Infrared microspectroscopic analysis of the chemical state and spatial distribution of hydrous species in minerals. *Geochem. J.*, **23**, 57–64.
5. Nagai, N., Imai, T., Terada, H., Seki, H., Okumura, H., Fujino, H., Yamamoto, T., Nishiyama, I. and Hatta, A. (2002) Depth profile analysis of ion-planted photoresist by infrared spectroscopy. *Surf. Interface Anal.*, **33**, 545–551.

17

Infrared Microspectroscopic Imaging

Shigeru Shimada
Bruker Optics K. K., Japan

17.1 Introduction

Over the past decade or so, the use of an array detector for spectroscopic measurements in the infrared region has been steadily increasing. An array detector has multiple infrared-sensitive pixels in either one or two dimensions. Trials to construct an image from a target sample based on its infrared absorption characteristics by combining an array detector with Fourier-transform infrared (FT-IR) microspectrometry began in the mid 1990s. Today, several FT-IR imaging systems equipped with either a one-dimensional or a two-dimensional array detector are commercially available. Since these systems can be used to quickly visualize the distribution of chemical components and the spatial difference of chemical structures in a sample, they are being utilized by many researchers, particularly those engaged in the characterization and analysis of industrial materials and biological systems. In this chapter, the principles and applications of FT-IR microscopic imaging systems using an array detector are described.

17.2 FT-IR Imaging Systems

17.2.1 Imaging by Sample Scanning (Mapping)

Mapping is one of the application techniques of FT-IR microspectrometry employed when using a single-element detector (see also Section 16.4). As illustrated schematically in Figure 17.1a, multiple infrared spectra are measured consecutively from small defined (x, y) coordinate masked areas of a sample mounted on the sample stage of the microscope, which can be moved stepwise along the x- and y-axis by computer control. The distribution

Introduction to Experimental Infrared Spectroscopy: Fundamentals and Practical Methods,
First Edition. Edited by Mitsuo Tasumi and Akira Sakamoto.
© 2015 John Wiley & Sons, Ltd. Published 2015 by John Wiley & Sons, Ltd.

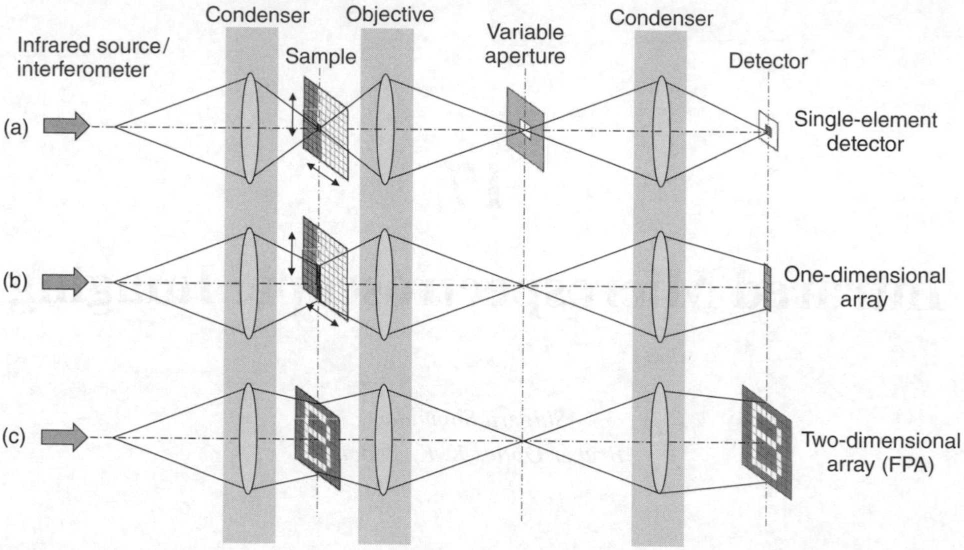

Figure 17.1 *Schematic illustration of FT-IR microspectroscopic imaging systems. (a) Mapping system using a single-element detector and an x–y motorized sample stage, (b) imaging system using a one-dimensional array detector and an x–y motorized sample stage, and (c) imaging system using a two-dimensional array detector.*

of the target component or chemical structure may then be visualized using the successively measured spectral information; that is, a "spectral map" may be constructed. In performing such a mapping measurement, the time needed for all the individual measurements necessary becomes increasingly longer as the number of small areas from which infrared spectra are measured increases. In practice, for a specific sample, measuring infrared spectra from a few thousand small areas is considered to be the upper limit reasonable for performing this type of measurement, although the practical limit will depend on the area of the target to be mapped, the lateral spatial resolution required, the wavenumber resolution and signal-to-noise (S/N) ratio of the spectra. When the overall measurement takes an excessively long time, the reliability of the mapping results will be influenced by the precision and stability of the measuring system, the environment for the measurement such as room temperature, and more importantly, any time-dependent changes in the sample itself.

17.2.2   Imaging by Using an Array Detector

The use of an array detector consisting of multiple infrared-sensitive pixels began as a means to overcome the limitations in the method of mapping using a single-element detector.

The principles of imaging using two types of array detector are schematically illustrated in Figure 17.1b,c, where the imaging methods using either a linear or a two-dimensional array detector are compared with the conventional mapping method using a single-element detector, shown in Figure 17.1a.

FT-IR imaging systems commercially available at present use two types of array detectors: a one-dimensional (linear) array in which the detection pixels are arranged linearly, or a two-dimensional array in which the detection pixels are laid out in a plane (often referred to as a *focal-plane array* or *FPA*).

Detection pixels used for an infrared array detector are usually made of a semiconductor such as HgCdTe (mercury cadmium telluride (MCT)), InSb (indium antimonide), or InGaAs (indium gallium arsenide), depending on the wavenumber region to be covered; MCT covers the mid-infrared region, while InSb or InGaAs is used in the near-infrared region.

A linear array detector used in the mid-infrared region typically consists of 16–32 pixels made of photoconductive MCT. The lower wavenumber limit (cut-off wavenumber) of mid-infrared radiation that it can detect is about $700 \, cm^{-1}$, similar to that of a single-element MCT detector. The advantage of a linear array detector is that it can be operated in the same way as a single-element detector. On the other hand, an FPA typically consists of 16×16 to 128×128 pixels each made of photovoltaic MCT, which can be laid out in a higher density than the photoconductive MCT. The lower cut-off wavenumber of this type of MCT is about $900 \, cm^{-1}$, which is considerably higher than that of the photoconductive MCT. However, in spite of the higher cut-off wavenumber of FPAs compared to that of linear array detectors, FPAs are widely used because of their tangible advantage of constructing an image of a larger area within a shorter time than linear array detectors.

As schematically illustrated in Figure 17.1b,c, when an array detector is used, interferogram data from small defined areas of a sample in either one dimension or two dimensions are simultaneously recorded. When a one-dimensional array detector is used, information measured by a scan of the interferometer is limited to one dimension as shown in Figure 17.1b. Therefore, the sample is scanned consecutively in x- and y-axis directions in a way similar to the mapping method with a single-element detector shown in Figure 17.1a. In this "hybrid" approach, adjacent regions of the sample are rapidly repositioned under the linear array detector until the full target area to be examined has been covered, and then a full region image is built up as a mosaic of the successive linear array images recorded. When an FPA is used, a spectral image of a field of view is obtained without scanning the sample, the size of which depends on the FPA used, and the magnification of the optical system (discussed in Section 17.3). Imaging over a larger area of a sample is also possible in a manner similar to the mapping or the "hybrid" approach described earlier, by appropriately "tiling" multiple images measured by the FPA.

As shown schematically in Figure 17.2, the advantage of using an array detector with an interferometer is to be able to record simultaneously (and in a shorter time) interferograms (in other words, infrared spectra) from individual small areas of the sample, the number of which is the same as that of the detector pixels of the array detector, by recording simultaneously the signals from each pixel of the array detector at all the optical path differences (OPDs) of the interferometer. This means that, in the same time span in which an interferogram of a single small area is measured in the conventional mapping approach with a single-element detector, the interferograms of small areas equivalent in number to the number of pixels in the array detector can be measured. As a consequence, efficiency is greatly increased, as the time of measuring an imaging data set from a sample can be improved by an order of magnitude up to a few orders of magnitude. In addition, when using an array detector, the need to place an aperture to mask a sample area, which is required in the conventional mapping method with a single-element

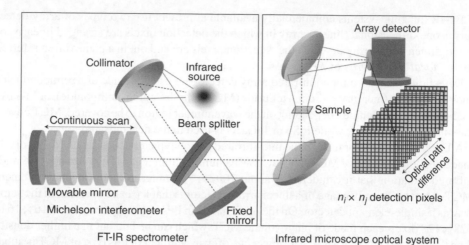

Figure 17.2 *Schematic illustration of an infrared microspectroscopic image measurement by using a combination of an FT-IR spectrometer and a microscope equipped with a two-dimensional array detector.*

detector in order to limit the area of the sample for measuring its image, becomes unnecessary.

The information obtained by the infrared spectroscopic imaging method by using an array detector is illustrated schematically in Figure 17.3. It is essentially the same as that

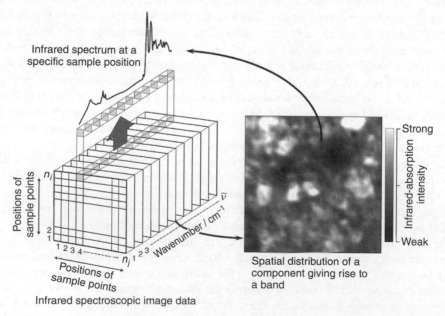

Figure 17.3 *Schematic illustration of an infrared microspectroscopic image data set after processing and the interrelationship between the image generated at a particular wavenumber and the infrared spectrum recorded at each pixel.*

obtained by the mapping method when using a single-element detector in that it consists of spectral information at defined positions (*x*- and *y*-coordinates) of small areas of the sample from which infrared absorption spectra characteristic of each of the individual small areas (pixels) can be extracted. By selecting an absorption band within this array and by using its normalized relative intensity, it is possible to construct an image showing the two-dimensional distribution of a chemical species or a chemical structure that gives rise to the selected absorption band. Additionally, it is also possible, by specifying a particular small area, to construct the infrared absorption spectrum of that small area and to use it for determining the chemical species existing within that small area.

17.2.3 Advantages of Using an Array Detector

Using an array detector offers a number of advantages to infrared spectroscopic measurements; the more important are the following:

1. *Reduction of the time needed for measuring an image*: As described already, the experimental time efficiency of imaging by using an array detector is greatly improved in comparison with the mapping approach using a single-element detector. This improvement has made it possible to follow in real time various time-dependent changes and to study physical and chemical reactions caused by external stimuli by observing changes in their infrared images. In other words, a new field of infrared spectroscopy has been developed by the introduction of a two-dimensional array detector which can collect spectroscopic imaging data from a sample in a single scan of the interferometer. This can be regarded as the most significant advantage of using a two-dimensional array detector.
2. *Improvement of lateral spatial resolution*: In the mapping method using a single-element detector, reduction of spatial resolution can be caused by diffraction due to the presence of a masking aperture which is needed to define the area from which the infrared spectrum should be measured. As already pointed out, no aperture is needed for a measurement with an array detector; as each pixel of the array detector is responsible for interrogating a different small area of a sample, diffraction effects are lessened.
3. *Enhancement of the spectral S/N ratio*: In a microspectroscopic measurement with a single-element detector, if the size of a sample is smaller than the surface area of the detector pixel, the size of the beam of radiation reaching the detector is limited by an aperture. Consequently, only a part of the infrared-sensitive surface of the detector is actively used for detecting the infrared radiation. For instance, if a detector has a square surface of $100\,\mu m \times 100\,\mu m$ and a sample's size defined by an aperture is $10\,\mu m \times 10\,\mu m$, only 1% of the surface of the detector is actively used, as the image of the sample projected onto the surface of the detector element is usually equal to the size of the sample. The intensity of the signal from the sample decreases with the decreasing size of the sample. By contrast, the noise produced by the detector element (or pixel) itself is always at a certain level as its noise level is proportional to its surface area. Consequently, the S/N ratio deteriorates greatly with decreasing sample size. On the other hand, with an imaging system using an array detector, the ratio between the surface area of a pixel of the array detector and the sample area corresponding to the detector pixel is always kept constant. As a consequence, there is much more efficient detection of the infrared radiation from the sample. Especially, as a two-dimensional array

detector can simultaneously receive almost the entire infrared beam from the sample, the imaging system using such a detector can take full advantage of the high through-put that is a characteristic benefit of FT-IR spectrometry (Jacquinot's Advantage; see Section 4.4.1.3).

17.3   Field of View and Spatial Resolution

Field of view (the sample area, the image of which is to be observed) and lateral spatial resolution are two important factors in infrared microspectroscopic imaging. These are given as follows:

$$\text{Field of view} = \frac{\text{size of a detector pixel} \times \text{number of detector pixels}}{\text{magnification of the optical system}} \qquad (17.1a)$$

$$\text{Spatial resolution} = \frac{\text{field of view}}{\text{number of detector pixels}} \qquad (17.1b)$$

It should be noted, however, that the spatial resolution given here is an approximate value that is simply derived from the geometry of the microspectroscopic system (called here *pixel resolution* for convenience); more theoretical considerations taking into account the diffraction phenomenon in the infrared region are needed to derive the correct spatial resolution of the system.

The theoretical spatial resolution is determined by the numerical aperture (NA) of the objective and the wavelength of radiation used for the imaging measurement. The theoretical spatial resolution R according to Rayleigh's criterion for a microscopic measurement is given as [1, 2]

$$R = \frac{0.61\lambda}{\text{NA}} \qquad (17.2)$$

In this equation, 0.61 is a constant derived from geometrical optics, λ is the wavelength of radiation used for the imaging measurement and NA is the numerical aperture of the microscope objective, which is defined as

$$\text{NA} = n \sin \theta \qquad (17.3)$$

where n is the refractive index of the medium (usually air; i.e., $n = 1$) and θ is the angle between the optical axis of the microscope and the outermost ray captured by the objective.

For example, if $\lambda = 10\,\mu\text{m}\,(1000\,\text{cm}^{-1})$ and $\text{NA} = 0.5$ (a typical value for an FT-IR microspectroscopic imaging system available commercially), $R = 12.2\,\mu\text{m}$. Based on such a consideration, it may be stated that the spatial resolution of most FT-IR microspectro-scopic imaging systems in the mid-infrared region is in the range of $3-20\,\mu\text{m}$, depending on the wavenumber used for the imaging. Therefore, it does not make sense to increase the magnification of the optical system excessively to obtain a higher "pixel resolution."

In the practical design of an FT-IR microspectroscopic imaging system, however, it is a usual practice to make the pixel resolution more than double that of the theoretical spatial resolution to be realized. This measure is the same in concept as that adopted between the interval of data points constituting a spectral band and spectral resolution; for example, in order to express a band having a width of $10\,cm^{-1}$, it is necessary to record data points at intervals of $5\,cm^{-1}$ or less. For this purpose, recent FT-IR microspectroscopic imaging systems are designed to have a pixel resolution of 2.5–10 µm.

In order to enhance spatial resolution, it is necessary to make the NA of the objective larger, as is clear from Equation (17.3); that is, either n or θ, or both of them should be increased. Due to the optical geometry of a microscope, there is an upper limit for θ. On the other hand, it is possible practically to increase n by introducing an attenuated total reflection (ATR) accessory into a microscope (see Chapter 13 for the ATR method; this is a frequently used accessory for many recent infrared microspectroscopy absorption measurements). The refractive index n of Ge, which is a commonly used material for an internal reflection element (IRE) in the ATR method, is about 4, and the NA when using a Ge IRE exceeds 2. This means that, if an ATR accessory with a Ge IRE is combined with a microscope, the theoretical spatial resolution is enhanced about four times that of the conventional reflection measurement. In fact, in the FT-IR microspectroscopic imaging measurement with a Ge ATR accessory, it has been confirmed that a spatial resolution comparable to the infrared wavelength used for the measurement is realized, and thus a higher spatial resolution may be attainable.

17.4 Methods of Measurement and Data Processing

17.4.1 Transmission Method

If a sample can be measured by the transmission method, it is often relatively easy to obtain reliable images as well as good-quality infrared spectra. However, because any pretreatment of the sample will likely greatly affect the quality of an image, it must be carried out very carefully. As in conventional infrared absorption measurements, due attention must be paid to make the thickness of the sample appropriate for such a measurement. For example, if the band of the target species used as the key band for generating an image has a high intensity (much >1 absorbance unit), an image of unreliable distribution of the target species is obtained, because quantitative accuracy is lost. This situation is the same as a quantitative analysis using an infrared absorption band; a very intense or saturated band cannot be used for such an analysis. In order to avoid saturation intensity of a key band, it is essential therefore to prepare a sample of appropriate thickness. If necessary, a thin film-like sample should be prepared by carefully sectioning the sample using a microtome. The appropriate thickness depends on the absorption coefficient of the key band, but usually a thickness of the order of a few micrometers is recommended.

Even when the absorption coefficient of a key band is low, it is not advisable to use simply a thick sample, as the refraction and diffuse reflection of infrared radiation that tend to occur within such a sample lead to deterioration of the spectral contrast and spatial resolution of an image.

In a measurement with a single-element detector, the field of view is limited; the spectral measurement is generated by focusing the infrared beam on a specified target area on the sample. On the other hand, in the microspectroscopic imaging measurement, homogeneous focusing over the entire field of view is necessary in order to obtain a good image. With increasing magnification of the microscope, the depth of field becomes shallower. As a result, the image obtained also becomes less clear because any wrinkles and gradients in the sample make its analysis more difficult. Therefore, if possible, pretreatment to remove such irregularities in the sample is needed.

17.4.2   Specular-Reflection Method

For an optically thick sample that does not transmit the infrared radiation or a sample on a highly reflecting substrate like a metal plate, the specular-reflection measurement (see Chapter 8) or "transflection"-type measurement (see Chapter 16) can be applied. In the same way as in the case of the transmission measurement, it is essential that the infrared beam is homogeneously focused over the entire field of view. If the surface is coarse, the contrast and spatial resolution of the image will deteriorate due to scattering of infrared radiation. For samples such as polymers with a flat surface and enough thickness, it is possible to measure specular-reflection spectra and obtain images based on the imaginary parts of the complex refractive calculated by using the Kramers–Kronig (KK) relations (see Sections 6.2.7 and 8.3). This method is worth trying, as it is sometimes useful for a qualitative analysis of the image obtained. It should be noted, however, that the values of reflectance in specular-reflection measurements of such a sample are generally low (typically, a small percentage in reflectance relative to that from a gold-covered mirror used as the background reference), and significant results can be expected by the calculation of KK relations only from high S/N reflectance spectra. Therefore, a prerequisite for undertaking this type of measurement is to be able to acquire a high S/N specular-reflection spectrum by performing sufficient accumulations of its interferograms.

17.4.3   Attenuated Total Reflection (ATR) Method

The FT-IR microscopic ATR sampling method is one of the measuring techniques that have come increasingly into wider use, especially since the middle of the 1990s. The main advantages of this method are

1. If the target sample can be brought into contact with the IRE of an ATR accessory, this method becomes applicable with a minimum (or even without) pretreatment of the sample.
2. As described in Section 17.3, when this method is used, a higher spatial resolution is expected to be obtained, when compared with that obtainable by the transmission method or the specular-reflection method. A sample of smaller size may also be handled by this method.

To exploit such advantages, combinations of the ATR method and microspectroscopic imaging methods have been developed, and in recent years ATR microscopic imaging is utilized much more frequently.

17.4.4   Data Processing

In data processing, an absorption band attributable to, for example, a specific chemical species or a particular functional group is selected, and a false-color or gray-scale spectroscopic image is constructed by using its normalized relative absorption intensity (either the peak intensity or integrated intensity). For calculating the band intensity, it is common to set a baseline in the vicinity of the band in order to minimize any effects arising from scattering by the sample itself and overlapping of bands from other species. This baseline setting construction should be made in an identical way for all spectra within the data set used for constructing an image. Alternatively, the ratios of the absorbance intensities of two bands may be used to construct an image. If a band to be used as the band for constructing an image shows a shift in its wavenumber maximum associated with a property of interest, it is possible to construct an image based on peak positions by automatically detecting the positions of band maxima, provided the measured spectra have satisfactorily high wavenumber resolution and S/N ratios for detecting the shifts. In either case, data on the difference in band intensities or shifts may be visualized as an image by using a contour map, steps in a gray scale, or color-coding. For some applications, it may also be possible to construct a three-dimensional image by additionally expressing band intensity or peak position changes in the z-axis.

Spectroscopic imaging can provide a great amount of visually realized information not available by simply looking at individual spectra recorded by the more conventional methods. However, attention should be paid to the potential for "misunderstanding" that might be caused by just looking at an image. For example, even if samples consist of equal components, the conditions of their surfaces may be different, thereby giving rise to different scattering or refraction of the infrared radiation. Such differences will likely cause changes in baselines and distortion of band shapes. When one examines spectroscopic images, one should be very aware that the physical conditions of samples can greatly affect the data set used for constructing images. If appropriate conditions are not met when constructing images, it is not possible to construct images that reliably reflect correct chemical information. In other words, even when imaging samples consisting of equal components, if the selection of a band that is to be used for constructing an image, the construction of the baseline, the effects of overlapping bands, the choice of scaling factor for the intensity axis, and so on, are not appropriately taken into account and processed, the resultant images will likely lead to an erroneous interpretation that the samples have different compositions. (For the baseline construction, see Section 3.4.3.) Consequently, it is very important to examine carefully the individual spectra from which an image has been constructed.

It is important to note, as mentioned in Section 17.3, that the spatial resolution of an image depends on the wavenumber of the band from which the image is constructed. For instance, when the lateral spatial distribution of a n-alkane compound is examined, images generated based on the CH_2 stretching bands in the region of $2925-2850\,cm^{-1}$ and the CH_2 bending band around $1450\,cm^{-1}$ are apparently different due to a difference in diffraction limits, as these depend on their wavelengths (wavenumbers). This point must be borne in mind when the lateral spatial distribution of a chemical species in a sample is the subject of a microspectroscopic imaging study.

17.5   Application Example of Microspectroscopic Imaging

As a representative application example, the analysis of a paint chip of an automobile by using an FT-IR microspectroscopic system equipped with a 64×64 two-dimensional MCT array detector is described. A paint chip embedded in an epoxy resin was sectioned using a microtome into a 10 μm-thick section, and a microspectroscopic imaging measurement from an area of this sample (345 μm × 345 μm) was performed with a pixel resolution of 2.7 μm by the transmission method. The time needed for the measurement was about 5 min.

A microphotograph taken to examine the layer structure of the sample is shown in Figure 17.4a. In this microphotograph, the sample appears to consist of four layers, each of which has a width of 40–60 μm. The infrared transmission microspectroscopic image generated from the same area is shown in Figure 17.4b, in which the relative magnitudes of the integrated absorption intensities are given using a gray scale. One result to be noticed in this figure is the fact that, as indicated by 2-1 and 2-2, layer 2 in Figure 17.4a is apparently two layers. Layer 2 in Figure 17.4a, which appears black over its entire region, would have led one to conclude that it consisted only of a single component; however, the infrared spectral image in Figure 17.4b indicates that layer 2 consists of two different layers, and the sample actually has a five-layer structure. A more detailed analysis made possible by comparing the infrared spectra of the five layers shown in Figure 17.4c

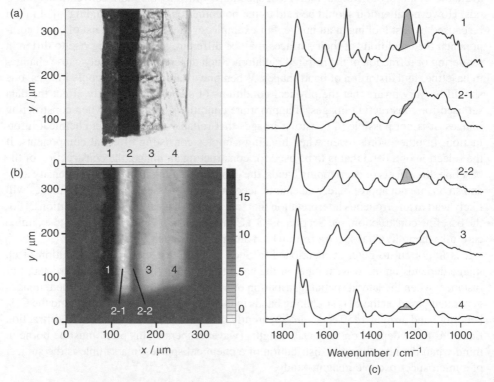

Figure 17.4 *Microscopic examination of a car paint chip. (a) Microphotograph of a cross section of the paint chip and (b) infrared microspectroscopic transmission image of the same sample, constructed by using the relative integrated intensities of a band in the region of 1277–1200 cm^{-1} (indicated by gray shading in the spectra). (c) Band intensities in each spectrum are normalized to the maximum intensity band. (Source: Figure supplied by Matthias Boese and Eric Klein of Bruker Optik GmbH with their permission.)*

clearly shows that the sample consists of five layers, and each layer has a component or composition specific to it.

Cars are coated with multiple layers of paint in order to weatherproof their exterior and make their appearance attractive. The types of paint used and the structure of paint layers vary with the car model. Since early in the development of FT-IR microspectroscopic systems, the analysis of car paint by using FT-IR microspectroscopy has been performed in forensic scientific investigation of crimes and accidents involving cars as a technique for identifying the car model or manufacturer.

The simple example described above demonstrates that infrared spectroscopic imaging using an FT-IR microspectroscopic system equipped with an array detector can reveal within a short time valuable information on a sample under investigation, which is difficult to obtain by traditional point-by-point analysis based on visual inspection using a non-imaging system. Today, many more complex sample systems are being subjected to study and analysis by FT-IR spectroscopic imaging techniques; one such field that is benefiting is studying samples for biomedical research investigations, such as disease-related human tissue.

References

1. Messerschmidt, R.G. (1995) Minimizing optical nonlinearities in infrared microspectroscopy, In: *Practical Guide to Infrared Microspectroscopy: Theory and Applications* (ed H.J. Humecki), Marcel Dekker, New York, pp. 3–39.
2. Born, M. and Wolf, E. (2009) *Principles of Optics*, 7th (expanded) edn, Cambridge University Press, Cambridge.

Further Reading

Bhargava, R. (2012) Infrared spectroscopic imaging: the next generation. *Appl. Spectrosc.*, **66**, 1091–1120.

Kidder, L.H., Haka, A.S. and Lewis, E.N. (2002) Instrumentation for FT-IR imaging, In: *Handbook of Vibrational Spectroscopy*, Vol. **2** (eds J.M. Chalmers and P.R. Griffiths), John Wiley & Sons, Ltd, Chichester, pp. 1386–1404.

Salzer, R. and Siesler, H.W. (eds) (2009) *Infrared and Raman Spectroscopic Imaging*, Wiley-VCH Verlag GmbH, Weinheim.

Sasic, S. and Ozaki, Y. (eds) (2010) *Raman, Infrared, and Near-Infrared Chemical Imaging*, John Wiley & Sons, Inc., Hoboken, NJ.

18

Near-Infrared Spectroscopy

Masao Takayanagi
United Graduate School of Agricultural Science, Tokyo University of Agriculture and Technology, Japan

18.1 Introduction

Transitions due to overtones and combination tones of molecular vibrations are mainly observed in the near-infrared (NIR) region (800 nm to 2.5 µm; i.e., $12\,500-4000\,cm^{-1}$), which is a continuation of the shorter-wavelength side of the mid-infrared (mid-IR) region where conventional fingerprint mid-IR spectra are observed. NIR spectroscopy is closely related with mid-IR spectroscopy, which deals mainly with the fundamental tones of molecular vibrations. NIR spectroscopy is commonly used, however, in a way quite different from mid-IR spectroscopy. In mid-IR spectroscopy, in many cases, assignments of IR bands observed from molecules under study are first considered for the purpose of obtaining information on the molecular structures present, chemical bonds, intermolecular interactions, and so on. In NIR spectroscopy, on the other hand, observed spectra are usually directly used for classification and quantitative analyses without paying much attention to the detailed assignments of the observed bands. Thus, the general use of spectra is greatly different between those of the NIR and mid-IR regions because of the different features of their spectra. In the first part of this chapter, the features of NIR spectra are briefly discussed. Then, spectrometers and techniques for measuring NIR spectra are described. Finally, some applications of NIR spectroscopy are discussed.

The abscissa of an NIR spectrum is given in units of either wavelength/nm or wavenumber/cm^{-1}. In many cases, wavelength/nm is preferred to wavenumber/cm^{-1}, because dispersive (non-Fourier-transform (non-FT)) spectrometers with a wavelength-scanning mechanism are often used for measurements in the NIR region, and wavelength/nm is more familiar to researchers and technicians working in the field of NIR spectroscopic analysis. Nevertheless, wavenumber/cm^{-1} is used for the abscissa in

Introduction to Experimental Infrared Spectroscopy: Fundamentals and Practical Methods,
First Edition. Edited by Mitsuo Tasumi and Akira Sakamoto.
© 2015 John Wiley & Sons, Ltd. Published 2015 by John Wiley & Sons, Ltd.

this chapter, because spectra shown in this format are more convenient for the assignments of observed bands in particular for their comparative study with mid-IR spectra.

18.2   Near-Infrared Spectroscopy

As is well known in quantum mechanics, transitions corresponding to overtones and combination tones of harmonic oscillators are optically forbidden and should therefore not be observed in absorption spectra. However, molecular vibrations are not completely harmonic; that is, vibrational potentials of real molecules have anharmonicities, which make the transitions of overtones and combination tones actually occur with probabilities smaller than those for fundamental transitions. Most absorptions due to such transitions are observed mainly within the NIR region. Therefore, NIR spectroscopy has the following characteristics:

1. NIR transitions have small absorption coefficients. Consequently, the NIR radiation used for NIR measurements penetrates into or transmits through samples more easily; it is possible to measure NIR spectra from the insides of samples nondestructively. On the other hand, it is difficult to observe NIR spectra from dilute or trace samples.
2. In NIR spectra, most of the observed absorptions are overtones and combinations related to the CH, OH, or NH stretching vibrations, which have relatively large anharmonicities. As a result, only a relatively few bands are actually observed in NIR spectra, even though a lot of overtones and combinations are expected to give absorptions in NIR spectra in view of the number of fundamental vibrations observed in mid-IR spectra. Based on these characteristics, NIR spectroscopy is applied mainly for two purposes.

Firstly, NIR spectroscopy is applied as a tool in analytical chemistry. Based on the observed spectra, not only qualitative analyses such as detection, identification, and discrimination of component materials in samples, but also quantitative analyses of components and correlation to physical properties are performed. In the analyses, detailed assignments of observed bands are not necessary. Chemometric routines (see Chapter 7) are used for discrimination of samples by abstracting spectral characteristics, and for examining the relationship between the spectra and the concentration of components or a sample's physical properties. Application of NIR spectroscopy to analytical chemistry in combination with chemometric techniques began about a half a century ago. The first reports in this field by Norris and coworkers [1, 2] on the analysis of water contents in wheat grains appeared in the 1960s. After these, the combination of NIR spectroscopy and chemometrics rapidly became a common practice. It might be said that the apparatus and technologies of NIR spectroscopy have been developed since then toward various analytical applications. In this chapter, the apparatus and technologies of NIR spectroscopy are reviewed mainly from the standpoint of analytical applications.

Secondly, information on structures, chemical bonds, intermolecular interactions, and so on, of target molecules is derived from observed NIR spectra by assigning observed bands to overtones and combinations of vibrational modes and analyzing vibrational potentials. This is considered an extension of the study of molecules by mid-IR spectroscopy.

More detailed information can be obtained by taking frequencies of not only fundamentals but also those of overtones and combinations into consideration. Although this type of study has been carried out since the early 1950s, it did not become popular because of difficulties in the analyses of NIR spectra. NIR spectra of samples in the solid and liquid states mostly consist of broad bands without structure, as the observed bands are formed by many overlapping component bands. Analyses of spectra by assigning bands are, therefore, very difficult. Samples in the gas phase or under low-temperature rare-gas matrix-isolated conditions, where the broadening of bands due to intermolecular interaction is nonexistent, provide spectra with narrow bands. It is very difficult practically, however, to observe weak transitions of essentially forbidden overtones and combinations of molecules in a low-pressure gas phase or under matrix-isolated conditions. Moreover, even if the complexity due to broadening is reduced, complexity due to Fermi resonances makes their detailed analyses practically impossible. Because of this, the physics of NIR spectroscopy has not been developed to a satisfactory degree in spite of its importance as a branch of basic science.

18.3 Near-Infrared Spectrometer

At present, FT spectrometers are almost exclusively used for mid-IR spectroscopy, except for some special purposes. On the other hand, dispersive spectrometers (two types exist as described below), spectrometers with optical filters, AOTF (acousto-optic tunable filter) spectrometers, inter alia, are used in NIR spectroscopy in addition to FT spectrometers. There are two types of dispersive spectrometers: the scanning type with a single-channel detector and the non-scanning type with a multichannel detector. The reason for the existence of such a variety of spectrometers in the NIR region is as follows.

First, both ultraviolet–visible (UV–vis) spectrometers, which are equipped with gratings, and FT-IR spectrometers can expand their functions into the NIR region. These two kinds of spectrometers are used almost equally in the NIR region.

Secondly, NIR spectrometers are often used for practical or routine purposes. Spectrometers developed for such purposes should, therefore, ideally be: (i) portable, small in size, and light in weight, (ii) capable of use under severe conditions such as at high or low temperatures, in dusty places, or with mechanical disturbances, and (iii) inexpensive. Since FT-NIR spectrometers cannot fully satisfy these conditions, various other types of spectrometers have been developed and used. A dispersive spectrometer with a multichannel detector can be used under the conditions with mechanical disturbances because it has no moving mechanism; such a spectrometer can be installed, for example, on a rotary blender of powdered drugs. On the other hand, spectrometers with optical filters and AOTF spectrometers, which are light in weight and inexpensive, are suitable for outdoor in-the-field type measurements.

Strictly speaking, FT spectrometers do not suit the differential processing often undertaken with NIR spectra. In differential processing, correlations between the measured values at neighboring wavelengths (or neighboring wavenumbers) are important. Since observed data are usually accumulated in FT spectrometers, the obtained value at each wavelength is a time average. On the other hand, the measured values at neighboring wavelengths in the spectra obtained by a UV–vis-NIR spectrometer correspond to the values at neighboring times. Spectra obtained by FT-NIR and UV–vis-NIR spectrometers are almost perfectly identical with each other when the sample does not change with time. However, time correlations between the values at neighboring wavelengths are lost when the spectra of the sample changing with time are measured by an FT-NIR spectrometer. The loss of time correlations may make a quantitative discussion of differential spectra impracticable. Careful considerations are needed to measure NIR spectra, because subtle spectral differences often have great influences in their analyses.

It is impossible to cover the whole target wavelength region with a single set of light source, detector, and optical elements in either a UV–vis-NIR spectrometer or an FT-NIR spectrometer. In many UV–vis-NIR spectrometers, these are automatically changed at appropriate wavelengths. On the other hand, it is often necessary to exchange manually both the light source and the beam-splitter of the interferometer in an FT-IR spectrometer for NIR use. The set of the beam-splitter made from potassium bromide (KBr) and the ceramic light source used in the mid-IR region is applicable only in the wavenumber region lower than $8000\,cm^{-1}$. The set of a beam-splitter made from calcium fluoride (CaF_2) and a halogen lamp for the NIR region cannot be used for the IR region lower than $2000\,cm^{-1}$ (see Figure 18.1). Some FT-NIR spectrometers recently marketed are equipped with automatic exchange systems for the beam-splitter and other optical elements. Otherwise, their exchange must be carried out by the user. Since the exchange and the concomitant optical alignment are time-consuming and require expertise, many FT spectrometers are actually used specifically only in either the NIR or mid-IR region.

In the rest of this section, spectrometers with optical filters and AOTF spectrometers, which are seldom used in the other wavelength regions, are briefly described.

18.3.1   Optical Filter Spectrometer

There are two types of spectrometers using optical filters. The first type is equipped with a number of optical filters transmitting NIR light of different wavelengths, which are used sequentially to transmit spectroscopic data. The other type applies wavelength-tunable filters such as a liquid crystal tunable filter (LCTF) or a Fabry–Perot interferometer.

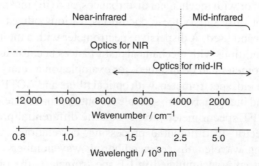

Figure 18.1 *Wavenumber regions typically covered by a single FT spectrometer without changing optics and source.*

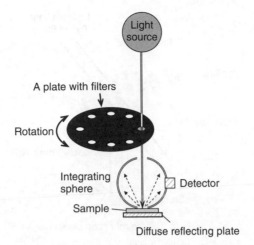

Figure 18.2 *Schematic illustration of a spectrometer using optical filters and an integrating sphere.*

In Figure 18.2, a schematic view of a spectrometer utilizing a number of optical filters is illustrated. Filters (up to about a maximum of 20) attached to a disk are used sequentially to select wavelengths for measurements by rotating the disk. Multiple regression analyses are often performed on the results of measurements with such filters. This spectrometer can be handled easily and is inexpensive, as neither an expensive optical element nor a high-precision mechanism is used. However, spectrometers of this type, because of their low spectral resolution and inability for continuous scan, have been replaced recently by dispersive or FT spectrometers. Only instruments marketed with a descriptive name such as "ingredient analyzer" are used now for specific analyses of food, cereals, tea, and so on.

Spectrometers using an LCTF as the dispersive element have not become popular yet. A charge-coupled device (CCD) camera, together with an LCTF with a large effective diameter, is used for the acquisition of the so-called "hyper spectra", which are images obtained simultaneously at different wavelengths.

18.3.2 AOTF Spectrometer

A schematic diagram of an AOTF spectrometer is shown in Figure 18.3. A crystal of tellurium oxide (TeO_2) works as a grating in which a longitudinal wave is generated with a piezoelectric oscillator. Since the wavelength of the longitudinal wave can be varied by changing the frequency of the piezoelectric oscillator, the output wavelength from the crystal can be changed. The spectrometer using this type of crystal for wavelength dispersion is called an *AOTF spectrometer*. Since the dispersion element is small and capable of high-speed sweeping, the spectrometer using the element is small in size and suitable for mobile use.

The wavelength resolution and reproducibility of current AOTF spectrometers seem to be inferior to those of FT spectrometers but if these could be improved technically, a high-performance AOTF spectrometer may be realized in the future.

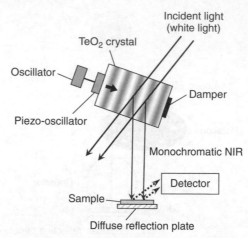

Figure 18.3   *Schematic illustration of an AOTF spectrometer in the arrangement for measuring diffuse reflection.*

18.4   Optical Elements in the Near-Infrared Region

NIR-transmitting materials used for windows, optical cells, and so on, in the NIR region are almost common with those in the visible region. Commonly used materials are summarized in Table 18.1; of these, optical glass (BK7) and synthetic quartz are those most commonly used. These materials do however have absorption bands related to OH stretching vibrations that occur around 1400 and 2220 nm (ca. 7143 and 4545 cm^{-1}, respectively), which interfere with the measurements of spectra. Since the transmittance of these materials decreases remarkably toward the longer wavelength side, they cannot cover the whole NIR and mid-IR regions. Although potassium bromide (KBr) and sodium chloride (NaCl) transmit radiation over a wide wavelength region from the UV to mid-IR, they are rarely used in the NIR region because they are so hygroscopic that they soon cloud in ambient conditions; consequently their transmittance in the NIR region falls rapidly.

Optical fibers are often used for NIR spectroscopy, as mentioned later. Synthetic quartz is mainly used as the material for these fibers. Since the NIR light for measurements is often required to travel a long distance through a fiber, special attention has to be paid to make

Table 18.1   *NIR-transmitting materials.*

Material	Longer-limit wavelength/nm
BK7 (optical glass)	2000
Silica (quartz)	2500
"Water-free" fused silica	3500
Calcium fluoride (CaF_2)	8000
Magnesium fluoride (MgF_2)	7000
Lithium fluoride (LiF)	8000

the absorption by the OH group as weak as possible when fibers for NIR spectroscopy are produced. Fibers with high transmittance made of tellurium oxide (TeO$_2$) or zirconium fluoride (ZrF$_4$) have been developed recently. Plastic fibers are also used for measurements in the wavelength region shorter than about 1700 nm.

18.5 Measurements Using NIR Spectroscopy

18.5.1 Methods of Measurements

NIR spectroscopy is applied to a wide variety of materials in various forms such as liquids, gels, crystals, tablets, powders, granules, and films. Various sources of samples become targets for these NIR measurements; included in these are not only farm products such as fruit and vegetables, food, and industrial products, but also samples from human bodies, animals, and plants. NIR spectra are often measured *in situ*, nondestructively, and without any prior processing of the sample. Various kinds of probes and sample cells are commercially available. By using them, NIR spectra of samples in various conditions are measured.

All the sampling techniques of transmission, diffuse reflection, transmission–reflection, and interactance, schematically shown in Figure 18.4, are commonly used for measuring NIR spectra. The transmission method shown in Figure 18.4a is applied to measurements on liquids and transparent materials such as small crystals, glass, and films which do not scatter NIR radiation seriously. The sample thickness appropriate for measuring NIR spectra depends strongly on the wavelength region, as described in the next section.

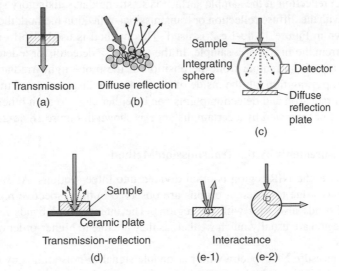

Figure 18.4 *Methods for measuring NIR absorption spectra. (a) Transmission, (b) diffuse reflection, (c) transmission and integration, (d) transmission and diffuse reflection, (e-1) and (e-2) interactance. Curves with arrows within the samples in (e-1) and (e-2) are meant to show an example of complicated paths through which NIR radiation transmits through the samples.*

Solid and suspension samples that hardly transmit NIR radiation are measured by the diffuse reflection method shown in Figure 18.4b, which is that used most frequently in NIR spectroscopy. In measurements by this method, an integrating sphere (see also Figure 18.2) is used. The integrating sphere is an optical element with a hollow sphere, whose inner wall has a highly reflecting coating. It has the function of making the NIR radiation introduced into it spatially homogeneous. The integrating sphere enables one to measure the intensity of NIR radiation scattered from the surface of samples without being affected by the directional differences of the scattering (diffuse reflection). The diffuse reflection method can be applied to almost any sample except those with smooth surfaces that exhibit specular reflection. Since NIR radiation has much higher transmittance for any samples compared with the ultraviolet, visible, and mid-IR radiation, as already mentioned, information relating to the inside of a sample can be sufficiently obtained even with the diffuse reflection method. Hence, the attenuated total reflection (ATR) method, whose application has increased considerably in the mid-IR region, is seldom used for measuring NIR spectra. The KBr-disk method, which is generally used in mid-IR spectroscopy, is not used either in the NIR region.

Samples such as suspensions and opaque films, which strongly scatter NIR radiation but transmit only a small portion, are handled in the following ways: the NIR radiation which has been transmitted by the sample is introduced into an integrating sphere as shown in Figure 18.4c, or the absorption is measured by the transmission–reflection method shown in Figure 18.4d. In the transmission–reflection method, a standard reflector such as a ceramic plate is placed at the back of the sample, and the ratio of intensities of the diffuse reflection measured with and without a sample is assumed to be the absorbance of the sample.

When direct reflection at the sample surface is so strong that satisfactory spectra cannot be obtained with the diffuse reflection or transmission–reflection method, the interactance method shown in Figure 18.4e-1,e-2 is used. This method is used also to record useful information from the inside of a sample. In the method, a detector (or a detection probe) is located at a position remote from the position of the probe-light irradiation; only the NIR radiation passing through the inside of the sample by dispersion and transmission is probed. The irradiation and detection points can be either close to each other as shown in Figure 18.4e-1 or separated by a certain distance as shown in Figure 18.4e-2.

18.5.2   Measurements by the Transmission Method

For convenience, the NIR region is often divided into three sections. As summarized in Table 18.2, bands due to different origins are observed in the respective regions and the intensities of bands also vary with the regions. The intensities of bands in the shorter-wavelength region are usually much weaker, as they are due to higher-order overtones and combinations.

In order to measure NIR spectra with reasonable signal-to-noise ratios by the transmission method, the optimal thickness (optical pathlength) of the sample should be chosen according to the wavelength region in which it is to be measured. It is necessary to use a cell with an appropriate pathlength for samples in the liquid or solution states. A cell with a pathlength of 1 mm or shorter should be used when the wavelength region closest to the mid-IR region is to be measured, because of the large absorption coefficients in this

Table 18.2 *Observed transitions in the three NIR regions, their intensities, and appropriate sample optical pathlengths.*

Wavelength/nm	Observed transitions	Transition intensities	Appropriate pathlengths
1800–2500	Combination tones First overtones	Large	Films of ~50 μm thick Cells with pathlengths <1 mm
1100–1800	First overtones Second overtones	Medium ~Small	Solids 0.1–1 mm thick Cells with pathlengths 5–20 mm
800–1100	Second overtones Higher overtones and combinations	Small ~Very small	Solids thicker than 1 cm Cells with pathlength 1–20 cm or longer

region. A cell with a pathlength of 1 cm can be used for measurements in 1100–1800 nm, the center of the NIR region. In the region near the visible, a cell with a pathlength of 10 cm or longer is needed in order to obtain sufficient signal intensities for bands with small absorption coefficients.

To measure a clear spectrum from a solid sample, it is also necessary to determine carefully its thickness. When the wavelength region to be measured is close to the mid-IR, and many bands are expected to exist, it is necessary to make a thin film of the sample. On the contrary, there are usually only two or three bands with appreciable intensities in the 700–900 nm region, which is close to the visible. In this region, samples with considerable thickness can be used for measuring reasonable NIR spectra. The insides of farm products, human bodies (e.g., oxygenated hemoglobin in human brains and infants), and other samples are investigated by measuring spectra in this wavelength region. Since the NIR radiation in this wavelength region easily transmits through living bodies, this region is sometimes called the *optical window in biological tissues.*

When NIR spectra of heterogeneous solid samples with internal structures, such as fruit and pharmaceutical tablets, are measured, information on the whole sample may not be obtained by the diffuse reflection method because only the NIR radiation scattered from near the surface of the sample will be observed. On the other hand, information on the whole sample can be obtained by the transmission or interactance method. Therefore, the transmission method is preferred for heterogeneous solid samples even if the absorbances of their bands become large, for example, >3.0 in absorbance units.

18.5.3 Optical Fibers and Various Probes

NIR spectroscopy is often used for *in situ* measurements, either on-line (on-site analyses of samples transported automatically to the spectrometer from an assembly-line), or in-line (on-site measurements performed for all samples directly in an assembly-line or process stream). In these measurements, for example, in factories, it is difficult in many cases to set the spectrometer close to the sample. In such cases, measurements using optical fibers are carried out widely, as highly permeable fiber materials are available in the NIR

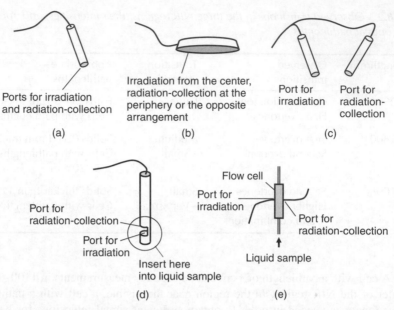

Figure 18.5   *Various types of probes used with optical fibers: (a) pen-type, (b) stethoscope–type, (c) linked probes, (d) liquid probe, and (e) flow cell.*

region. Application of optical fibers enables measurements for samples in various states and hazardous conditions and/or dangerous measurements at sites where analysts cannot approach or it is unsafe to do so.

Various probes using optical fibers have been employed in many and varying cases for irradiating and condensing NIR radiation. Representative examples are shown in Figure 18.5. Figure 18.5a shows a pen-type probe having both the irradiating and radiation-collection ports at an end. Figure 18.5b shows a stethoscope-type probe which is designed to collect diffuse reflection from a wide area. Figure 18.5c shows a pair of probes, one for irradiation and the other for radiation collection; they are set independently at different positions on a sample. The probe shown in Figure 18.5d, which has a set of irradiation and radiation-collection ports facing each other with a gap of several millimeters, is placed into a liquid sample to measure spectra with an optical pathlength corresponding to the gap. A combination of optical fibers and a flow cell is sometimes used as shown in Figure 18.5e; a flow cell equipped with the irradiation and collection ports on the opposing sides is used for measuring the NIR spectrum of a liquid sample flowing through a cell.

18.5.4   Accessories for NIR Spectrometers

Various accessories and sample holders, which are designed to measure NIR spectra in various conditions, are commercially available for most NIR spectrometers. Their proper use is necessary in order to obtain useful results by NIR spectroscopic analysis. A careful study of their manufacturers' catalogs and instruction manuals is often helpful for an appropriate choice. Some examples are given below.

The accessories usually attached to commercially available NIR spectrometers are considered first. As NIR spectra depend sensitively on temperature, thermostats to stabilize the temperature of samples, mainly liquid samples in an optical cell, are usually provided. An integrating sphere is necessary for measurements of diffuse reflection. Attachments for measuring transmission spectra from samples outside the body of the spectrometer are used for samples that cannot be put into the sample compartment of the spectrometer, or are used with an automatic sampler. Ports for the optical fibers are of especial practical importance, because optical fibers of many kinds are used to measure NIR spectra of samples in various shapes and states, including high pressure and high temperatures; an appropriate optical fiber should be used for the sample to be studied.

Cells and vessels of the Petri-dish type are generally used for measurements of NIR spectra. The latter vessels are often simply called *sample cups*. Cells not only to be used for transmission measurements for liquid samples but also cells used for diffuse reflection from powder and granule samples are available. A cell fit for the intended purpose must be used. For example, large cells with long optical paths and wide openings for putting in and taking out samples are often used for large granule type samples, because an increase in the vacant space in the cell and a concomitant decrease in sample density inevitably occur as the size of granules increases.

When the diffuse reflection method is used, a sample cup made of either glass or metal is used, depending on the direction of the incident NIR radiation; if the sample is irradiated from the bottom, a sample cup made of glass should be used, and if the sample is irradiated downward, it is better to use a sample cup made of metal. Cups used specifically for a powder, granule, paste, or suspension sample are provided individually. In order to measure an averaged spectrum from an inhomogeneous sample, some spectrometers are equipped with a mechanism of rotating the sample cup during the measurement. Thus, the optimal spectrum of a specific sample can be measured by using a cup most appropriate for the sample. Moreover, some spectrometers are provided with a means to measure the spectrum of a sample placed in a beaker, a vial, or a sample bag, and also a mechanism whereby transmission and diffuse reflection spectra can be recorded simultaneously.

18.6   Issues Still to be Solved in NIR Spectral Analysis

In this section, some problems which are desirable to be solved for more effective analyses of NIR spectra are summarized.

Although assignments of observed bands to vibrational modes are essential for reliable spectral analyses, only a limited amount of information is currently available in the literature [3, 4]. Quantum chemical calculations, especially DFT (density functional theory) calculations, have become very powerful as an aid for the analyses of mid-IR spectra. In the NIR region, calculations of the frequencies of overtones and combinations by taking the effects of vibrational anharmonicities into account have become possible by using at least a commercially available program package. Developments of more useful program packages are strongly hoped for at the time of preparing this chapter.

Chemometrics usually applied for the analyses of NIR spectra is described in detail in Chapter 7 and in other literature [4–8]. The most important and essential points remaining in this field seem to be the generalization and the systematic maintenance of calibration models. In traditional optical spectroscopic analysis, a concentration of an analyte is calculated by using Lambert–Beer's law from the intensity of the band observed at a specific wavelength and assigned to the analyte; a spectral band and the concentration of the analyte have a definite relation via the molar absorption coefficient. Therefore, quantitative analyses of an analyte with different spectrometers and under different circumstances are possible. In the multivariate data analysis of NIR spectra, on the other hand, a model to abstract the information of the concentration of an analyte from a spectrum is constructed. Since the features of spectra depend on the spectrometer and on the circumstances of the measurements such as the wavelength interval of data points, the model for quantitative analyses should be constructed case by case. Since the NIR spectra measured *in situ* or nondestructively depend highly on the concentrations of other species coexisting with the analyte in the sample at the time the model was generated, the model once constructed for an analyte will not be applicable for the same analyte if there is any variation of coexisting species. When a specific component in a complex mixture consisting of many components like agricultural products is analyzed quantitatively, the model used for the analyses should be revised or updated periodically according to the variation of concentrations of coexisting components. The need of revising or updating a model is a great obstacle in NIR spectroscopic analysis. A method which enables the easy management of models for quantitative analyses is strongly desired. The best solution for this issue is the development of a technique for using a common model on different spectrometers and under different circumstances.

18.7   Some Examples of Near-Infrared Spectroscopic Analyses

In this section, some examples of NIR spectroscopic analyses are shown.

In Figure 18.6, the NIR absorption spectra of fabrics of six different materials, cotton, hemp, wool, silk, polyester, and acrylic fibers are shown. The main component of cotton and hemp is cellulose. Wool and silk are mainly composed of proteins of animal origins. The other two are made of synthetic polymers. Identification of the materials of fabrics is of great importance because the materials mainly determine the characteristics of the fabrics. Spectra in Figure 18.6 were obtained by performing the Kubelka–Munk conversion (see Section 12.2) for the diffuse reflection spectra obtained with an integrating sphere. In the NIR absorption spectra, only a small number of broad bands are usually observed. All the spectra in Figure 18.6 show a tendency to increase in intensity on going to the right (longer wavelength), that is, the absorption becomes stronger at lower wavenumbers. This tendency is commonly seen in NIR absorption spectra.

Although materials of fabrics can be identified by finding key bands for each material in the observed NIR spectra, the quantitative analyses of spectra by chemometrics are more useful. Quantitative comparison of the observed spectra with the standard spectra by evaluating the degree of similarity reveals the material of sample fabrics. Score plots based on the principal component analyses have been found effective for the identification of fabrics. Use of appropriate wavelength regions for analyses has enabled even the distinction of cotton and hemp, the spectra from which closely resemble each other.

NIR spectroscopy is also useful for the quantitative analyses of blended fabrics. Figure 18.7 shows the NIR absorption spectra recorded from blended fabrics of cotton and polyester at various blend ratios. The spectra of the blended fabrics are an overlap of

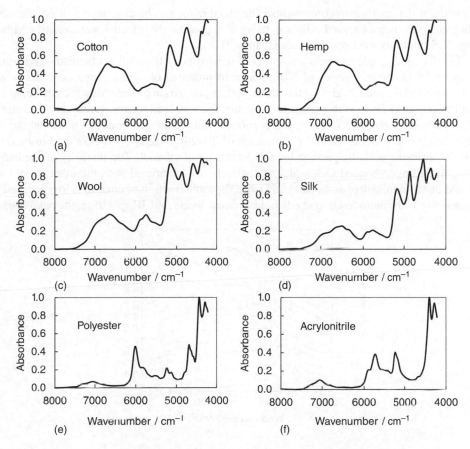

Figure 18.6 *(a–f) NIR absorption spectra of fabrics made from six kinds of fibers.*

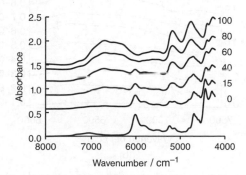

Figure 18.7 *NIR absorption spectra of blended fabrics of cotton and polyester with various blend ratios. Numbers to the right indicate the blend ratios of cotton.*

spectra of the constituent components. The blend ratios can be estimated to a satisfactory degree from their observed NIR spectra over appropriately selected wavelength regions by PLS (partially least squares) (see Chapter 7).

Finally, an example of application of NIR spectroscopy to solution chemistry is shown. Figure 18.8a shows a series of NIR spectra of mixtures of cyclohexane and ethanol at various mixing ratios. A glass cell with the optical pathlength of 1 cm for which the temperature was set at 30 °C with a temperature controller was used for measurements. A narrow band around 7080 cm^{-1} is due to the overtone of non-hydrogen-bonded OH stretching of ethanol, while a broad feature in the region of 7000–6000 cm^{-1} is considered to be due to the overtone of hydrogen-bonded OH stretching of ethanol. The intensity of the band due to hydrogen-bonded OH, which is relatively low compared with the intensity of the band due to non-hydrogen-bonded OH when the mole fraction of ethanol is low, increases markedly as the mole fraction of ethanol increases. In the mid-IR spectrum, the band due to

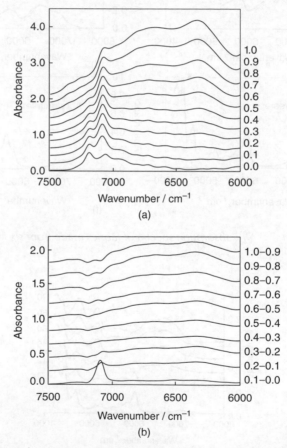

Figure 18.8 (a) NIR absorption spectra of cyclohexane–ethanol binary mixtures with various mixing ratios. Numbers to the right indicate the mixing ratios of ethanol. (b) Difference spectra between the spectra for neighboring mole fractions; for example, spectrum (0.9 − 0.8) was obtained by subtracting the spectrum of ethanol mole fraction 0.8 from the spectrum of ethanol mole fraction 0.9. The spectra in both (a) and (b) have been offset to aid clarity.

non-hydrogen-bonded OH is often hidden by the intense hydrogen-bonded OH; the variation of the degree of hydrogen bonding upon the mixing ratio is therefore difficult to follow by mid-IR spectroscopy. On the other hand, both bands due to hydrogen-bonded and non-hydrogen-bonded OH are observed distinctly as different bands in NIR spectra. Therefore, NIR spectroscopy seems to have a possibility of providing a technique useful for the study of hydrogen bonding in solvents and solvent mixtures. In Figure 18.8b, the difference spectra between the spectra measured at neighboring mole fractions are shown, where the variation of spectrum against the mole fraction is clearly observed. In fact, detailed information on the intermolecular interaction and the microscopic structures in the solvent mixtures has been reported, based on difference spectra of this kind [9, 10].

References

1. Hart, J.R., Norris, K.H. and Golumbie, C.I. (1962) Determination of the moisture content of seeds by near-infrared spectrophotometry of their methanol extracts. *Cereal Chem.*, **39**, 94–99.
2. Massie, D.R. and Norris, K.H. (1965) Spectral reflectance and transmittance properties of grain in the visible and near infrared. *Trans. ASAE*, **8**, 598–600.
3. Workman, J. Jr., and Weyer, L. (2012) *Practical Guide and Spectral Atlas for Interpretive Near-Infrared Spectroscopy*, 2nd edn, CRC Press, Boca Raton, FL.
4. Siesler, H.W., Ozaki, Y., Kawata, S. and Heise, H.M. (eds) (2002) *Near-Infrared Spectroscopy: Principles, Instruments, Applications*, Wiley-VCH Verlag GmbH, Weinheim.
5. Burns, D.A. and Ciurczak, E.W. (2007) *Handbook of Near-Infrared Analysis*, 3rd edn, CRC Press, Boca Raton, FL.
6. Ciurczak, E.W. and Drennen, J.K. III, (2002) *Pharmaceutical and Medical Applications of Near-Infrared Spectroscopy*, CRC Press, Boca Raton, fl.
7. Ozaki, Y., McClure, W.F. and Christy, A.A. (eds) (2006) *Near-infrared Spectroscopy in Food Science and Technology*, John Wiley & Sons, Inc., Hoboken, NJ.
8. Ozaki, Y. (2012) Near-infrared spectroscopy – its versatility in analytical chemistry. *Anal. Sci.*, **28**, 545–563.
9. Li, Q., Wang, N., Zhou, Q., Sun, S. and Yu, Z. (2008) Excess infrared absorption spectroscopy and its applications in the studies of hydrogen bonds in alcohol-containing binary mixtures. *Appl. Spectrosc.*, **62**, 166–170.
10. Koga, Y., Sebe, F., Minami, T., Otake, K., Saitow, K. and Nishikawa, K. (2009) Spectrum of excess partial molar absorptivity. I. Near infrared spectroscopic study of aqueous acetonitrile and acetone. *J. Phys. Chem. B*, **113**, 11928–11935.

Reference

19

Far-Infrared Spectroscopy and Terahertz Time-Domain Spectroscopy

Seizi Nishizawa
Research Center for Development of Far-Infrared Region, University of Fukui,
Japan

19.1 Introduction

As described in Section 1.2.2.1, it is common to call the wavelength region from 25 μm to 1 mm the far-infrared region; in wavenumber, this wavelength region corresponds to the region from 400 to 10 cm^{-1}. Far-infrared spectroscopy, which has a history nearly as long as that of mid-infrared spectroscopy, began in the early 1900s. Randall [1], a pioneer in this field, constructed a high-performance far-infrared spectrometer in the 1930s, which could measure a rotational spectrum of water vapor over a wide wavenumber region from 555 to 75 cm^{-1} at a wavenumber resolution of 0.5 cm^{-1} or better. The quality of this spectrum appears to be comparable to that of a spectrum that might be measured currently by a modern Fourier transform infrared (FT-IR) spectrometer set up for measuring far-infrared spectra at an equivalent wavenumber resolution. Presently, the Martin–Puplett interferometer [2, 3] configuration, as described later, is better suited for a far-infrared spectrometer, particularly in the region from about 150 to 10 cm^{-1}, than a conventional Michelson interferometer.

By comparison with the phrase "far-infrared spectroscopy," the term *terahertz spectroscopy* or *terahertz time-domain spectroscopy* is a relatively recent one, which probably began to be used in the early 1990s. As was already described in Section 1.2.1, the wavenumber region of 400 to 10 cm^{-1} corresponds to the frequency region of $(12–0.3) \times 10^{12}$ Hz

Introduction to Experimental Infrared Spectroscopy: Fundamentals and Practical Methods,
First Edition. Edited by Mitsuo Tasumi and Akira Sakamoto.
© 2015 John Wiley & Sons, Ltd. Published 2015 by John Wiley & Sons, Ltd.

or 12–0.3 THz. This means that the far-infrared region mostly coincides with the terahertz frequency region. For this reason, currently, the term *terahertz spectroscopy* or *terahertz time-domain spectroscopy* is often used instead of far-infrared spectroscopy. However, the term *far-infrared spectroscopy* is a better designation of the field because of its consistency with other fields of optical spectroscopies. The term *terahertz spectroscopy* or *terahertz time-domain spectroscopy* is used to emphasize intentionally that the mechanism of measurement in studies using such a title is completely different from that used in the more traditional far-infrared spectroscopy instrumentation.

In this chapter, the terms *terahertz time-domain spectroscopy* and *terahertz time-domain spectrometry* are used; both of them are abbreviated as THz-TDS. Since the averaged intensity of terahertz pulses used in THz-TDS is much higher than that emitted in the far-infrared region from a Globar or a high-pressure mercury lamp source, THz-TDS is easier to perform, at least in principle, compared with the more traditional far-infrared spectrometry methods of measurement. For this reason, THz-TDS is not only used currently for basic research studies in molecular science, solid-state physics, materials science, and biological science but also has proven applications or possibilities of application in various other fields including the detection of hazardous materials, the testing of pharmaceuticals, food, and agricultural products, medical examination (biopsy), and the quality control of a variety of manufactured products.

19.2   FT-IR Spectrometry in the Far-Infrared Region

19.2.1   Use of a Conventional FT-IR Spectrometer in the Far-Infrared Region

Because the lower wavenumber limit covered by a conventional FT-IR spectrometer operating in the mid-infrared region is typically $400\,\mathrm{cm}^{-1}$, it is necessary, in order to measure far-infrared spectra, to exchange a few of the main optical components of an FT-IR interferometer; these are described here:

1. *The source of radiation*: A Globar, which is usually used as the source for the mid-infrared radiation, can still be used in the wavenumber region of $400–150\,\mathrm{cm}^{-1}$. However, in order to measure the wavenumber region lower than $150\,\mathrm{cm}^{-1}$, the Globar needs to be replaced by a high-pressure mercury lamp (see Section 5.2.1.2).
2. *The beamsplitter*: As was mentioned in Section 5.2.1.1, it has been common to use a tightly held Mylar® film as the beamsplitter for the far-infrared beam; *Mylar*, the name by which such a far-infrared beamsplitter material has become commonly referred to, is actually a registered trade name of one manufacturer of polyester (poly(ethylene terephthalate) (PET) film). The PET film thickness may be selected in order to avoid the effect of interference arising from multireflections between the inner surfaces of the film (practically, for example, films of the following thickness are used: $6\,\mu\mathrm{m}$ for $400–50\,\mathrm{cm}^{-1}$, $12\,\mu\mathrm{m}$ for $250–30\,\mathrm{cm}^{-1}$, and $25\,\mu\mathrm{m}$ for $125–20\,\mathrm{cm}^{-1}$). A PET film on which Ge is vapor-deposited can be more useful, as it can cover a wider wavenumber region of $680–30\,\mathrm{cm}^{-1}$. A thin silicon plate can also be used for extending the lower wavenumber limit to $10\,\mathrm{cm}^{-1}$ (the upper limit being $600\,\mathrm{cm}^{-1}$), although a limitation may occur for wavenumber resolution when this beamsplitter is used.

3. *The detector*: A pyroelectric detector using deuterated triglycine sulfate (DTGS) or deuterated L-alanine-doped triglycine sulfate (DLATGS) as an element for detecting the far-infrared radiation, fitted with a polyethylene window, can be used at room temperature. The lower wavenumber limit of detection by DTGS is about $30\,cm^{-1}$ and that by DLATGS is about $50\,cm^{-1}$. A "bolometer," an extremely sensitive thermometer using a semiconductor (e.g., Ge, Si) as an element for detecting the far-infrared radiation, operating at cryogenic temperatures (liquid He temperatures), is superior in performance to either a DTGS or DLATGS detector in the following: (i) its sensitivity ($D*$ in more accurate terminology; see Section 5.2.1.3) is orders of magnitude higher than that of either a DTGS or a DLATGS detector and (ii) it maintains its sensitivity over a lower wavenumber region; for example, a bolometer using composite Si as an element for detecting far-infrared radiation can cover a region of about $600-10\,cm^{-1}$. A modification of the bolometer called the *hot-electron detector* using an InSb element is sometimes used.

In practice, it is not convenient for a user to routinely exchange the above-mentioned components. Therefore, if measurements of far infrared spectra need to be carried out constantly, it is advisable to have an FT-IR spectrometer that is set up specifically and used exclusively for measuring far-infrared spectra.

19.2.2 Use of an FT-IR Spectrometer with a Martin–Puplett Interferometer

It has been established that an interferometer design first proposed by Martin and Puplett [2] is the most effective FT-IR spectrometer configuration for far-infrared measurements, particularly in the wavenumber region lower than $150\,cm^{-1}$ [3].

An example of a Martin–Puplett interferometer is schematically illustrated in Figure 19.1. The important characteristic of the Martin–Puplett interferometer lies in its use of a wire-grid polarizer (WGP) as the beamsplitter; this is free from the effects of interference arising from multiple internal reflections occurring inside a PET film. The WGP used for the Martin–Puplett interferometer is an array of parallel metallic wires, each having a diameter of about $10\,\mu m$ and separated by intervals of about $12.5\,\mu m$. A beam of far-infrared radiation incident on the WGP is divided into two orthogonal polarized components; that is, a component with its plane of polarization parallel to the wires, which is reflected, and a component with its plane of polarization perpendicular to the wires, which is transmitted. Thus, the WGP performs the role of a beamsplitter.

The Martin–Puplett interferometer illustrated in Figure 19.1 has two further WGPs (P_{in} and P_{out}) in addition to that used as the beamsplitter (B). The plane of P_{in} located before B is set to be perpendicular to the x-axis, while the plane of B is set at an angle of $45°$ to both the x- and y-axes. P_{in} has its wires parallel to the y-axis, so that only x-polarized radiation is transmitted from P_{in} and advances toward B. The wires in B are arranged in a direction rotated clockwise by $45°$ from the x-axis, when viewed from the side of P_{in}, so that the component of the beam of radiation incident on B with its plane of polarization rotated clockwise by $45°$ from the x-axis is reflected by B, and advances toward the fixed mirror R_f along the positive direction of the y-axis. It should be noted that the plane of polarization of this beam, when viewed in the direction toward R_f, is rotated anticlockwise

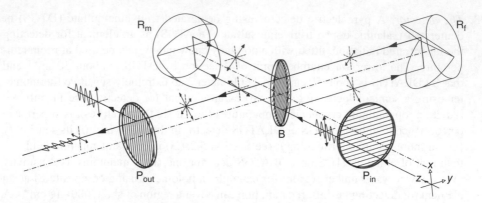

Figure 19.1   *Schematic illustration of a Martin–Puplett interferometer. For details, see text.*

by 45° from the x-axis. This beam is reflected by R_f and returns toward B. Because R_f is a roof mirror, the beam of radiation reflected by R_f has its plane of polarization rotated anticlockwise by 45° from the x-axis, when viewed in the direction toward B. This means the plane of polarization of this beam is perpendicular to the wires of B. Accordingly, this beam is transmitted by B and advances toward P_{out}.

On the other hand, the component of the beam of radiation incident on B from P_{in} polarized perpendicularly to the wires of B is transmitted by B, and advances toward the movable mirror R_m along the positive direction of the z-axis. The plane of polarization of this beam, when viewed toward R_m is rotated anticlockwise by 45° from the x-axis. This beam is reflected by R_m and returns to B. The plane of polarization of this beam, when viewed toward B, is rotated anticlockwise by 45° from the x-axis. The direction of the wires of B, when viewed from the side of R_m, is rotated anticlockwise by 45° from the x-axis. The plane of polarization of the beam returning from R_m is therefore parallel to the wires of B, so that this beam is reflected by B along the y-axis toward P_{out}.

Thus, the two beams reflected by R_f and R_m are mixed at B, although their planes of polarizations are orthogonal to each other, and advance to P_{out}. The plane of P_{out} is rotated anticlockwise by 45° about the x-axis, and its wires are parallel to the x-axis. Therefore, only the z-polarized component is transmitted from P_{out} and advances to the detector. Interference between the two beams reflected by R_f and R_m occurs during this step. The spectroscopic measurement process after this is exactly the same as that of a conventional FT-IR spectrometer. The x-polarized component reflected by P_{out} along the z-axis is not utilized for any purpose.

The advantage of the Martin–Puplett interferometer is that, as already mentioned, it is free from the troublesome effects arising from the multireflections inside a PET film. However, the Marin–Puplett interferometer, as illustrated in Figure 11.1, has drawbacks. The drawbacks are: (i) the quantity of radiation energy is theoretically decreased to 1/4 of the original radiation energy before entering P_{in} because there are two WGPs (P_{in} and P_{out}), and (ii) the use of such an interferometer is difficult in the wavenumber region higher than about $300\,cm^{-1}$ because of difficulties in producing WGPs having high mechanical precision.

19.2.3 Spectral Measurements in the Far-Infrared Region

Because the intensities of absorption bands in the far-infrared region are generally weaker than those in the mid-infrared region, the amount of sample required for measuring a far-infrared spectrum tends to be larger.

The window materials that can be used in the far-infrared region are limited (see Table 2.1). A practically useful material not listed in Table 2.1 is spectroscopic-grade polyethylene; high-density polyethylene plates of a few millimeters thickness can be used as cell windows in the far-infrared region. Polyethylene has two weak bands in the vicinity of $100\,cm^{-1}$, but these bands do not seriously disturb measurements of far-infrared spectra. Quartz (z-cut) may be used as windows in the low-wavenumber region (below about $250\,cm^{-1}$). Synthetic diamond and sapphire crystal windows may also be used.

A disk using spectroscopic-grade polyethylene powder (commercially available), instead of the KBr powder used for mid-infrared sample preparations, may be formed with some solid samples. Whether such a disk is firmly and usefully formed depends on the affinity between the sample and the polyethylene powders.

A solid sample that can be finely pulverized and does not change its property when heated up to about $150°$ may be completely mixed with heated, molten low-density polyethylene, and made into a sample-containing polyethylene film of an appropriate thickness after cooling. This method usually works for most inorganic compounds and some coordination compounds and some stable organic compounds as well. As mentioned in Section 6.2.7, the specular-reflection method is particularly useful for measuring the far-infrared spectra from inorganic crystals [4].

As the absorption bands caused by water vapor (its rotational spectrum) are intense throughout the far-infrared region, it is important to purge efficiently the inside of the spectrometer with dried air or nitrogen. Some commercial spectrometers can be evacuated, but usually their sample compartment needs to be purged with dried air or nitrogen. This is commonly needed also in terahertz time-domain spectrometry, which is described in the following section.

19.3 Terahertz Time-Domain Spectrometry

As described later in more detail, light pulses with an extremely short width from a femtosecond laser irradiating a terahertz-radiation generation element can give rise to terahertz pulses covering the entire terahertz frequency region. The terahertz pulses are used as the source of radiation for THz-TDS. The strength of the electric field of this terahertz radiation $E_{THz}(t)$ is measured as time-series signals by using as a detector a device similar to the terahertz-radiation generation element. By the Fourier transform of $E_{THz}(t)$, the distribution of the complex electric field $\widehat{E}(\omega)$ in the frequency axis (ω, angular frequency) is obtained, and $\widehat{E}(\omega)$ can be easily converted to the complex electric field $\widehat{E}(\tilde{\nu})$ along the wavenumber axis, from which the far-infrared spectrum is obtained. It is worth emphasizing here that, in THz-TDS, not only is the intensity of the far-infrared spectrum obtained but the strength of the electric field $E_{THz}(t)$ and the phase of the terahertz radiation $\phi(\omega)$ are also determined.

THz-TDS was initiated by the pioneering research on the emission of terahertz radiation by Auston *et al.* [5] and subsequent studies on the coherent detection of the terahertz radiation by van Exeter and Grischkowski [6], Arjavalingram *et al.* [7], and Nuss *et al.* [8]. Since then, THz-TDS, enabled by the advent of femtosecond laser technology, has become a new spectrometric method covering the region from the millimeter wavelength to the far-infrared.

19.3.1   Basic Description of a THz-TDS Instrument

In Figure 19.2, an instrument for THz-TDS is illustrated schematically. A beam of femtosecond pulses generated by a laser system, shown located at the lower left corner of this figure, is divided into two beams by a beamsplitter; one beam is used to generate the terahertz radiation, while the other is used for detecting the terahertz radiation, and also for processing the detected signals. The laser pulses going to the components illustrated in the upper left corner of this figure are focused onto a photoconductive thin film antenna (labeled as PCA; see Section 19.3.3) supported on a semi-insulating substrate, and the laser photons incident on the PCA generate terahertz radiation within the electrode gap of the PCA.

The image of the generated terahertz radiation is then focused onto a sample through an optical system consisting of a Si hyper-spherical lens, a plane mirror, and an ellipsoid mirror. Although a sample is shown set at the position for a transmission measurement in Figure 19.2, other methods of spectral measurement are also possible, at least in principle.

The terahertz radiation that has passed through the sample is detected with an optical system having essentially the same structure as the optical system for generating the terahertz radiation and focusing it onto the sample, as depicted in Figure 19.2. The terahertz radiation that has passed through the sample is focused onto the terahertz-radiation detection

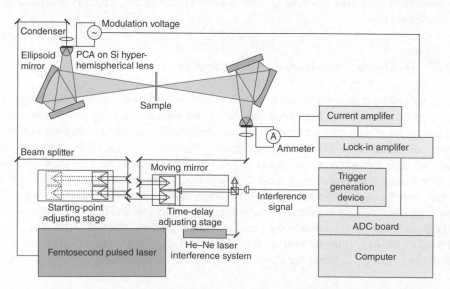

Figure 19.2   *Schematic illustration of a THz-TDS instrument. For details, see text.*

element, together with the laser beam that has passed through the optical stages for adjusting the starting point and time delays. This process generates an electric current in the detection system, and the resultant current is detected, amplified, digitized, and sent to a computer.

More detailed explanations of the main components and their functions are given in Sections 19.3.2 and 19.3.3.

19.3.2   Types and Efficiencies of Femtosecond Pulsed Lasers

Two types of pulsed lasers are mainly used for THz-TDS: a femtosecond synchronous Ti:sapphire laser and a femtosecond fiber laser. A Ti:sapphire laser is superior in generating stable femtosecond pulses. To excite the photoconductive antenna element consisting of low-temperature-grown gallium arsenide (LT-GaAs), the commonly used Ti:sapphire laser has the following properties: central wavelength, 780 nm; pulse width, 100–45 fs; average power, several hundred mW; repetition rate, 80–40 MHz. If a pulse-compression system is added to this laser, the values cited become as follows: pulse width, 10–7 fs, average power, 300–200 mW; repletion rate, 80 MHz.

Among femtosecond fiber lasers, the Er(erbium)-doped fiber amplifier (EDFA), generating femtosecond pulses at about 1.55 μm, is that most commonly used for THz-TDS. The ytterbium fiber laser has a wide band and a high quantum effect. Development of other lasers is also in progress.

The output pulse width of the EDFA is narrower than that of the Ti–sapphire laser, but it is difficult to obtain high output power from an EDFA. Furthermore, as the wavelength of the fundamental line is in the vicinity of 1.55 μm, it is necessary to obtain the second harmonic in order to excite the LT-GaAs photoconductive antenna element. This reduces the average power needed for the excitation. However, an EDFA has many practical advantages: that is, it is small in size, light weight, has a simple and stable operation, low power consumption, and is low cost. In practice, the EDFA used in the instrument for THz-TDS usually has the following properties: central wavelength (second harmonic output), 780 nm; pulse width, 120–75 fs; average power, 30 mW; and repetition rate, 40 MHz.

19.3.3   Generation and Detection of the Terahertz Radiation Using a Photoconductive Antenna Element

19.3.3.1   Generation of the Terahertz Radiation

Several different types of wide-band terahertz-radiation generation element are available: these include a photoconductive antenna element, a nonlinear optical effective element (NOE), and a surface outgoing radiational-type semiconductor device. In this subsection, brief descriptions of the most commonly used PCAs are given.

A typical PCA is schematically shown in Figure 19.3. The PCA consists of a thin film of LT-GaAs grown on a substrate of semi-insulating gallium arsenide (SI-GaAs), Au transmission lines, and emission electrodes that extend from the transmission lines. The transmission lines and emission electrodes are etched onto the LT-GaAs film. The distance between the transmission lines is 30 μm, and pulsed modulation voltage of ±50 Vpp is applied between the transmission lines including the emission electrodes. The distance d of the gap between the emission electrodes is 5 μm. When a bias voltage V_b is applied between

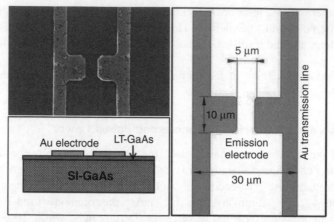

Figure 19.3 *Schematic illustration of the photoconductive antenna element (PCA). For details, see text.*

these electrodes, electric charges of $\pm Q_0$ are stored in the electrodes, and an electric dipole $P_0 = Q_0 d$ is formed between the gap.

The femtosecond laser pulses are focused onto the gap between the emission electrodes, while a bias voltage V_b is applied to the emission electrodes. Then, a photocurrent $J_{op}(t)$ due to photoexcitation occurs between the emission electrodes. This photocurrent causes the V_b-induced electric dipole to disappear instantaneously. The dipole radiation arising from this extremely rapid dipole change covers the entire terahertz region, and is used for THz-TDS.

The process described above may be stated also in the following way. When a PCA is irradiated with a laser pulse, carriers (electrons) of electric charge q_e and positive holes are formed between the electrodes as a result of photoexcitation. As the positive holes are formed in the vicinity of the cathode, they disappear instantaneously by coupling with the electrons accumulated on the cathode. On the other hand, the carriers move to the anode side accelerated by the electric field, and decay with a time constant τ_c. The magnitude of the electric dipole may be expressed as a function of time $P(t)$ expressed as

$$P(t) = q_e e^{-\frac{t}{\tau_c}} \left(d - \frac{\mu V_b}{d} t \right) + \left(Q_0 - q_e e^{-\frac{t}{\tau_c}} \right) d \qquad (19.1a)$$

where μ is the mobility of the carriers generated by photoexcitation. Actually, the time-dependent part of Equation (19.1a), denoted by $P_{op}(t)$, may be reduced to a much simpler form expressed as

$$P_{op}(t) = -q_e e^{-\frac{t}{\tau_c}} \frac{\mu V_b}{d} t \qquad (19.1b)$$

According to the physical meaning of $P_{op}(t)$, $P_{op}(t) = 0$ for $t \leq 0$. When the laser pulse is cut off, the carriers decay with the time constant τ_c and finally disappear (τ_c is 0.1–1 ps in an LT-GaAs installed in a typical instrument). Then, the transmission lines and the emission electrodes become insulated.

The electric field of the terahertz radiation $E_{THz}(t)$ emitted from the PCA in the direction at right angles to the emission electrodes and at a distance r from the electrodes is expressed approximately as [9]

$$E_{\text{THz}}(t) \simeq -\frac{1}{4\pi\varepsilon_0 rc^2}\frac{\partial^2 P_{\text{op}}(t-r/c)}{\partial t^2} \tag{19.2}$$

where ε_0 is the electric constant (permittivity of vacuum) and c is the speed of light.

19.3.3.2 Detection of the Terahertz Radiation

A PCA is effective also as a detector for the terahertz radiation. As shown in Figure 19.2, the terahertz radiation that has passed through the sample is focused onto the PCA in the detection system. The femtosecond laser pulse is synchronously superimposed onto the terahertz radiation at the same PCA. Since the density of the electric current $j_{\text{op}}(t)$ occurring between the electrodes is proportional to the number of carriers formed by photoexcitation $N_{\text{exc}}(t)$, the current formed is also proportional to the complex electric field $\hat{E}_{\text{THz}}(t)$ of the terahertz radiation. Therefore, by measuring the electric current between the electrodes as illustrated in Figure 19.2, it is possible to determine the electric field of the terahertz radiation $E_{\text{THz}}(t)$ and its phase $\phi(t)$.

However, it is not possible to determine directly $E_{\text{THz}}(t)$ and $\phi(t)$ of the terahertz radiation, because the terahertz radiation is emitted as an extremely short pulse. For this reason, as illustrated schematically in Figure 19.4, time delays are instigated between the laser pulses for generating the terahertz radiation and those for detecting the terahertz radiation. The shape of the terahertz radiation on the time axis is determined by scanning the delay time of the laser pulses.

As shown in Figure 19.2, the femtosecond laser pulses used for detecting the terahertz radiation are separated from those used to generate the terahertz radiation by a beamsplitter, and brought to the two optical stages, that is, the starting-point adjusting stage and the time-delay adjusting stage. These optical stages consist of multiple cube-corner retroreflectors and mechanisms for driving them. The former adjusts the starting point of the detection, while the latter controls time delays for femtosecond pulses, which are superimposed on the terahertz radiation at the PCA in the detection system in order to determine the shape of the terahertz radiation. By these optical stages, the distance (optical path difference) between the beamsplitter, shown located in the left middle of Figure 19.2, and the PCA of the detection system can be changed either stepwise or continuously by the movement of the axial drive in the time-delay adjusting stage. The movement of the axial drive is monitored and controlled by a He–Ne laser interference reference system.

The procedures for measuring the electric current corresponding to the electric field of the terahertz radiation are schematically illustrated in Figure 19.4. The time-dependent electric field of the terahertz radiation $E_{\text{THz}}(t)$ and the measured time-dependent electric current $I_m(t)$ (subscript m indicates the numbering of the time-delayed femtosecond laser pulses for detecting the terahertz radiation) are depicted, respectively, at the top and at the bottom of this figure; in between, the femtosecond laser pulses that irradiate the PCA in the detection system at delayed time τ_m are shown. As the gate for detecting the terahertz radiation is opened with the irradiation of the femtosecond laser pulse, $I_m(t)$ on the time axis is measured with increasing τ_m.

$I_m(t)$ usually consists of very narrow flat steps; the flat step Δt corresponds to there being an optical path difference existing between the femtosecond laser pulses for the generation

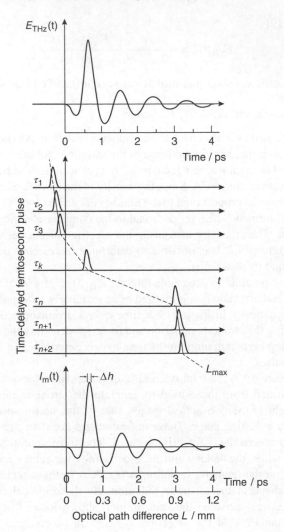

Figure 19.4 *Detection scheme for the terahertz radiation. For details, see text.*

of the terahertz radiation and that for the detection of the terahertz radiation. This optical path difference is an integral multiple of the wavelength of the He–Ne laser light, so that the minimum optical path difference $L_{min} = n\lambda_{He-Ne}$, where n is an integer, and λ_{He-Ne} is the wavelength of the He–Ne laser light. Then, Δt corresponding to time needed for light to pass through L_{min} may be given as $\Delta t = L_{min}/c = n\lambda_{He-Ne}/c$. In Figure 19.4, the optical path difference L is also shown as an abscissa axis in parallel with the corresponding time axis.

The Fourier transform of $I_m(t)$ gives a spectrum in the terahertz region (on the frequency scale). The maximum wavenumber of this spectrum $\tilde{\nu}_{max}$ is estimated as about $1/2L_{min}$ (see Section 4.4.2). For example, if n is 4, $\tilde{\nu}_{max} \approx 2000$ cm^{-1}, which means the spectrum covers the entire terahertz region.

The wavenumber resolution $\delta\tilde{\nu}$ of the measured terahertz spectrum is given as $\delta\tilde{\nu} \approx 1/L_{max}$ (see Section 4.4.1). For example, if the maximum optical path difference is 30 cm, the wavenumber resolution of the measured spectrum if about 0.03 cm^{-1}. In practice, as a result of the effect of apodization, the wavenumber resolution may not be as high as this value, but in principle the relationship described in Sections 3.2.2 and 4.4.1 holds also in THz-TDS.

If the mechanism for the detection of the terahertz radiation is described in more detail, it is necessary to consider the relationship between the density of current $j_{op}(t)$ induced in the transmission lines of the PCA in the detection system and the electric field of the terahertz radiation $E_{THz}(t)$ as well as the number of photoexcited carriers $N_{exc}(t)$. The relationship is expressed as a convolution of the latter two quantities as [9]

$$j_{op}(t) \propto e\mu \int_{-\infty}^{\infty} E_{THz}(\tau)N_{exc}(t - \tau)d\tau \qquad (19.3)$$

$N_{exc}(t)$ is also expressed as a convolution of the intensity of the femtosecond laser pulse for detection $P_L^d(t)$ and the response function of LT GaAs in the PCA $F_{GaAs}(t)$ as

$$N_{exc}(t) \propto \int_{-\infty}^{\infty} P_L^d(\tau) F_{GaAs}(t - \tau)d\tau \qquad (19.4)$$

These relationships mean that, in order to expand the wavenumber region of the terahertz radiation, it is necessary to make the width of the laser pulse for generating the terahertz radiation narrower and the response of the photogenerated carrier faster in order to have a shorter lifetime. At present, LT-GaAs is widely used for the PCA as a material appropriate for the above-mentioned requirement.

19.3.3.3 Data Analysis

As described in Section 19.3.3.2, the strength of the electric field of the terahertz radiation is detected as a time-series signal $E_{THz}(t)$. These signals are actually measured as $I_m(t)$ signals that are proportional to $ш_{\Delta t}(t)E_{THz}(t)$, where $ш_{\Delta t}(t)$ indicates the Dirac delta comb (see Section D.3.3). $E_{THz}(t)$ signals are converted to equivalent complex electric field $\hat{E}(\omega)$ values on the frequency axis as

$$\hat{E}(\omega) = \frac{1}{2\pi} \int_{-\infty}^{\infty} E_{THz}(t) \exp(-i\omega t)\, dt \qquad (19.5)$$

where ω is the angular frequency; that is, $\omega = 2\pi c\tilde{\nu}$.

As $\hat{E}(\omega)$ is a complex quantity, it may be expressed by using the amplitude $E(\omega)$ and the phase factor $\phi(\omega)$ as $\hat{E}(\omega) = E(\omega)\exp[-i\phi(\omega)]$. It is a common practice to express $E(\omega)$ and $\phi(\omega)$ of a sample as the quantities relative to those of a reference material. By denoting the electric fields of the sample and the reference material by $E_{sam}(\omega)$ and $E_{ref}(\omega)$, respectively, the transmittance spectrum $T(\omega)$ is obtained as $T(\omega) = |E_{sam}(\omega)/E_{ref}(\omega)|^2$. By denoting the phases of the sample and the reference material by $\phi_{sam}(\omega)$ and $\phi_{ref}(\omega)$, respectively, the phase delay $\Delta\phi(\omega)$ is defined as $\Delta\phi(\omega) = \phi_{sam}(\omega) - \phi_{ref}(\omega)$. $T(\omega)$ and $\Delta\phi(\omega)$ are depicted in spectral forms as independent quantities.

$T(\omega)$ and $\Delta\phi(\omega)$ for a sample, which are determined by taking the signals for vacuum (or dried air) as the reference, are related with the complex refractive index of the sample

$\hat{n}(\omega)$ in the following way. (As described in Section 1.2.4, $\hat{n}(\omega) = n(\omega) + ik(\omega)$, where the real part $n(\omega)$ means the conventional refractive index, and the imaginary part $k(\omega)$ is the absorption index.)

$$\frac{\ln\left[1/T\left(\omega\right)\right]}{d} = \frac{2\omega}{c}\, k(\omega) \tag{19.6}$$

$$\Delta\phi(\omega) = \frac{\omega}{c}\left[n\left(\omega\right) - 1\right]d \tag{19.7}$$

where c is the speed of light and d is the sample thickness.

Equations (19.6) and (19.7) mean that the complex refractive index can be determined from THz-TDS measurements without resorting to the Kramers–Kronig relations used in specular-reflection measurements (see Section 8.3). The complex refractive index $\hat{n}(\omega)[= n(\omega) + ik(\omega)]$ is related to the complex permittivity $\hat{\varepsilon}(\omega)[= \varepsilon'(\omega) + i\varepsilon''(\omega)]$ by the following relations: $\hat{\varepsilon}(\omega) = \hat{n}(\omega) \cdot \hat{n}(\omega)$, $\varepsilon'(\omega) = n^2(\omega) - k^2(\omega)$, and $\varepsilon''(\omega) = 2n(\omega)k(\omega)$. Thus, it is possible to calculate the real and imaginary part of the complex permittivity from a THz-TDS measurement.

Not only transmission measurements but also other types of spectral measurements, including the specular-reflection measurement method, are possible by THz-TDS. By utilizing the result of the specular-reflection measurement, it is possible to determine $n(\omega)$ and $k(\omega)$. In practice, this knowledge is important for analyzing the measured spectrum, which often exhibits a pronounced interference fringe effect arising from multiple reflections between the inner surfaces of a sample, as shown in Section 19.3.5. As the procedure of the analysis of such a measured spectrum is performed by a curve-fitting routine based on complicated formulae, these are not described here; readers interested in the curve-fitting procedure are advised to refer to Ref. [10].

From a practical standpoint, users of commercial THz-TDS instruments usually do not have any difficulty in using various methods of measurement, because accessories for those methods of measurement and the software needed for analyzing the measured results are provided by the manufacturer of the instrument used.

19.3.3.4   Spectral Measurements in THz-TDS

Methods used for spectral measurements and the points to note in THz-TDS are essentially the same as those described in Section 19.2.3.

As the terahertz radiation can usually pass through thick samples, powder samples may often be formed into self-supporting disks of appropriate thickness without any other coexisting diluent material, and the disks formed may be used as samples for THz-TDS. This method can be applied to many kinds of organic compounds including pharmaceuticals.

19.3.4   Performance of Time-Domain Terahertz Spectrometry

In Figure 19.5, a high-resolution absorption spectrum (rotational spectrum) of water vapor in the atmosphere at a reduced pressure of 60 Pa is shown, together with the measured THz-TDS signal shown in (a). The abscissa range shown in (b) and (c) covers $0-150\,\text{cm}^{-1}$, although the spectrum actually measured covers a region from about 1.3 to $230\,\text{cm}^{-1}$. The phase-delay spectrum is also shown in (c), but it does not have any particular analytically useful information. The instrument used for this spectral measurement was manufactured

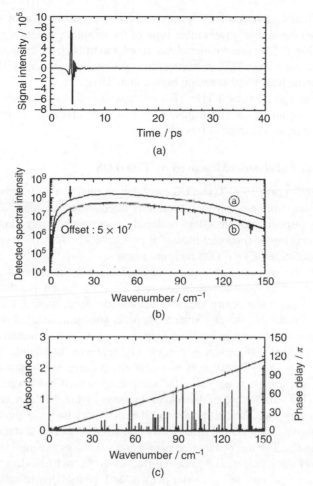

Figure 19.5 *THz-TDS measurement of the rotational spectrum of water vapor. (a) Measured interference signal intensity in arbitrary unit, (b) transmission spectra of ⓐ vacuum and ⓑ water vapor in the atmosphere at a reduced pressure of 60 Pa, and (c) absorbance and phase-delay spectra of water vapor in the atmosphere at a reduced pressure of 60 Pa.*

by a company in which the author of this chapter was deeply involved. The following performance was achieved: a wavenumber resolution of about $0.05 \, \text{cm}^{-1}$, a dynamic range $>75 \, \text{dB}$ (which corresponds to more than 7.5 in absorbance units), and a signal-to-noise ratio of about 2500 (per one scan of 1 cm of the optical path difference). When a weak band at about $64 \, \text{cm}^{-1}$ in Figure 19.5 was scale-expanded in order to see its fine structure, it became clear that the band actually consists of two bands at 63.950 and $63.979 \, \text{cm}^{-1}$, and the full width at half maximum of the former band is $0.02 \, \text{cm}^{-1}$, while that of the latter band is slightly narrower.

In theory, the full wavenumber region of the terahertz radiation generated by the PCA is inversely proportional to the pulse width of the exciting femtosecond pulse, so that it would be possible for this THz-TDS instrument to cover an extended wavenumber region higher than $230\,cm^{-1}$ also. Actually, however, the phonon absorption of LT-GaAs in the PCA seems to set the higher wavenumber limit of the terahertz radiation generated by the PCA at about $230\,cm^{-1}$. At present, therefore, in order to fully cover the far-infrared region, it is necessary to use a THz-TDS instrument in combination with an FT-IR spectrometer, which can cover the wavenumber region higher than $230\,cm^{-1}$.

A definite advantage of using a THz-TDS instrument is that its dynamic range is much greater than that of an FT-IR spectrometer, so that it is advisable to use THz-TDS for purposes that can make the most of this advantage.

19.3.5   Examples of Results Measured by THz-TDS

As mentioned in Section 19.1, THz-TDS has been applied not only to basic research in molecular science, solid-state physics, materials science, and biomedical science but is also used for the inspection of hazardous materials, noncontact testing of pharmaceuticals, and on-site monitoring of food and industrial products. The following are a few typical examples of application of THz-TDS measurements.

In Figure 19.6, the spectra of absorption, phase delay, and the complex refractive index of three types of Si (ⓐ, ⓑ, and ⓒ) in the very low THz region are shown [11]. In the spectra shown in (a), many sharp bandlike shapes are seen; these are due to the multiple reflections inside the sample; such a shape is sometimes called the *Fabry–Pérot effect*. The phase-delay spectrum becomes stepwise when the transmittance spectrum has the Fabry–Pérot effect. This effect is particularly apparent for sample ⓐ, which has a high resistivity ($\rho = 284.5\,\Omega\,cm$); it is weak for sample ⓑ, which has a low resistivity ($\rho = 10\,\Omega\,cm$), and almost nonexistent for sample ⓒ, which has a very low resistivity ($\rho = 1.1\,\Omega\,cm$). For sample ⓒ, as the reflection arising from free carriers increases, the transmittance is very low. The real part n and the imaginary part k of the complex refractive index of sample ⓒ, calculated by using the transmittance in (a) and the phase delay in (b), are shown in (c). The marked increases of n and k with decreasing frequency indicate that the resistivity decreases with decreasing frequency. From these data, it is possible to evaluate the complex conductivity, carrier density, and the mobility of carriers.

The THz-TDS method is superior to the conventional electrical impedance method of four-terminal sensing in that the THz-TDS method is faster in determining the required physical quantities in a noncontact condition without a magnetic field.

In Figure 19.7, the absorption and phase-delay spectra of bismuth titanate ($Bi_4Ti_3O_{12}$; abbreviated as BIT) in the wavenumber region of $0–60\,cm^{-1}$ are shown, together with the spectra of the real and imaginary parts (ε' and ε'') of its complex permittivity calculated by using the measured spectral data [12, 13]. BIT is an important ferroelectric material. The transmission measurement was made for a very thin film of monoclinic crystal with the terahertz radiation advancing along the c-axis of the crystal with the plane of its polarization parallel to the a-axis. The solid curves are the results of calculation using the phonon wavenumber of $28.3\,cm^{-1}$ observed by Raman spectroscopic measurement and a damping constant of $3\,cm^{-1}$ [12, 13]. The values of ε' and ε'' obtained by THz-TDS are in good agreement with the results of calculations based on the Raman data.

The determination of ε' and ε'' within a short time is important from an industrial viewpoint, and the use of the THz-TDS method at a production site of ferroelectric materials is in progress.

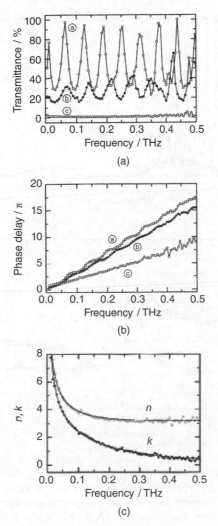

Figure 19.6 *THz-TDS measurements for three types of Si. (a) Transmission spectrum, (b) phase-delay spectrum, and (c) real (n) and imaginary (k) parts of the complex refractive index versus frequency spectrum. (Source: Adapted from Ref. [11] with permission from Springer Verlag.)*

In Figure 19.8, the transmittance spectra of six pharmaceuticals in two groups, each of which has three pharmaceuticals, in the region of $0-100 \, cm^{-1}$ are shown. These spectra were measured from the tablets of these pharmaceuticals. A useful characteristic of THz-TDS is that such measurements can be made directly, so that it is hopeful that THz-TDS will be used in the future as an in-line inspection tool at the site of manufacturing pharmaceutical tablets.

It has been argued for a long time that crystal polymorphs of pharmaceuticals affect their activities. It has been confirmed that THz-TDS is useful for discriminating crystal

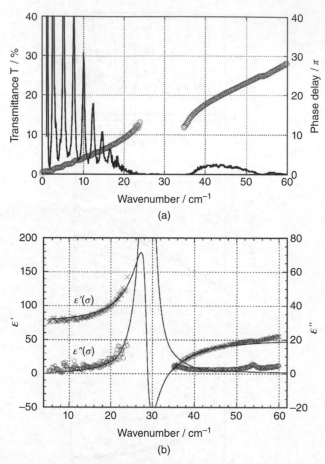

Figure 19.7   *THz-TDS measurements for bismuth titanate ($Bi_4Ti_3O_{12}$). (a) Transmittance and phase-delay spectra and (b) real (ε') and imaginary (ε'') parts of the complex permittivity versus frequency spectra. (Source: Reproduced with permission from [12] © Institute of Physics and Engineering in Medicine. Publishing on behalf of IPEM by IOP Publishing Ltd. All rights reserved.)*

polymorphs by their transmittance spectra. In fact, the crystal polymorphs A and B of famotidine, a well-known pharmaceutical for stomach disorder, show clearly different transmittance spectra in the region of $0-150\,cm^{-1}$. (It has been claimed that only one of the two crystal polymorphs is effective for the disorder.) This example provides THz-TDS with another useful application possibility for use in the pharmaceutical industry.

19.3.6   Future of THz-TDS

THz-TDS is a relatively new technology, and it is still in progress and development. Two beneficial and important characteristics are that (i) it has a large dynamic range and (ii) the complex refractive index can be determined without using the Kramers–Kronig relations.

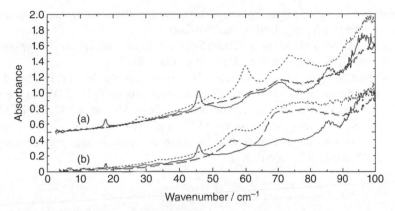

Figure 19.8 *Transmittance spectra of six pharmaceuticals. (a) Three medicines for a cold produced by three different companies and (b) three medicines for the digestive system produced by three different companies.*

These should be developed further; then, many more applications will be found in various fields of study.

References

1. Randall, H.M. (1938) The spectroscopy of the far infra-red. *Rev. Mod. Phys.*, **10**, 72–85.
2. Martin, D.H. and Puplett, E. (1970) Polarised interferometric spectrometry for the millimeter and submillimetre spectrum. *Infrared Phys.*, **10**, 105–109.
3. Lambert, D.K. and Richards, P.L. (1978) Martin-Puplett interferometer: an analysis. *Appl. Opt.*, **17**, 1595–1602.
4. Yamamoto, K. and Masui, A. (1995) Complex refractive index determination of bulk materials from infrared reflection spectra. *Appl. Spectrosc.*, **49**, 639–644.
5. Auston, D.H., Cheung, K.P. and Smith, P.R. (1984) Picosecond photoconducting Hertzian dipoles. *Appl. Phys. Lett.*, **45**, 284–286.
6. van Exeter, M. and Grischkowski, D. (1990) Carrier dynamics of electrons and holes in moderately doped silicon. *Phys. Rev. B*, **41**, 12140–12149.
7. Arjavalingram, G., Theophilou, N., Pastol, Y., Kopcsay, G.V. and Angelopoulos, M. (1990) Anisotropic conductivity in stretch oriented polymers measures with coherent microwave transient spectroscopy. *J. Chem. Phys.*, **93**, 6–9.
8. Nuss, M.C., Goossen, K.W., Gordon, J.P. , Mankiewich, P.M., O'Malley, M.L. and Bhushan, M. (1991) Terahertz time-domain measurement of the conductivity and superconductivity band gap in niobium. *J. Appl. Phys.*, **70**, 2238–2241.
9. Tani, M., Herrmann, M. and Sakai, K. (2002) Generation and detection of terahertz pulsed radiation with photoconductive antennas and its application to imaging. *Meas. Sci. Technol.*, **13**, 1739–1745.

10. Nishizawa, S. (2005) Principles of THz-TDS, In: *Terahertz Optoelectronics* (ed K. Sakai), Springer-Verlag, Berlin, pp. 204–206.
11. Hangyo, M. and Nagashima, T. (2005) Semiconductors, In: *Terahertz Optoelectronics* (ed K. Sakai), Springer-Verlag, Berlin, pp. 2006–2012.
12. Nishizawa, S., Tsumura, N., Kitahara, H. , Wada Takeda, M. and Kojima, S. (2002) New application of terahertz time-domain spectroscopy (THz-TDS) to the phonon-polariton observation on ferroelectric crystals. *Phys. Med. Biol.*, **47**, 3771–3776.
13. Kojima, S., Tsumura, M., Takeda, M.W. and Nishizawa, S. (2003) Far-infrared phonon-polariton dispersion probed by terahertz time-domain spectroscopy. *Phys. Rev. B*, **67**, 035102(1)–035102(5).

20

Time-Resolved Infrared Absorption Measurements

Akira Sakamoto
Department of Chemistry and Biological Science, College of Science and Engineering,
Aoyama Gakuin University, Japan

20.1 Introduction

Compared with other time-resolved spectroscopic methods for measuring electronic transitions (electronic absorption, fluorescence, etc.) in molecules and molecular systems, time-resolved vibrational spectroscopy, which measures vibrational transitions, has the potential for simultaneously providing information on both the molecular structure (and vibrational force field) and the dynamics of a short-lived transient molecular species. Time-resolved vibrational spectroscopy is therefore considered to be one of the most effective means for monitoring changes in molecular structure over a wide time scale, particularly for condensed-phase studies.

Infrared and Raman spectroscopies are the two main representative methods of vibrational spectroscopy. Although development of time-resolved infrared spectroscopy began earlier than that of time-resolved Raman spectroscopy, the advent of time-resolved infrared spectroscopy over a short time scale (particularly the nanosecond to femtosecond range) was delayed in comparison with time-resolved Raman spectroscopy developments for a similar time scale. This is attributed to the following reasons: (i) generation of short laser pulses is more difficult in the mid-infrared region than in the ultraviolet to visible region, (ii) the detector sensitivity and high-speed response capability in the mid-infrared region are inferior to those in the ultraviolet to visible region, and (iii) while the resonance Raman effect is utilized effectively in time-resolved Raman measurements of molecular species that have an electronic absorption in resonance with the Raman excitation laser light, there is no comparable effect for enhancing the intensity of infrared absorption.

Introduction to Experimental Infrared Spectroscopy: Fundamentals and Practical Methods,
First Edition. Edited by Mitsuo Tasumi and Akira Sakamoto.
© 2015 John Wiley & Sons, Ltd. Published 2015 by John Wiley & Sons, Ltd.

For many molecules in their ground electronic state, infrared absorption and Raman scattering spectroscopic measurements are complementary to each other. In general, for molecules having polar functional groups, infrared spectroscopy usually gives the more useful information on the associated molecular structure, because this is not as easily obtained from Raman spectroscopy. This situation is expected to be similar also in time-resolved spectroscopy. In addition, compared with Raman spectroscopy, infrared spectroscopy is not hampered by "fluorescence," which is thought to arise from either the sample itself or from coexisting extremely small amounts of impurities, and often hinders Raman spectral measurements. All these considerations lead to a conclusion that time-resolved infrared measurements are likely to yield some significant and important results, even if they are difficult to perform from an experimental standpoint.

In this chapter, millisecond time-resolved infrared measurements are first described in Section 20.2; for this time scale, time resolution is set by the time needed to measure (scan) a spectrum. Then, microsecond to nanosecond time-resolved measurements, which are limited by the detector response time are described in Section 20.3, and finally, picosecond to femtosecond time-resolved measurements, the time resolution for which is determined by the width of the laser pulse used for the measurement, are described in Section 20.4.

20.2   Millisecond Time-Resolved Infrared Absorption Measurements

If the lifetime of a transient phenomenon is sufficiently longer than (typically over 10 times that of) the time needed for one scan of the moving mirror of the interferometer in a Fourier-transform infrared (FT-IR) spectrometer used for the measurement, the interferogram measured by each scan of the interferometer is considered to correspond to the transient state during which the interferogram has been measured. This means that, if multiple scans of the interferometer are performed while a transient phenomenon is in progress, time-resolved interferograms are measured by these scans. From the Fourier-transform calculations for these interferograms, corresponding time-resolved infrared spectra are obtained [1].

In practical measurements, an instrument (e.g., a pulsed laser, a function generator) that gives an external stimulus to the sample is used in combination with an FT-IR spectrometer. The procedures of spectral measurements are essentially the same as those for normal FT-IR spectroscopic measurements.

Millisecond time-resolved infrared absorption measurements can often be applied to transient phenomena that are not easily repeatable. The lifetime of the transient phenomenon to be studied by time-resolved measurements is limited by the time needed for one scan, namely, the speed of the moving mirror of the interferometer; in the case of a conventional FT-IR spectrometer equipped with a high-sensitivity mercury cadmium telluride (MCT) detector, a phenomenon that can be the target of such measurements must have a lifetime longer than a few hundred milliseconds. If an ultra-rapid-scanning FT-IR interferometer equipped with a rotating wedge-shape mirror and a cube-corner retroreflector is used, a transient phenomenon with a lifetime of about 1 ms can be studied [2, 3]. By using an ultra-rapid-scanning spectrometer of this type in combination with a stopped-flow apparatus, a reaction of $CpCo(CO)_2$ (Cp, cyclopentadienyl ligand) with ClNO in CH_2Cl_2 was traced; the results of the measurements are shown in Figure 20.1 [3]. After mixing the solutions of the two reactants, the bands of $CpCo(CO)_2$ at 2026 and 1961 cm^{-1} disappeared

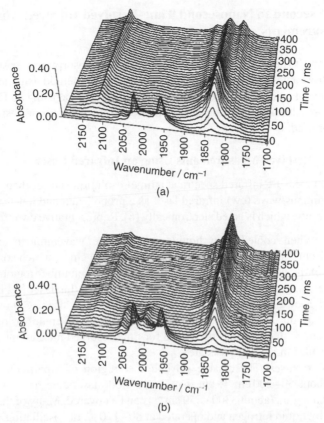

Figure 20.1 *Millisecond time-resolved infrared spectra measured during a reaction of CpCo(CO)₂ with CINO [3]. (a) Spectra at room temperature and (b) spectra at 10°C. Cp, cyclopentadienyl. (Source: Reproduced with permission from [3]. Copyright © 2010, SAS.)*

rapidly, and the bands at 2060, 2008, and 1780 cm^{-1}, which are assignable to an intermediate species [CpCo(CO)$_2$(NO)]$^{+}$, arose within 10 ms. After a short period, these bands also decayed rapidly, and the band at 1836 cm^{-1} due to the second intermediate species CpCo(NO)Cl began to be observed. These reactions occurred faster at room temperature than at 10 °C, as is clear from a comparison of Figure 20.1a,b. As can be readily seen, at room temperature, the intensity of the 1836 cm^{-1} band decays gradually, and is accompanied by a gradual increase in the intensity of the band of the reaction product Co(NO)$_2$Cl$_2$ at 1799 cm^{-1}.

An infrared spectrometer system using a dispersive polychromator in combination with a two-dimensional MCT detector has been reported to be able to perform fast spectral measurements in the region of 2000–975 cm^{-1} [4]. In this type of spectrometer also, the time needed for the measurement of one spectrum limits its highest time resolution. It was reported that an absorbance change on the order of 2.4×10^{-4} was detectable at a time resolution of 8.7 ms [4]. Measurements with time resolution <100 µs were also reported, although the signal-to-noise (S/N) ratio of these was much poorer [5].

20.3   Microsecond to Nanosecond Time-Resolved Infrared Absorption Measurements

For microsecond to nanosecond time-resolved infrared absorption measurements, three types of spectroscopic methods have been developed: (i) a method using an infrared laser, (ii) a method using a dispersive spectrometer, and (iii) a method using an FT-IR spectrometer. The time resolution of each of these is limited to the fastest time-response capability of the detector used.

20.3.1   Time-Resolved Measurements Using an Infrared Laser

It is possible to measure infrared spectra at a time resolution shorter than 100 ns by using a tunable continuous wave (cw) infrared laser as a probe beam and a detector with a fast-response capability, which is gated electronically [6]. Representative cw infrared lasers are:

1. *CO laser*: When cooled with liquid nitrogen, the wavenumber region of about $2100-1550 \, cm^{-1}$ can be covered at intervals of about $4 \, cm^{-1}$, which corresponds to the separation of the neighboring laser lines. Since this wavenumber region coincides with the CO stretching region of metal carbonyl compounds, this laser has been often used for the studies of these compounds as well as for the study of the CO molecule itself.
2. *CO_2 laser*: Being tunable over a relatively narrow wavenumber region of about $1090-920 \, cm^{-1}$, this laser has been used for time-resolved measurements of relatively small molecules in the gaseous state.
3. *Infrared diode laser*: The tunable wavenumber region of a particular diode laser is typically about $50-200 \, cm^{-1}$, but if multiple diode lasers are used, nearly the entire mid-infrared region (about $3300-350 \, cm^{-1}$) can be covered. Many of these diode lasers are cooled by liquid nitrogen and operated at $80-120 \, K$; the oscillating wavenumber of a diode laser can be continuously tuned by controlling its temperature and the operating electric current. The output power of a diode laser ($0.1-10 \, mW$) is much smaller than that of the CO or CO_2 laser, and the laser beam propagates over a large solid angle. In addition, as multimode laser oscillation often occurs, a condenser and a monochromator are usually needed for undertaking time-resolved measurements. Infrared diode lasers are frequently used for high-resolution time-resolved studies of photochemical reactions of small molecules in the gaseous state as well as for time-resolved measurements of the condensed state.
4. *Color-center laser*: The tunable wavenumber region is about $12\,500-2500 \, cm^{-1}$. As the laser wavenumber can be very accurately controlled, this laser is used for high-resolution, time-resolved measurements of the gaseous state.
5. *Difference-frequency laser* (cw) (see Section 20.4.2.1 for the generation of difference frequency): By using various nonlinear crystals, this laser can cover almost the entire mid-infrared region (the lower limit is about $550 \, cm^{-1}$). Since the width of this laser line is very narrow, this laser is important for performing high-resolution time-resolved measurements in the gaseous state.
6. *Quantum cascade laser*: Each laser of this type has its laser oscillation center somewhere in the region of about $2400-900 \, cm^{-1}$ and its tunable wavenumber region is over about $60-250 \, cm^{-1}$. Application of such lasers to time-resolved infrared spectroscopic measurements is fairly recent and will likely increase in future [7].

The advantages of using infrared lasers are that: (i) their intensities are much higher than those of a conventional infrared source such as a globar, and (ii) they are better suited for high-resolution measurements. However, they have a drawback that the tunable wavenumber region is generally narrow.

20.3.2 Time-Resolved Measurements Using a Dispersive Infrared Spectrometer

It has been reported that, by combining a dispersive spectrometer (with a continuous broadband infrared source), a fast-response high-sensitivity MCT detector, and a time-gating electronic circuit, infrared spectral measurements at about 50 ns time resolution are attainable [8]. In this spectrometer, alternating current (AC) components of signals from the detector, which occur as a consequence of an external stimulus, are extracted and processed by an amplifier with a high gain and a fast-response capability. Time-resolved measurements are performed by electronic gating using either a digital sampling oscilloscope or a boxcar integrator. The spectrum directly obtained by this method is a difference spectrum. If a sampling oscilloscope is used for accumulation of signals, it is possible to follow changes of signals at a wavenumber position over all time ranges. If the dispersive monochromator is scanned step by step, and the time-resolved data are obtained at each step and processed by the sampling oscilloscope, information along the wavenumber axis as well as that on the time axis is obtained.

By this approach, detection of very small absorbance changes has been reported for time-resolved measurements in both the nanosecond and microsecond time ranges [8].

20.3.3 Time-Resolved Measurements Using a Fourier Transform Infrared (FT-IR) Spectrometer

The features of FT-IR spectrometry, that is, the multiplex advantage (Fellgett's advantage), the optical throughput advantage (Jacquinot's advantage), and the accuracy of measured wavenumbers (the Connes advantage) (see Chapter 4 for these advantages) are, in principle, effective also in time-resolved measurements. In time-resolved measurements, time-resolved interferograms are obtained by time-gating analog interferograms from the interferometer in a gate circuit at certain delay times after giving an external stimulus to the sample. In the more conventional use of an FT-IR spectrometer, according to its basic operation principles, an analog interferogram from the interferometer is digitized by an analog-to-digital (AD) converter at certain equidistant intervals along the optical path difference of the interferometer. In the digital interferogram obtained, data points exist at these predetermined intervals. Therefore, in order to perform time-resolved measurements, it is necessary to render the timing of two kinds of signal processing compatible; that is, the timing of the time resolving must be compatible with that of the sampling by the AD converter of the FT-IR spectrometer.

20.3.3.1 Time-Resolved Measurements with a Step-Scan FT-IR Spectrometer

In time-resolved measurements with a step-scan FT-IR spectrometer (see Section 5.2.1.1 for description of a step-scan FT-IR spectrometer), the timing of the two types of signal processing described above is rendered compatible by scanning the movable mirror of the interferometer step by step and sampling the interferogram signals at each step [1, 9]. The

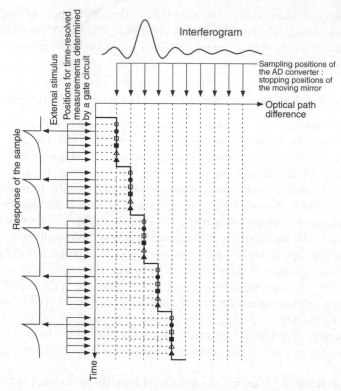

Figure 20.2 *Schematic illustration of the relationship between the interferometer scan and data acquisition in time-resolved measurements using a step-scan FT-IR spectrometer.*

total system consists of a step-scan FT-IR spectrometer, a detector with fast-response capability, a generator of an external stimulus, a pulse generator for controlling the timing of signal processing (some commercial FT-IR spectrometers have a built-in pulse generator), and an AD converter for fast data acquisition (instead of this, a transient recorder or digital-storage oscilloscope may be used).

In Figure 20.2, the relationship between the scan of the movable mirror and data acquisition in time-resolved measurements using a step-scan FT-IR spectrometer is illustrated schematically [1]. The horizontal direction represents the optical path difference of the interferometer, and the vertical dotted lines drawn at equal intervals indicate the sampling positions at which the movable mirror stops and the magnitudes of the interferogram are detected. The ordinate indicates time; the transient phenomena repeatedly caused by an external stimulus are schematically illustrated on the left-hand side of this axis. While the movable mirror is at rest at a point indicated on the abscissa, the interferogram signals at some predefined delay times (indicated by horizontal dotted lines) from the generation of an external stimulus are sent to the AD converter, and data for the time-resolved measurements are recorded. Then, the movable mirror shifts to the next sampling point, and similar measurements are repeated. After collecting the necessary data at all the sampling points, the data measured at the same delay times from the generation of the external stimulus (shown in Figure 20.2 by different marks such as ○, ●, and so on) are collated, and time-resolved interferograms are produced in the computer. As interference signals at all delay times at each sampling point are obtained by this procedure, time-resolved interferograms (in other words, time-resolved spectra) at each delay time are obtained during one complete scan of the movable mirror of the interferometer.

As described, by this method, the same transient phenomenon is repeatedly caused, and time-resolved interferograms are produced by collating the signals of the different transient phenomena. If the stability of repeated transient phenomena is not maintained for reasons of, for example, damage to the sample caused by the external stimulus or unstable operation of a pulsed laser for exciting the sample, the resultant interferogram will be distorted, and, as a consequence, time-resolved spectra may exhibit false bands and/or low S/N ratios. By this method, however, it is relatively easy to cause the same transient phenomenon repeatedly at each sampling point and to take an average of the accumulated signals. This will contribute to enhancing the S/N ratio of the resultant time-resolved spectra. If fluctuation of the sampling position occurs, it will become the origin of additional noise and thereby adversely affect the result.

Time resolution is limited by the response speed of the detector and amplifier, and the speed of taking signals into the AD converter. If the pulse width of the external stimulus is short enough; at a time resolution of about 100 ns, an absorbance change of smaller than 10^{-4} is detectable [1]. In measurements using a photovoltaic MCT detector capable of producing output in both AC and direct current (DC) and a high-speed preamplifier [1, 9], a time resolution of about 10 ns has been achieved.

This time-resolved measurement method can be applicable to relatively slow transient phenomena, as its time-resolved measurements are undertaken while the movable mirror is at rest. The number of applications of step-scan FT-IR spectrometry to time-resolved measurements currently is more than that by any other method, and it has been applied to various studies in many fields such as studies of biomolecules, liquid crystals, polymers, photochemical reactions in zeolites, oxidation–reduction reactions on electrode surfaces, and excited electronic states of inorganic complexes. Further, this method has been applied to time-resolved measurements in combination with attenuated total reflection (ATR) (see Chapter 13), surface-enhanced infrared absorption (see Section 13.2.2) [10, 11], infrared microscopic measurements (see Chapter 16) [12], and infrared spectroscopic imaging (see Chapter 17) [13].

As an example of a time-resolved measurement covering a time scale from nanoseconds to seconds, time-resolved difference spectra of PYP (photoactive yellow protein) photoexcited by nanosecond laser pulses at 396 nm are shown in Figure 20.3 [14]. The ordinate axis indicates the absorbance change. The infrared difference spectrum measured first after 50 ns from the excitation by the laser pulse, which is shown by a black curve at the front of the figure, corresponds to the difference between the first intermediate species I_1 and PYP in the ground state. The spectral changes after the first spectrum follow the process that I_1 changes to the second intermediate species I_2, and finally returns to the ground state. On the basis of these results, the structural changes and dynamics of the chromophore and the protein were discussed [14].

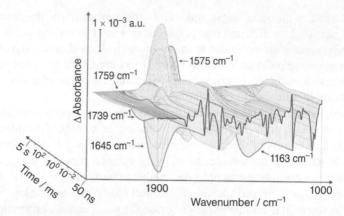

Figure 20.3   *Time-resolved infrared difference spectra of photoexcited PYP (photoactive yellow protein) measured using a step-scan FT-IR spectrometer [14]. Time is indicated on a common logarithmic scale after 50 ns from photoexcitation. ΔAbsorbance, absorbance difference; a.u., absorbance unit. (Source: Reprinted by permission from Macmillan Publishers Ltd: Nature Structural and Molecular Biology [14]. Copyright © 2001.)*

20.3.3.2   Synchronous Time-Resolved Measurements Using a Continuous-Scan FT-IR Spectrometer

In time-resolved measurements with a continuous-scan FT-IR spectrometer, the timing of giving an external stimulus to the sample and that of time resolving are synchronized with the timing of the sampling of an interferogram of the FT-IR spectrometer. In this way, the time-resolving process is made compatible with the sampling timing of the interferogram [1]. The measuring system consists of a continuous-scan FT-IR spectrometer and a detector with a fast-response capability, together with a means of generating the external stimulus.

In time-resolved measurements made with a continuous-scan FT-IR spectrometer, a typical relationship between the scan of the movable mirror of the interferometer and data acquisition in time-resolved measurements is illustrated schematically in Figure 20.4 [1]. During the first scan of the movable mirror, an external stimulus is applied to the sample while at a certain sampling position, and the interferogram signals before and after the applied external stimulus are taken at the sampling positions of the interferometer by the AD converter. During the second scan of the movable mirror, an external stimulus is applied to the sample at a sampling position next to that at which the external stimulus was applied to the sample in the first scan of the movable mirror, and the interferogram signals before and after the external stimulus are taken at the sampling positions. This process is repeated until the external stimulus has been applied to the sample at all the sampling positions, and the interferogram signals before and after the external stimulus have been recorded at the sampling positions; the interferogram signals taken at the same delay times after the external stimulus (discriminated in Figure 20.4 by the marks o, •, etc.) are collected, collated, and processed by using software for this specific purpose in order to generate time-resolved interferograms in the computer. Time-resolved infrared spectra are obtained by performing the Fourier transform of the time-resolved interferograms obtained. Time resolution is decided by the sampling interval (or the scanning speed of the movable mirror), and its shortest limit is typically a few microseconds. Transient phenomena occurring within

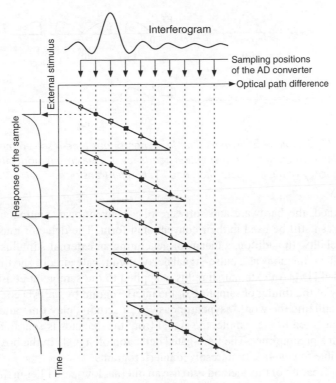

Figure 20.4 *Schematic illustration of the relationship between the interferometer scan and data acquisition in synchronous time-resolved measurements using a continuous-scan FT-IR spectrometer.*

a time range from several hundred milliseconds to several tens of microseconds can be measured by this method.

In a similar way as for time-resolved measurements by a step-scan FT-IR spectrometer, signals collected from different transient phenomena are processed to form time-resolved interferograms. If the repeatability of transient phenomena is not maintained, the quality of the time-resolved spectra will deteriorate rapidly. The sampling interval must be always the same, that is, the speed of the moving mirror must be constant. Unstable scans of the movable mirror will not only affect the accuracy of time resolution but also increase the noise level.

20.3.3.3 Asynchronous Time-Resolved Measurements Using a Continuous-Scan FT-IR Spectrometer

In time-resolved measurements described in Sections 20.3.3.1 and 20.3.3.2, the time-resolving process is synchronized with the timing of sampling of the interferogram. In this respect, both of them may be called *synchronous time-resolved measurements*. By contrast, a method (called *asynchronous time-resolved measurements*) that does not need the synchronization of these two timings although still using a continuous-scan FT-IR spectrometer has been reported [15].

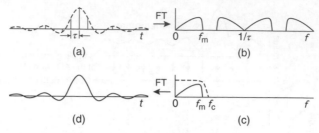

Figure 20.5 *(a–d) Role of a low-pass filter in asynchronous time-resolved measurements. t, time; f, modulation frequency; $1/\tau$, repetition frequency; f_m, maximum modulation frequency; f_c, cut-off frequency of the low-pass filter; and FT, Fourier transform.*

In this method, the hardware and software of a conventional continuous-scan FT-IR spectrometer can still be used in their conventional form. The detector must have a fast-response capability. In addition, a means of generating an external stimulus and a boxcar integrator (which functions as a pulse-signal delay circuit, gate circuit, and the main amplifier) are needed [15]. An external stimulus is applied to the sample at constant intervals, independently of the timing of sampling of the interferogram by the AD converter. This is the origin of utilizing the word "asynchronous." At a certain delay time after the external stimulus was applied to the sample, the signal from the detector is sent to a boxcar integrator through a preamplifier. The time-gated detection of signals in the boxcar integrator makes it possible to generate time-resolved interferograms.

An important feature of this method is to use an internal low-pass filter in the continuous-scan FT-IR spectrometer as a link between the gate circuit of the boxcar integrator and the AD converter of the FT-IR spectrometer (see Figure 20.5). The low-pass filter removes high-frequency components (Figure 20.5c) from a discrete interferogram time-resolved by the gate circuit (Figure 20.5a), and converts it to an analog interferogram (Figure 20.5d). This analog interferogram in Figure 20.5d is free from the timing of time resolving by the gate circuit, so that it is then possible to convert this analog interferogram into a digital interferogram by the AD converter with its own sampling timing. Further, it is possible to accumulate interferograms from scans of the interferometer, as their sampling is synchronized.

The time resolution of the measuring system using a conventional infrared source of continuous broadband radiation is determined by the response speed of the detector and its preamplifier. If a conventional MCT detector is used, the highest time resolution is about 100 ns. However, if pulsed infrared radiation could be used, the time resolution would be determined by the pulse width, and it would become higher [1, 16]

The repeating frequency of the external stimulus to the sample $(1/\tau)$ must be set higher than twice the highest modulation frequency f_m the spectra originally have, in order to avoid overlapping of the measured spectra with their harmonics (the Nyquist sampling theorem; see Figure 20.5b and Section 4.4.2). According to this requirement, the period between external stimuli to the sample should normally be a few milliseconds, when a conventional continuous-scan FT-IR spectrometer is used; that is, systems to be measured by this method are limited to phenomena that can be repeated quickly.

20.4 Picosecond to Femtosecond Time-Resolved Infrared Absorption Measurements

Many photochemical and photophysical phenomena occur on a time scale shorter than a nanosecond. In order to follow such fast phenomena by infrared spectroscopy, picosecond to femtosecond time-resolved infrared measurements are required. Since time resolving in this time range cannot be performed by utilizing the fast-response capability of a detector and the time-resolving power of an electronic circuit (gate circuit, etc.), the following optical methods are mainly used: (i) a method based on the upconversion (optical gating) process, and (ii) a method which detects pulsed infrared radiation itself. At present, the latter method is commonly used for picosecond to femtosecond time-resolved measurements.

20.4.1 The Method Based on the Upconversion (Optical Gating) Process

In this method, a sample under study is excited by a pump pulsed light, and the excited sample is subjected to infrared irradiation of frequency ν_{IR}. Then, the infrared radiation of ν_{IR}, which has passed through the sample, is combined with a time-gating pulsed light of frequency ν_{GATE}, which is delayed, by a certain time from the exciting pump pulse, in a nonlinear crystal ($LiIO_3$, $AgGaS_2$, etc.) to generate a pulse of sum frequency $\nu_{IR} + \nu_{GATE}$. Since the intensity of the sum-frequency pulse is proportional to that of the infrared radiation at the time when the time-gating pulsed light enters the nonlinear crystal, the infrared spectrum at that time is converted to the sum-frequency spectrum. Usually, the time-gating pulsed light is in the near-infrared to visible region, so that the sum-frequency spectrum is obtained as a spectrum in the visible region where the detector sensitivity is high.

Although infrared radiation of either cw or pulsed operation (see Section 20.4.2.1 for pulsed infrared radiation) may be used for this method, a cw infrared laser (see Section 20.3.1) is the more commonly used for high-sensitivity measurements because it is more stable. The measuring system with a cw infrared laser has a simpler structure, and the scan of the delay time is also simpler with a cw infrared laser. The drawback of using a cw infrared laser is that these do not however have a wide tunable range (see Section 20.3.1). Furthermore, because the phase-matching conditions for sum-frequency generation change with the frequency of a cw infrared laser when it is scanned, these lasers are not suited for measuring time-resolved infrared spectra over a wide wavenumber region. The time resolution of the measurement with a cw infrared laser is mainly decided by the width of the time-gating pulsed light, and the wavenumber resolution is determined by the width of the infrared laser line.

20.4.2 The Method Based on the Direct Detection of Infrared Pulses

20.4.2.1 Generation of Infrared Pulses

Infrared pulses used for ultrafast time-resolved infrared spectroscopic measurements have been generated mainly by the following three methods:

1. By irradiating the vapors of metals such as Ba and Cs with ultrashort laser pulses, infrared pulses are generated by the process of stimulated electronic Raman scattering.

2. Infrared pulses can be generated by optical parametric generation (OPG) and/or optical parametric amplification (OPA) induced by ultrafast laser pulses. Both OPG and OPA are nonlinear optical processes that generate two pulses at frequencies v_S and v_I from an exciting pulse at frequency v_P (i.e., $v_P = v_S + v_I$). Pulses at v_S and v_I ($v_S > v_I$) are called, respectively, the *signal output* and the *idler output*. Examples of nonlinear crystals used for OPG and OPA are $LiNbO_3$ and β-barium borate (BBO; $β$-BaB_2O_4).

3. Infrared pulses at frequency v_D are generated as the difference frequency between two ultrashort pulses at frequencies v_1 and v_2 (i.e., $v_D = v_1 - v_2$; $v_1 > v_2$). Examples of non-linear crystals used for the generation of a difference frequency are $LiIO_3$ and $AgGaS_2$.

Recently, the third method has become the most commonly used; the second method may also be combined with the third method, that is, the difference frequency between the signal and idler frequencies $v_S - v_I$, which is in the infrared region, is generated by the third method.

20.4.2.2   The Pump–Probe Method

In ultrafast, time-resolved infrared absorption measurements by the pump–probe method, the sample is first excited by an ultrashort pump pulse, and then irradiated by an ultrashort infrared pulse (probe pulse) after a certain delay time from the excitation by the pump pulse. The delay time of the probe pulse from the pump pulse is usually changed by the difference in the optical path lengths of the pump and probe pulses (a delay time of 1 ps arises from a path difference of about 0.3 mm). When the infrared spectrum of a molecule in an excited electronic state is measured, pulses in the ultraviolet to visible region are used for the pump purpose, and pulses in the infrared region are used for the probe purpose. When a vibrationally excited molecule is the target of such a measurement, pulses in the infrared region are used for both the pump and probe purposes. The transient (or time-resolved) infrared absorption spectra by this method are usually measured as the difference in absorption intensities for the probe pulses between the measurements with the pump pulses and those without the pump pulses.

The pump and probe pulses are usually obtained from the same ultrashort pulse by wavelength conversion, so that the jitter between the pump and probe pulses is kept small, and the time resolution is determined by the width of these pulses. It is a common practice to divide the probe pulses into two; one is used for the time-resolved measurements and the other as a reference signal in order to correct for effects of fluctuation of intensity from pulse to pulse. In practice, the probe pulses for the sample are focused on the same point in the sample as are the pump pulses, and the reference pulses are focused onto a different point in the sample. The probe pulses that have passed through the sample and the reference pulses are detected by different infrared detectors (or by the upper and lower pixels of the same array detector). In practice, the measurement is often performed by focusing every other pump pulse on the sample by reducing the repetition frequency of the pump pulses by half with a synchronous optical chopper, and to then obtain the transient infrared absorption intensities by dividing the intensities of the probe pulses when the sample is excited by the pump pulses by those when the sample is not excited. This is also effective in reducing the effects of any long-term fluctuations in the intensities of the probe pulses.

The time resolution in such measurements is decided by the mutual correlation time between the pump and probe pulses, which corresponds to the full width at half maximum

(FWHM) of the mutual correlation function representing the temporal overlap of the pump and probe pulses. The wavenumber resolution in such measurements depends on the following. If the probe pulse having a narrow spectral bandshape is directly detected by an infrared detector such as an MCT detector, the wavenumber resolution is decided by the bandwidth of such probe pulses. If the probe pulses cover a wide infrared region, and it is necessary to use an infrared polychromator and to detect the monochromatic radiation by an infrared array detector such as an MCT array detector, the wavenumber resolution of such a measurement system is determined by both the wavenumber resolution of the polychromator used and the pixel resolution of the array detector.

20.4.2.3 Time-Resolved Spectroscopic Measurements with Narrow-Band Ultrashort Infrared Pulses and Applications

Picosecond to femtosecond time-resolved infrared absorption measurements were initiated in the middle of the 1980s. In 1984, Heilweil *et al.* [17] studied the dynamics of vibrational relaxation by using picosecond infrared pulses obtained from an OPA ($LiNbO_3$) excited by a mode-locked Nd:YAG laser.

The transient infrared absorption spectra of molecules in excited electronic states were measured by Elsaesser and Kaiser [18]. The probe pulses used were: (i) picosecond infrared pulses in the wavenumber region of $3500-3000 \, cm^{-1}$ generated by an OPA in $LiNbO_3$ excited by the Nd:YAG laser line, and (ii) picosecond infrared pulses in the region of $3000-1400 \, cm^{-1}$ obtained by the difference-frequency generation in $AgGaS_2$ between the Nd:YAG laser line and near-infrared dye-laser lines excited by the same Nd:YAG laser line.

About 10 years later, Okamoto and Tasumi [19] measured the transient infrared spectrum in the finger-print region ($1700-900 \, cm^{-1}$) of an organic molecule in an excited electronic state. The picosecond pulses in the finger-print region were obtained by using the difference-frequency generation twice: (i) the output from an amplified dye laser and the fundamental output at 1053 nm from the Nd:YLF regenerative amplifier were combined, and a tunable difference-frequency beam (wavelength $1.15-1.3 \, \mu m$) was generated in a BBO crystal and (ii) the tunable near-infrared and the 1053 nm beams were difference-frequency mixed in an $AgGaS_2$ (AGS) crystal to generate a tunable mid-infrared beam between 1700 and $800 \, cm^{-1}$. Further, a method of high-sensitivity measurement of picosecond transient infrared spectra was developed by utilizing the anisotropy of absorption (optically heterodyned detection of absorption anisotropy) [20].

A spectrometer system for picosecond time-resolved measurements was developed using the picosecond infrared probe pulses obtained by difference-frequency generation in $AgGaS_2$ between the signal output and the idler output from an OPA excited by a Ti:sapphire regenerative amplifier [21]. Using this spectrometer system, the photoexcited states of conjugated materials were studied. In Figure 20.6, a transient infrared absorption spectrum of photoexcited regioregular poly(3-hexylthiophene) (*RR*-P3HT) is shown [22]. The sample was in the form of a spin-coated film. The transient spectrum in Figure 20.6a measured at about 100 ps after photoexcitation is similar to a difference spectrum (the difference between the spectrum of $FeCl_3$-doped *RR*-P3HT and that of the intact polymer) shown in Figure 20.6b, which is attributed to the polaron (a radical cation produced in the polymer because of the doping) [23]. It was inferred that the polaron produced at about 100 ps after photoexcitation is the carrier of photoconduction.

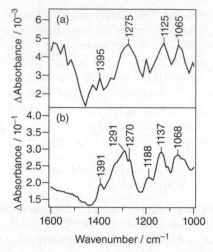

Figure 20.6 *(a) Picosecond time-resolved infrared spectrum of RR-P3HT (regioregular poly (3-hexylthiophene)) at about 100 ps delay time after photoexcitation [22] and (b) infrared difference spectrum of RR-P3HT doped with $FeCl_3$ [23]. ΔAbsorbance, absorbance difference. (Source: Reproduced with permission from [22] and [23]. Copyright (2009) and (2003), Elsevier.)*

20.4.2.4   Time-Resolved Spectroscopic Measurements with Broadband Ultrashort Infrared Pulses and Applications

In the middle of the 1990s, a new type of an ultrafast time-resolved measurement system was developed by Hamm *et al.* [24]; femtosecond infrared pulses over a broadband were dispersed by a polychromator and detected by a multichannel infrared detector. They obtained broadband infrared pulses tunable over a wide frequency range (pulse width 400 fs, spectral bandwidth (FWHM) 65 cm^{-1}), and divided these pulses into two: one was used for probing the sample and the other for reference. The infrared pulses passing through the sample and the reference pulses were detected separately by two MCT array detectors each with 10 pixels. By this system, time-resolved infrared spectra of photoexcited molecules were measured [24].

Arrivo *et al.* [25] obtained broadband infrared pulses (spectral width 500 cm^{-1} in the wavenumber region centered at about 3300 cm^{-1}) by difference-frequency generation in $LiIO_3$ between the second harmonic of the Nd:YAG laser line (wavelength 532 nm, repetition rate 20 Hz) and a dye laser excited by this second harmonic line. They used these infrared pulses for the study of the dynamics of hydrogen bonding by exciting the target molecule by the infrared pulses, and detecting the changes in infrared absorption spectra after the excitation with two-dimensional InSb and MCT detectors (both having 256×256 pixels) [25].

Since about the end of the 1990s, generation of ultrashort pulses has become easier due to the progress of laser technology, and, as a consequence, a measuring method based on femtosecond Ti:sapphire regenerative amplifier with a kilohertz repetition rate has become the mainstream of fast time-resolved infrared absorption measurements.

Snee *et al.* [26] obtained broadband femtosecond infrared pulses (FWHM about 200 cm^{-1}) by the difference-frequency generation in $LiIO_3$ between the fundamental output of a femtosecond Ti:sapphire laser (wavelength 810 nm, repetition rate 30 Hz) and a white light generated in a sapphire crystal by focusing the fundamental output of the laser. These infrared pulses were divided into probe and reference pulses, which were detected separately by the upper and lower pixels of a two-dimensional MCT detector.

Hamm *et al.* [27] and Wang *et al.* [28] excited an OPA by an output of a femtosecond Ti:sapphire regenerative amplifier (repetition rate 1 kHz), and produced broadband femtosecond infrared probe pulses by the difference-frequency generation in $AgGaS_2$ between the signal output and the idler output from the OPA. The two groups used these infrared probe pulses for time-resolved measurements using an MCT array detector with 32 pixels. They did not however correct the effects of fluctuation of pulse energies by measuring the intensities of reference pulses. In 2003, Towrie *et al.* [29] constructed a type of spectrometer with two polychromators (focal length 25 cm) each equipped with MCT array (64 pixels) detectors, one for detecting the probe pulses and the other for the reference pulses. With this spectrometer, any fluctuations of pulse energies were canceled, and it was reported that detection of absorbance changes on the order of 10^{-4} to 10^{-5} was possible in a measuring time of 1 min.

In 2005, Towrie *et al.* [30] developed another time-resolved infrared spectrometer capable of performing femtosecond to microsecond time-resolution measurements, by adding to their spectrometer described in Ref. [29] a sub-nanosecond Q-switch $Nd:YVO_4$ laser (wavelength 1064 nm, pulse width 0.6 ns). The pulses generated by this laser were electronically synchronized with the probe pulses with about 0.3 ns jitter, and the harmonics of pulses from this laser were used as the pump pulses.

Using the above-described time-resolved spectrometer [29], Kuimova *et al.* [31] measured the picosecond time-resolved infrared spectra of a nucleotide (2′-deoxyguanosine 5′-monophosphate (5′-dGMP)) in an excited state generated by irradiation of 200-nm laser pulses as a model for the process of photoinduced damage to DNA. The results obtained are shown in Figure 20.7 [31]. The band at 1662 cm^{-1} in spectrum (a) arises from the ground state, while in spectrum (b), measured at a delay time of 5 ps from the photoexcitation, this band shows photoinduced fading and a new band appears at 1638 cm^{-1}, which is assignable to a vibrationally excited state in the ground electronic state formed via very fast internal conversion from the excited electronic state. In spectrum (c), an averaged spectrum for delay times over a range from 30 to 500 ps is shown. A new band observed at 1702 cm^{-1} in spectrum (c) is assigned to the radical cation of 5′-dGMP, which has been formed via photoionization. Similar measurements were also performed for polynucleotides. Measurements for longer delay times were also made by using the spectrometer described in Ref. [30].

In 2010, Greetham *et al.* [32] reported on a femtosecond infrared probe and reference pulses over a wavenumber region tunable between 4000 and 800 cm^{-1} by the difference-frequency generation in $AgGaS_2$ between the signal output and the idler output obtained by pulses from a higher repetition rate femtosecond Ti:sapphire amplifier (fundamental wavelength at 800 nm, pulse width 40–80 fs, repetition rate 10 kHz) [32]. Actually, as illustrated schematically in Figure 20.8, the Ti:sapphire amplifier excites two OPAs; one is used for generating the pump pulses in a certain wavelength region, the other is used for generating the probe and reference pulses in the infrared region. The probe pulses pass through two different polychromators and are detected by two different MCT array detectors each with 128 pixels, each of which covers a different wavenumber region, and the reference

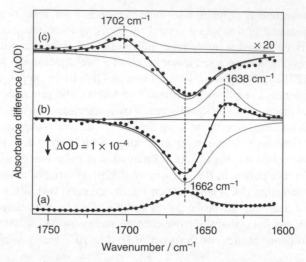

Figure 20.7 *(a) Infrared absorption spectrum of 5'-dGMP in the ground state in H₂O, (b) time-resolved infrared spectrum of 5'-dGMP at 5-ps delay time after photoexcitation, and (c) an average of time-resolved infrared spectra measured during delay times from 30 to 500 ps after photoexcitation [31]. Thin solid curves indicate the results of band fitting using Lorentzian band shapes. (Source: Reproduced from [31]. Copyright © 2006, National Academy of Sciences, USA.)*

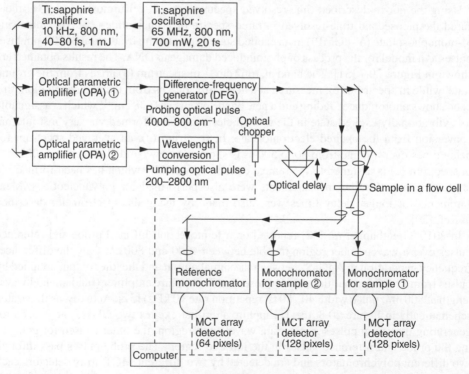

Figure 20.8 *Explanatory schematic illustration of the femtosecond time-resolved infrared spectrometer reported in Ref. [32].*

pulses pass through a different polychromator and are detected by an MCT array detector with 64 pixels. With this spectrometer system, it is possible to measure, using the two MCT array detectors simultaneously, a transient infrared spectrum over a range of $500\,cm^{-1}$, and it was reported that an absorbance change of about 10^{-5} could be detected in 1 s.

20.5   Points to Note Common to Time-Resolved Measurements

Time-resolved measurements can be made for a sample in any state. When the sample is a solution, attention should be paid to infrared absorption bands of the solvent, as these bands will likely hinder the detection of weak absorption bands of a transient species. It is important to use a solvent that does not have strong infrared absorption bands in the region in which the absorption bands of a transient species might be observed. It is of course important to ensure that there is no particular molecular interaction between the solvent and sample, unless the purpose of the study is the molecular interaction between them. It is also necessary to have a high enough concentration of the target molecule in order to minimize the effects of solvent absorptions. It is desirable also to use a flow cell system in order to avoid the attachment of any sample burnt by the pump pulses to the window materials.

In time-resolved infrared measurement systems, gold-coated mirrors should be used in the optical paths of infrared laser lights, as the reflectance from a gold-coated mirror in the infrared region is higher than that from an aluminum-coated mirror. For transmitting infrared pulses, use of infrared-transparent materials such as BaF_2, CaF_2, ZnSe, and Ge, etc. is advised. To focus an infrared beam onto a sample, either an off-axis parabolic mirror or a ZnSe lens is usually used.

References

1. Smith, G.D. and Palmer, R.A. (2002) Fast time-resolved mid-infrared spectroscopy using an interferometer, In: *Handbook of Vibrational Spectroscopy*, Vol. **1** (eds J.M. Chalmers and P.R. Griffiths), John Wiley & Sons, Ltd, Chichester, pp. 625–640.
2. Griffiths, P.R., Hirsche, B.L. and Manning, C.J. (1999) Ultra-rapid-scanning Fourier transform infrared spectrometry. *Vib. Spectrosc.*, **19**, 165–176.
3. Reback, M.L., Roske, C.W., Bitterwolf, T.E., Griffiths, P.R. and Manning, C.J. (2010) Stopped-flow ultra-rapid-scanning Fourier transform infrared spectroscopy on the millisecond time scale. *Appl. Spectrosc.*, **64**, 907–911.
4. Pellerin, C., Snively, C.M., Chase, D.B. and Rabolt, J.F. (2004) Performance and application of a new planar array infrared spectrograph operating in the mid-infrared ($2000-975\,cm^{-1}$) fingerprint region. *Appl. Spectrosc.*, **58**, 639–646.
5. Snively, C.M., Pellerin, C., Rabolt, J.F. and Chase, D.B. (2004) Acquisition of mid-infrared spectra from nonrepeatable events with sub-100-μs temporal resolution using planar array infrared spectroscopy. *Anal. Chem.*, **76**, 1811–1816.
6. Grills, D.C. and George, M.W. (2002) Fast and ultrafast time-resolved mid-infrared spectrometry using lasers, In: *Handbook of Vibrational Spectroscopy*, Vol. **1** (eds J.M. Chalmers and P.R. Griffiths), John Wiley & Sons, Ltd, Chichester, pp. 677–692.
7. Grills, D.C., Cook, A.R., Fujita, E., George, M.W., Presses, J.M. and Wishart, J.F. (2010) Application of external-cavity quantum cascade infrared lasers to nanosecond

time-resolved infrared spectroscopy of condensed-phase samples following pulse radiolysis. *Appl. Spectrosc.*, **64**, 563–570.

8. Hashimoto, M., Yuzawa, T., Kato, C., Iwata, K. and Hamaguchi, H. (2002) Fast time-resolved mid-infrared spectroscopy using grating spectrometers, In: *Handbook of Vibrational Spectroscopy*, Vol. **1** (eds J.M. Chalmers and P.R. Griffiths), John Wiley & Sons, Ltd, Chichester, pp. 666–676.

9. Rödig, C. and Siebert, F. (2002) Instrumental aspects of time-resolved spectra generated using step-scan interferometers, In: *Handbook of Vibrational Spectroscopy*, Vol. **1** (eds J.M. Chalmers and P.R. Griffiths), John Wiley & Sons, Ltd, Chichester, pp. 641–654.

10. Pronkin, S. and Wandlowski, Th. (2003) Time-resolved in situ ATR-SEIRAS study of adsorption and 2D phase formation of uracil on gold electrodes. *J. Electroanal. Chem.*, **550–551**, 131–147.

11. Wisitruangsakul, N., Zebger, I., Ly, K.H., Murgida, D.H., Ekgasit, S. and Hildebrandt, P. (2008) Redox-linked protein dynamics of cytochrome c probed by time-resolved surface enhanced infrared absorption spectroscopy. *Phys. Chem. Chem. Phys.*, **10**, 5276–5286.

12. Zhou, Z.-Y., Lin, S.-C., Chen, S.-P. and Sun, S.-G. (2005) In situ step-scan time-resolved microscope FTIR spectroscopy working with a thin-layer cell. *Electrochem. Commun.*, **7**, 490–495.

13. Bhargava, R. and Levin, I.W. (2003) Time-resolved Fourier transform infrared spectroscopic imaging. *Appl. Spectrosc.*, **57**, 357–366.

14. Brudler, R., Rammelsberg, R., Woo, T.T., Getzoff, E.D. and Gerwert, K. (2001) Structure of the I_1 early intermediate of photoactive yellow protein by FTIR spectroscopy. *Nat. Struct. Biol.*, **8**, 265–270.

15. Masutani, K. (2002) Time-resolved mid-infrared spectrometry using an asynchronous Fourier transform infrared spectrometer, In: *Handbook of Vibrational Spectroscopy*, Vol. **1** (eds J.M. Chalmers and P.R. Griffiths), John Wiley & Sons, Ltd, Chichester, pp. 655–665.

16. Sakamoto, A., Okamoto, H. and Tasumi, M. (1998) Observation of picosecond transient Raman spectra by asynchronous Fourier transform Raman spectroscopy. *Appl. Spectrosc.*, **52**, 76–81.

17. Heilweil, E.J., Casassa, M.P., Cavanagh, R.R. and Stephenson, J.C. (1984) Picosecond vibrational energy relaxation of surface hydroxyl groups on colloidal silica. *J. Chem. Phys.*, **81**, 2856–2858.

18. Elsaesser, T. and Kaiser, W. (1986) Visible and infrared spectroscopy of intramolecular proton transfer using picosecond laser pulses. *Chem. Phys. Lett.*, **128**, 231–237.

19. Okamoto, H. and Tasumi, M. (1996) Picosecond transient infrared spectroscopy of electronically excited 4-dimethylamino-4′-nitrostilbene in the fingerprint region (1640–940 cm^{-1}). *Chem. Phys. Lett.*, **256**, 502–508.

20. Okamoto, H. (1998) High-sensitivity measurement of ultrafast transient infrared spectra based on optically heterodyned detection of absorption anisotropy. *Chem. Phys. Lett.*, **283**, 33–38.

21. Sakamoto, A., Nakamura, O., Yoshimoto, G. and Tasumi, M. (2000) Picosecond time-resolved infrared absorption studies on the photoexcited states of poly(*p*-phenylenevinylene). *J. Phys. Chem. A*, **104**, 4198–4202.

22. Sakamoto, A. and Takezawa, M. (2009) Picosecond time-resolved infrared absorption study on photoexcited dynamics of regioregular poly(3-hexylthiophene). *Synth. Met.*, **159**, 809–812.

23. Furukawa, Y., Takao, H., Yamamoto, J. and Furukawa, S. (2003) Infrared absorption induced by field-effect doping from poly(3-alkylthiophene)s. *Synth. Met.*, **135–136**, 341–342.

24. Hamm, P., Wiemann, S., Zurek, M. and Zinth, W. (1994) Highly sensitive multichannel spectrometer for subpicosecond spectroscopy in the midinfrared. *Opt. Lett.*, **19**, 1642–1644.

25. Arrivo, S.M., Kleiman, V.D., Dougherty, T.P. and Heilweil, E.L. (1997) Broadband femtosecond transient infrared spectroscopy using a 256 × 256 element indium antimonide focal-plane detector. *Opt. Lett.*, **22**, 1488–1490.

26. Snee, P.T., Yang, H., Kotz, K.T., Payne, C.K. and Harris, C.B. (1999) Ultrafast infrared studies of the reaction mechanism of silicon–hydrogen bond activation by η^5-CpV(CO)$_4$. *J. Phys. Chem. A*, **103**, 10426–10432.

27. Hamm, P., Lim, M. and Hochstrasser, R.M. (1997) Vibrational energy relaxation of the cyanide ion in water. *J. Chem. Phys.*, **107**, 10523–10531.

28. Wang, Y., Asbury, J.B. and Lian, T. (2000) Ultrafast excited-state dynamics of Re(CO)$_3$Cl(dcbpy) in solution and on nanocrystalline TiO$_2$ and ZrO$_2$ thin films. *J. Phys. Chem. A*, **104**, 4291–4299.

29. Towrie, M., Grills, D.C., Dyer, L., Weinstein, J.A., Matousek, P., Barton, R., Bailey, P.D., Subramaniam, N., Kwok, W.M., Ma, C., Phillips, D., Parker, A.W. and George, M.W. (2003) Development of a broadband picosecond infrared spectrometer and its incorporation into an existing ultrafast time-resolved resonance Raman, UV/visible, and fluorescence spectroscopic apparatus. *Appl. Spectrosc.*, **57**, 367–380.

30. Towrie, M., Gabrielsson, A., Matousek, P., Parker, A.W., Rodriguez, A.M.B. and Vlček, Jr., A. (2005) A high-sensitivity femtosecond to microsecond time-resolved infrared vibrational spectrometer. *Appl. Spectrosc.*, **59**, 467–473.

31. Kuimova, M.K., Cowan, A.J., Matousek, P., Parker, A.W., Sun, X.Z., Towrie, M. and George, M.W. (2006) Monitoring the direct and indirect damage of DNA bases and polynucleotides by using time-resolved infrared spectroscopy. *Proc. Natl. Acad. Sci. U.S.A.*, **103**, 2150–2153.

32. Greetham, G.M., Burgos, P., Cao, Q., Clark, I.P., Codd, P.S., Farrow, R.C., George, M.W., Kogimtzis, M., Matousek, P., Parker, A.W., Pollard, M.R., Robinson, D.A., Xin, Z.-J. and Towrie, M. (2010) ULTRA: a unique instrument for time-resolved spectroscopy. *Appl. Spectrosc.*, **64**, 1311–1319.

21

Two-Dimensional Correlation Spectroscopy

Shigeaki Morita[1], Hideyuki Shinzawa[2], Isao Noda[3], and Yukihiro Ozaki[4]

[1]*Department of Engineering Science, Osaka Electro-Communication University, Japan*
[2]*National Institute of Advanced Industrial Science and Technology (AIST), Japan*
[3]*Department of Materials Science and Engineering, University of Delaware, USA*
[4]*Department of Chemistry, School of Science and Technology, Kwansei-Gakuin University, Japan*

21.1 Introduction

Two-dimensional (2D) correlation spectroscopy [1–4] is a useful analytical technique orig-inally developed for infrared (IR) spectroscopy to probe submolecular-level dynamics and interactions of a system observed under the influence of some externally applied physical perturbation, such as mechanical deformation or electric field. Such a perturbation induces systematic variation of spectral features, reflecting the changes in the state or population of system constituents. In 2D correlation spectroscopy, spectral maps defined by two inde-pendent spectral variable axes, such as wavenumber axes, are generated by applying a special form of correlation analysis to a set of dynamically varying spectral intensities. Such perturbation-induced variations of spectral signals contain surprisingly rich informa tion about the unique behavior of individual submolecular moieties comprising the system. Some of the notable features of 2D correlation spectra are:

- simplification of complex spectra consisting of many overlapped peaks,
- apparent enhancement of spectral resolution by spreading peaks over the second dimension,

- establishment of unambiguous band assignment through correlation of bands selectively coupled by various interaction mechanisms, and
- determination of the sequential order of events depicted by asynchronous changes in spectral intensities.

This chapter is organized to serve as an introduction to the basic concept of 2D correlation spectroscopy. The central theoretical background of 2D correlation spectroscopy is provided first. The rest of this chapter covers some pertinent examples of 2D correlation spectroscopy derived from experimentally obtained IR spectral data sets.

21.2   2D Correlation Spectra

Detailed description of the mathematical background of 2D correlation theory is provided in Appendix F of this book. Here we only briefly go over the correlation treatment of a set of discretely observed spectral data commonly encountered in practice. Let us assume spectral intensity $x(v, u)$ described as a function of two separate variables: spectral index variable "v" of the probe and additional variable "u" reflecting the effect of the applied perturbation. Typically, spectral intensity x is sampled and stored as a function of variables v and u at finite and often constant increments. For convenience, here, we refer to the variable v as wavenumber $\tilde{v}$ and the variable u as time t. For a set of m spectral data $x(\tilde{v}, t_i)$ with $i = 1, 2, \ldots, m$, observed during the period between $t_{\min}$ and $t_{\max}$, we define the dynamic spectrum $y(\tilde{v}, t_i)$ as

$$y(\tilde{v}, t_i) = x(\tilde{v}, t_i) - \bar{x}(\tilde{v}) \tag{21.1}$$

It is customary to select $\bar{x}(\tilde{v})$ as the spectrum averaged over the observation interval given by

$$\bar{x}(\tilde{v}) = \frac{1}{m} \sum_{i=1}^{m} x(\tilde{v}, t_i) \tag{21.2}$$

The functional similarity or dissimilarity of intensity variation patterns between a pair of spectral bands measured under a common perturbation at two different spectral variables, $\tilde{v}_1$ and $\tilde{v}_2$, can be mathematically characterized in terms of synchronous and asynchronous correlations. Synchronous and asynchronous 2D correlation spectrum pair, $\Phi(\tilde{v}_1, \tilde{v}_2)$ and $\Psi(\tilde{v}_1, \tilde{v}_2)$, are given by

$$\Phi(\tilde{v}_1, \tilde{v}_2) = \frac{1}{m-1} \sum_{i=1}^{m} y(\tilde{v}_1, t_i) \cdot y(\tilde{v}_2, t_i) \tag{21.3}$$

$$\Psi(\tilde{v}_1, \tilde{v}_2) = \frac{1}{m-1} \sum_{i=1}^{m} y(\tilde{v}_1, t_i) \cdot h(\tilde{v}_2, t_i) \tag{21.4}$$

The Hilbert transform $h(\tilde{v}_2, t_i)$ of the dynamic spectrum $y(\tilde{v}_2, t_i)$ is given by

$$h(\tilde{v}_2, t_i) = \sum_{j=1}^{m} N_{ij} \cdot y(\tilde{v}_2, t_j) \tag{21.5}$$

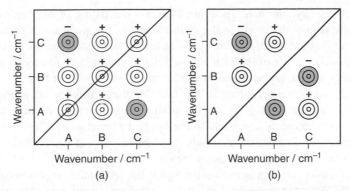

Figure 21.1 *Schematic illustrations of (a) synchronous and (b) asynchronous 2D correlation spectra. White and gray areas in the contour maps represent positive and negative correlation intensities, respectively.*

with N_{ij} being the ith row and jth column element of the "Hilbert–Noda transformation matrix" N given by [5].

$$N_{ij} = \begin{cases} 0 & \text{if } i = j \\ \dfrac{1}{\pi(j-i)} & \text{otherwise} \end{cases} \tag{21.6}$$

The synchronous spectrum $\Phi(\tilde{v}_1, \tilde{v}_2)$ represents the pattern similarity or in-phase nature of spectral intensity changes observed at $\tilde{v}_1$ and $\tilde{v}_2$ along the perturbation variable t. On the other hand, the asynchronous spectrum $\Psi(\tilde{v}_1, \tilde{v}_2)$ represents the dissimilarity or out-phase nature of intensity variation patterns. Although the procedure described is a highly simplified one, it should be applicable to many practical cases of spectral analysis. More general treatment of 2D correlation analysis, including the case for unevenly sampled spectral data, is discussed in Appendix F.

It is now instructive to see how a pair of 2D correlation spectra can be interpreted by using a schematic example illustrated in Figure 21.1. The horizontal and vertical axes in this Figure correspond to spectral variables $\tilde{v}_1$ and $\tilde{v}_2$, respectively. The synchronous spectrum $\Phi(\tilde{v}_1, \tilde{v}_2)$ is a symmetric map with respect to the main diagonal representing the position along $\tilde{v}_1 = \tilde{v}_2$. Positive correlation peaks on the diagonal are called *autopeaks*. They represent the extent of perturbation-induced spectral intensity variations at a given wavenumber. Cross peaks located at off-diagonal positions of a synchronous spectrum represent the coincidental or simultaneous changes of spectral intensities observed at two different spectral variables. A positive synchronous cross peak indicates that spectral intensities of two bands are either increasing or decreasing together. In contrast, a negative peak, indicated by the shaded regions, means that the intensity of one band is increasing while the other band is decreasing in a coordinated manner under the same perturbation. For example, in this synchronous correlation spectrum, a cross peak observed for band pairs A and C reveals that bands A and C are synchronously correlated. In addition, the negative sign of the cross peak indicates the intensity of one band (let us assume the band A) is increasing while that of the other band B is decreasing. Consequently, assuming that the intensity of the band A is increasing, the interpretation of the correlation patterns in the synchronous spectra indicates: A ↑, B ↓, and C↓ or A ↓, B ↑, and C↑.

In contrast, the asynchronous spectrum $\Psi(\tilde{v}_1, \tilde{v}_2)$ consisting of only cross peaks is an antisymmetric map with respect to the main diagonal. The presence of an asynchronous peak indicates that the spectral intensity variations at two wavenumbers corresponding to the cross peak coordinate of the map are not fully synchronized. In other words, correlation peaks appear in an asynchronous spectrum only if the patterns of the spectral intensity changes at $\tilde{v}_1$ and $\tilde{v}_2$ are not identical. The fact that they can vary independently of each other implies that spectral signals may be arising from different constituents of the system, which are responding separately to the applied perturbation. This feature becomes especially useful in distinguishing closely overlapped spectral bands, which may indeed respond asynchronously to a given perturbation.

The sign of cross peaks can be used to determine the sequential order of spectral intensity changes. If the signs of the synchronous and asynchronous cross peaks located at the same coordinate are the same, that is, either both peaks are color-coded gray or both white, the spectral intensity change at spectral variable $\tilde{v}_1$ occurs predominantly before that at $\tilde{v}_2$ during the course of applied perturbation. If the signs are different, the order is reversed. In the example asynchronous correlation spectrum, the presence of cross peaks between A and B indicates that intensity changes at band A occur before the changes at B. Likewise, from the cross peak between bands B and C, one can determine that intensity changes at band B occur before the changes at C. Consequently, the sequential order of intensity variations is deduced to be A → B → C.

This seemingly simple sequential rule has been used quite successfully to probe the mechanistic understanding of various process dynamics in a number of chemical, physical, and biological systems [4, 6, 7].

21.3 Illustrative Application Examples

Some pertinent examples for 2D correlation spectral analyses are provided here to show how certain useful information can be effectively extracted for different types of applications of the 2D correlation technique.

21.3.1 Transient Evaporation of Ethanol from a Solution Mixture with Oleic Acid

Let us start with a very simple example to demonstrate how straightforward the use of 2D correlation spectroscopy is to extract significant results from spectral data. Oleic acid (OA) tends to form dimers in the pure liquid state owing to the strong hydrogen bonding of its carboxyl group. In addition, possible interaction between the dimers induces spontaneous assembling of the dimers to develop a highly ordered liquid crystalline structure. It is believed that such association dynamics is essentially controlled by the hydrogen bonding of the carboxyl group and segmental motion of the alkyl chain [8, 9]. Time-dependent attenuated total reflection (ATR) spectra of a binary mixture solution of OA and ethanol undergoing a spontaneous evaporation process were measured by a Fourier transform-infrared (FT-IR) spectrometer equipped with a mercury cadmium telluride (MCT) detector. The initial mole fraction of the OA in the mixture solution was 0.2. The sample solution was analyzed by depositing it onto a horizontal IRE (internal reflection element: ZnSe plate in this case). The sample was exposed to open atmosphere at room temperature, and a set

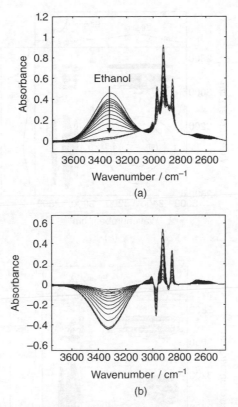

Figure 21.2   *(a) Time-dependent IR spectra of a binary solution mixture of OA and ethanol and (b) corresponding difference spectra obtained by subtracting the initial spectrum.*

of IR spectra were collected at intervals of approximately 3 s for the duration of 60 s. The obtained time-dependent IR spectra are shown in Figure 21.2a. Once the solution mixture was exposed to air, ethanol started evaporating. Eventually, ethanol was completely removed from the system, and only pure OA remained. Consequently, the IR spectra substantially reflect the transient variations of OA and ethanol during the observation period. For example, the peak observed around 3350 cm^{-1}, assignable to the stretching vibration mode of the OH group associated with ethanol, shows a gradual decrease with time. Such a variation of the spectral intensity can obviously be interpreted as the gradual evaporation of ethanol from the system.

It is often useful to explore the variation of a spectral feature by subtracting a specific spectrum from the entire spectral data. Figure 21.2b represents the absorbance difference spectra obtained by subtracting the initial IR spectrum from the individual transient spectra. One can find that ethanol undergoing evaporation provides negative peaks, while OA generates positive peaks in the spectra. It is, thus, very likely that spectral features observed in Figure 21.2 reveals that gradual evaporation of ethanol and the decrease in ethanol is compensated by the relative increase in OA concentration.

Now it is interesting to point out that the application of 2D correlation analysis provides an additional opportunity to sort out subtle but important patterns of the spectral

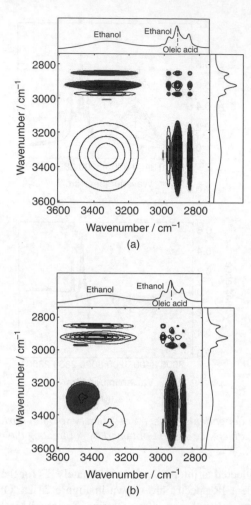

Figure 21.3 *(a) Synchronous and (b) asynchronous correlation spectra directly calculated from time-dependent IR spectra of a binary solution mixture of OA and ethanol.*

intensity variations that are not readily observable in Figure 21.2. Synchronous and asynchronous correlation spectra calculated from the time-dependent IR spectra are shown in Figure 21.3a,b. It is noted that the distinct negative cross correlation peak between 3350 and 2820 cm^{-1} appears in the synchronous correlation spectrum. Since the spectral intensity at 3350 cm^{-1} represents the gradual decrease of ethanol during the evaporation, the intensity variation at 3350 cm^{-1} can be interpreted to mean that the alternation occurs in the opposite direction to the ethanol concentration decrease. Consequently, the intensity variation at 3350 cm^{-1} is assignable to the OA.

More importantly, the asynchronous correlation spectrum reveals in a surprisingly simple manner the substantial interaction between OA and ethanol. For example, the development of the negative correlation peak at 3450 and 3300 cm^{-1} indicates that the variation in the spectral intensity at 3300 cm^{-1} predominantly occurs before that at 3450 cm^{-1}. If spectral intensities simply change simultaneously, reflecting the concentration shift of a purely binary system without any sequential or successive changes in the constituent, it should only show synchronous correlation and no asynchronous correlation. In other words, the appearance of the asynchronous correlation peak reveals the presence of a different type of ethanol species having a somewhat different structure. In fact, such a delay in the evaporation can be interpreted to mean that some portion of the ethanol molecules form a complex with OA, and the peak observed at $3450\,cm^{-1}$ can be assigned to the complex between OA and ethanol [8, 9]. In general, OA molecules form dimers in the pure liquid state. Ethanol essentially induces dissociation of the OA dimer, and it in turn develops the complex between OA and ethanol. Such a molecular interaction between OA and ethanol may enable the ethanol to hold onto the OA, and it subsequently generates the delay in the evaporation, which is not readily detected in the spectra shown in Figure 21.2.

The system at first glance may seem a simple binary mixture of OA and ethanol. The onset of the possible complex formulation is not readily detected from the 1D spectra. In contrast, the 2D correlation analysis of the spectra revealed that subtle differences in the variations of IR spectral intensities were assignable in fact to different species. The nature of the surprisingly intricate spectral variation can be effectively elucidated, providing strong evidence for the existence of possible interaction between OA and ethanol molecules. Consequently, this example demonstrates the deconvolution capability of 2D correlation analysis of highly overlapped bands arising from different moieties and their variations during a physicochemical process.

21.3.2 Partial Fusion of Crystalline Lamellae of High-Density Polyethylene

Polymers have been extensively studied by 2D IR correlation spectroscopy. A number of temperature-, pressure-, time-, and composition-dependent IR spectral variations of polymers have been subjected to the 2D correlation analysis. We showcase here the study of a seemingly simple premelting of a high-density polyethylene (HDPE) film as an illustrative example of the application of 2D correlation spectroscopy in polymer science.

HDPE is composed of very complex supermolecular structures consisting of crystalline lamellae embedded within an amorphous matrix. The crystalline structure undergoes temperature-induced variation well below its melt temperature (T_m). The resulting small level of spectral intensity variation may not be readily identified by the simple observation of a stack plot of 1D spectra. The application of 2D correlation analysis to such spectral data provides an interesting opportunity to elucidate the meaningful information related to perturbation-induced variation of the system. Figure 21.4 represents temperature-dependent IR spectra recorded from an HDPE film. An HDPE film prepared from the melt was placed on a horizontal ZnSe IRE by applying 5 MPa pressure. A set of spectra of the film was measured under increasing temperature from 20 to 100 °C. The sharp IR bands at 1463 and $1473\,cm^{-1}$ are both assignable to the contributions from the crystalline component of HDPE [10, 11]. Although it is not readily observed, the amorphous component also provides spectral intensity change around at 1465 and $1455\,cm^{-1}$ [10, 11]. In Figure 21.4, it is noted that the increase in the temperature resulted in a very small level of change in the spectral intensity of the crystalline bands at $1473\,cm^{-1}$. On the other hand, the variation of the amorphous components is unclear because of the overlapping of the IR bands.

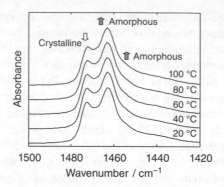

Figure 21.4   *Temperature-dependent IR spectra of high-density polyethylene (HDPE) film.*

Synchronous and asynchronous correlation spectra derived from the temperature-dependent IR spectra of HDPE are presented in Figure 21.5. The synchronous correlation spectrum shows the negative cross peaks at the coordinate (1473, 1465) and (1473, 1455), which suggest that the spectral intensity changes of the crystalline and amorphous bands occur in the opposite direction upon heating. This result can be interpreted that the decrease in the crystalline band intensities is compensated for by the rising intensity of the amorphous bands. Consequently, crystalline and amorphous components of HDPE must be undergoing the population changes similar to those observed for the melting observed above its T_m [12]. Asynchronous correlation spectrum, on the other hand, shows cross peaks at the coordinate (1473, 1465) and (1473, 1455), indicating the presence of a specific delay in the spectral intensity changes between the crystalline and amorphous bands. For example, the correlation peak between 1473 and 1465 cm^{-1} here reveals that variation of the spectral intensity of the crystalline components predominantly occurs before that of the amorphous component. In other words, the generation of the fully amorphous component occurs at a higher temperature range than that for the disappearance of the crystalline component. Apparently, the decrease in the crystalline component does not result in the simultaneous increase in the amorphous component, suggesting the possibility of the existence of an intermediate state between the highly ordered crystalline state and the totally liquid-like amorphous state. It can be speculated that, upon heating, some portion of crystals (probably less ordered small crystals) undergoes a transition similar to those observed for the regular melting process. This is quickly followed by the development of the partially fused structure. Further heating results in the generation of totally liquid-like amorphous structures.

Application of 2D correlation analysis effectively sorted out the spectral intensity variation of HDPE induced even well below its T_m. A sequence of the event induced by the increasing temperature was detected to show an apparent additional step in the transformation between the crystals and the amorphous component, namely, the possible partial fusion of the less ordered small crystals of HDPE induced below its T_m was suggested from the patterns appearing in the 2D correlation maps.

21.3.3   pH-Dependent 2D Correlation Study of Human Serum Albumin

The basis of 2D correlation is so flexible and general that the application is not limited to any particular research field, such as classical physical chemistry of polymers or liquid

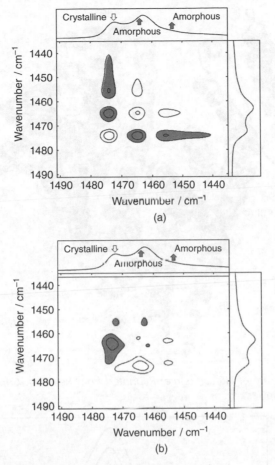

Figure 21.5 *(a) Synchronous and (b) asynchronous correlation spectra directly calculated from temperature-dependent IR spectra of HDPE.*

crystals. It has, for example, been applied to the analysis of IR spectra of proteins. An interesting testament to the versatility of 2D correlation spectroscopy was demonstrated by Murayama *et al.* [13], in which the basic approach of 2D correlation was applied to sort out the pH-dependent structural variation of human serum albumin (HSA).

The first three-dimensional crystal structure determination of HSA at a resolution of 2.8 Å was obtained by He and Carter [14]. Figure 21.6 illustrates the structure of HSA [15]. It is characterized by a repeating pattern of three α-helical homologous domains, derived from genic multiplication and numbered I, II, and III consisting of 585 amino acid residues. It contains 98 acidic side chains with 36 aspartic (Asp) acid and 62 glutamic (Glu) residues and 100 basic side chains based on aromatic amino acid residues. The secondary structure and hydrogen bonding of the side chains in aqueous solutions are sensitive to the environmental variations, such as pH and temperature.

Figure 21.7 shows representative pH-dependent IR spectra of HSA in the buffer solutions (2.0 wt%). At each pH, the spectrum of the buffer was subtracted from the

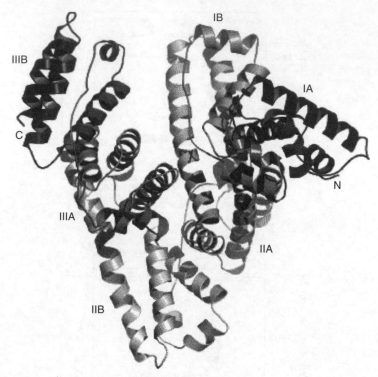

Figure 21.6 *Structure of HSA. (Source: Reprinted from Ref. [15]. Copyright © 2013, with permission from Elsevier.)*

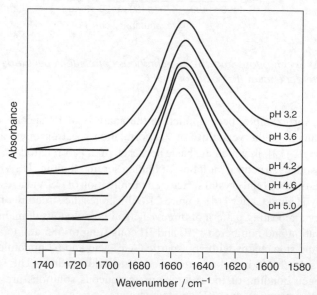

Figure 21.7 *Representative pH-dependent IR spectra of HSA in the buffer solutions (2.0 wt%). (Source: Reprinted with permission from Ref. [13]. Copyright © 2001, American Chemical Society.)*

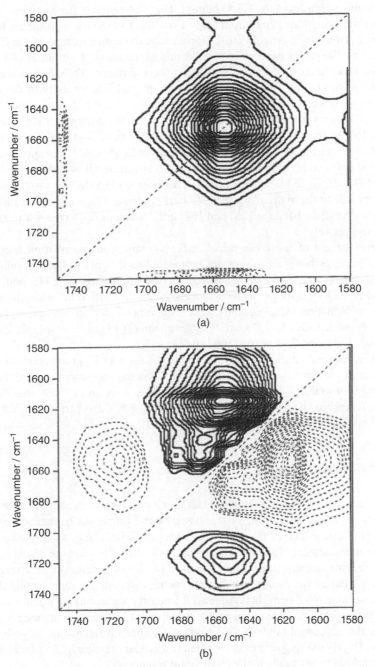

Figure 21.8 (a) Synchronous and (b) asynchronous 2D correlation spectra constructed from the pH-dependent IR spectra. Solid lines in the contour maps represent positive correlation intensity and dashed lines mean negative correlation intensity. (Source: Reprinted with permission from Ref. [13]. Copyright © 2001, American Chemical Society.)

corresponding spectrum of the HSA solution. The HSA in the buffer solution substantially undergoes the structural change from the N-isometric form to the F-isometric form, which takes place primarily in domain III. Unfortunately, the entire feature of the IR spectra is dominated by the major band at $1654\,cm^{-1}$ due to the amide I vibration of HSA, which makes the identification of some minor bands very difficult. Thus, it becomes useful to highlight the subtle but significant variation of spectral intensities with the help of 2D correlation analysis.

Figure 21.8 shows the 2D IR correlation spectra of the N-isomeric form of HSA in a buffer solution, derived from the pH-dependent (pH 5.0, 4.8, 4.6, and 4.4) spectral variations. The synchronous spectrum shows a major autopeak around $1654\,cm^{-1}$, assigned to the α-helix. The analysis of several peaks in the asynchronous spectrum allows one to identify several bands at 1715, 1667, 1654, and $1641\,cm^{-1}$. The band at $1715\,cm^{-1}$ is assignable to a C=O stretching mode of the hydrogen-bonded COOH groups of Glu and Asp residuals of HSA. On the other hand, the bands at 1667 and $1641\,cm^{-1}$ are assigned to the β-turn and β-strand of HSA, respectively.

The development of the cross peaks in the asynchronous correlation spectrum indicates that some carboxylate groups are protonated in this pH range depending on their structures. For example, the broad feature of the cross peak at 1715 and $1654\,cm^{-1}$ reveals that COOH groups with hydrogen bonds of different strength are formed upon the protonation. Consequently, the sequential order of the event becomes as hydrogen-bonded COOH $(1715\,cm^{-1})$, side chain $(1614\,cm^{-1}) \rightarrow \alpha$-helix $(1654\,cm^{-1})$ $\rightarrow \beta$-strand $(1641\,cm^{-1}) \rightarrow \beta$-turn $(1667\,cm^{-1})$

The given sequence indicates that protonation of the COO^- groups and microenvironmental changes in the side chains predominantly precede the secondary structural change of α-helix, followed by those of β-strand and β-turn. It is, thus, likely that the protonation of some carboxylic groups at relatively low pH (4.5–5.0) is triggering the secondary structural changes (i.e., N–F transition) in domain III.

21.4   Concluding Remarks

The basic concept and some examples of 2D correlation spectroscopy have been covered in this chapter. 2D correlation analysis is based on the simple mathematical treatment of a set of spectral data collected from a system under the influence of an applied perturbation during the measurement. This perturbation can take different forms of changes, including temperature, concentration, or pH, and the like. The set of spectra is then converted to the synchronous and asynchronous correlation spectra, respectively, representing the similarity and dissimilarity of perturbation-induced intensity variations between wavenumbers. 2D correlation peaks provide easier access to pertinent information, making it possible to determine the sequential order of the variations of spectral intensities, as well as relative directions. Highly overlapped peaks are often resolved more clearly. This technique can be a useful addition to the toolbox of experimental scientists.

References

1. Noda, I. (1989) Two-dimensional infrared spectroscopy. *J. Am. Chem. Soc.*, **111**, 8116–8118.

2. Noda, I. (1990) Two-dimensional infrared (2D IR) spectroscopy: theory and applications. *Appl. Spectrosc.*, **44**, 550–561.
3. Noda, I. (1993) Generalized two-dimensional correlation method applicable to infrared, Raman, and other types of spectroscopy. *Appl. Spectrosc.*, **47**, 1329–1336.
4. Noda, I. and Ozaki, Y. (2004) *Two-Dimensional Correlation Spectroscopy: Applications in Vibrational and Optical Spectroscopy*, John Wiley & Sons, Ltd, Chichester.
5. Noda, I. (2000) Determination of two-dimensional correlation spectra using the Hilbert transform. *Appl. Spectrosc.*, **54**, 994–999.
6. Ozaki, Y. (2002) 2D correlation spectroscopy in vibrational spectroscopy, In: *Handbook of Vibrational Spectroscopy* (eds J.M. Chalmers and P. Griffiths), John Wiley & Sons, Ltd, Chichester, pp. 2135–2172.
7. Noda, I. (2010) Two-dimensional correlation spectroscopy – Biannual survey 2007–2009. *J. Mol. Struct.*, **974**, 3–24.
8. Shinzawa, H., Iwahashi, M., Noda, I. and Ozaki, Y. (2008) Asynchronous kernel analysis for binary mixture solutions of ethanol and carboxylic acids. *J. Mol. Struct.*, **883**, 27–30.
9. Pi, F.W., Shinzawa, H., Czarnecki, M.A., Iwahashi, M., Suzuki, M. and Ozaki, Y. (2010) Self-assembling of oleic acid (cis-9-octadecenoic acid) and linoleic acid (cis-9, cis-12-octadecadienoic acid) in ethanol studied by time-dependent attenuated total reflectance (ATR) infrared (IR) and two-dimensional (2D) correlation spectroscopy. *J. Mol. Struct.*, **974**, 40–45.
10. Schachtschneider, J.H. and Snyder, R.G. (1964) Normal coordinate calculations of large hydrocarbon molecules and polymers. *J. Polym. Sci. Part C*, **7**, 99–124.
11. Tasumi, M. and Shimanouchi, T. (1965) Crystal vibrations and intermolecular forces of polymethylene crystals. *J. Chem. Phys.*, **43**, 1245–1258.
12. Tashiro, K. and Sasaki, S. (2003) Structural changes in the ordering process of polymers as studied by an organized combination of the various measurement techniques. *Prog. Polym. Sci.*, **28**, 451–519.
13. Murayama, K., Wu, Y., Czarnik-Matusewicz, B. and Ozaki, Y. (2001) Two-dimensional/attenuated total reflection infrared correlation spectroscopy studies on secondary structural changes in human serum albumin in aqueous solutions: pH-dependent structural changes in the secondary structures and in the hydrogen bondings of side chains. *J. Phys. Chem. B*, **105**, 4763–4769.
14. He, X.M. and Carter, D.C. (1992) Atomic structure and chemistry of human serum albumin. *Nature*, **358**, 209–215.
15. Artali, R., Bombieri, G., Calabi, L. and Del Pra, A. (2005) A molecular dynamics study of human serum albumin binding sites. *IL Farmaco*, **60**, 485–495.

22

Vibrational Circular Dichroism

Yoshiaki Hamada
Studies of Nature and Environment, The Open University of Japan, Japan

22.1   Introduction

The pioneering research of detecting vibrational optical activity (VOA) of chiral molecules stemmed from experimental studies undertaken in the 1970s [1–3], and this was followed by theoretical developments during the period beginning from the middle of the 1980s to the first half of the 1990s [4–8]. Over the same period, the practical method of measurements made rapid progress [9], and theoretical calculations of VOA spectra also became possible because of the evolution and developments in computer hardware and software [10, 11]. Nowadays, fairly accurate predictions of VOA spectra are possible on a personal computer by using commercially available software [12].

Traditionally, electronic circular dichroism (ECD) associated with molecular electronic transitions has been used for the experimental study of chiral molecules. ECD has much higher sensitivity and produces simpler patterns than VOA. This point is sometimes an advantage as extremely small differences in the ECD spectra of a pair of chiral molecules can be used to differentiate between the molecules. ECD spectra have a disadvantage as less information on molecular structure can be obtained from their simpler patterns. Another disadvantage of ECD is that a chromophore giving rise to an electronic absorption in the visible or accessible UV region must exist in the target molecule.

In VOA spectroscopy, differences between vibrational spectra for left-circularly polarized radiation and right-circularly polarized radiation are detected. By the early 2000s, spectrometers for measuring VOA had been developed and they became commercially

Introduction to Experimental Infrared Spectroscopy: Fundamentals and Practical Methods,
First Edition. Edited by Mitsuo Tasumi and Akira Sakamoto.

available. Vibrational spectroscopy has two complementary methods, that is, infrared and Raman spectroscopies; likewise, VOA spectroscopy has two complementary methods. Vibrational circular dichroism (VCD) detects the difference between infrared absorption spectra for left- and right-circularly polarized infrared radiation, while Raman optical activity (ROA) detects the difference between Raman scattering spectra for left- and right-circularly polarized radiation of the laser used for Raman excitation. In view of growing interest in biological systems and chiral materials, both VCD and ROA have diverse applications and potential in many fields of study. Because both methods need to be able to accurately determine a difference of the order of 10^{-4} to 10^{-5} of the original spectral intensities measured by left- and right-circularly polarized radiation, VCD and ROA instrumentation must have high precision and sensitivity. Special attention needs to be paid to measuring conditions and spectrometer operation, and sample preparations. This chapter discusses in brief the principles of VCD, the points to note in measuring VCD spectra, and how to interpret measured spectra.

22.2   Theoretical Explanation of VCD

A molecule that has two equivalent three-dimensional structures corresponding to mirror images of each other, like a person's right and left hands, is called a *chiral molecule* or an *optically active molecule*, and each of the two structures is called an *enantiomer*. The VCD spectrum of a chiral molecule gives information on the absolute three-dimensional structure of the molecule, together with its ordinary infrared absorption spectrum. Thus, VCD spectroscopy provides a useful diagnostic tool for studying peptides, proteins, nucleic acids, saccharides, natural products, and pharmaceuticals, which are to differing extents associated with chirality.

The VCD signal is defined as

$$\Delta A = A_{\mathrm{L}} - A_{\mathrm{R}} \tag{22.1}$$

where A_{L} and A_{R} indicate, respectively, the absorbance (absorption intensity) for left- and right-circularly polarized infrared radiation, respectively. The quantum-mechanical explanation of the transition between energy levels of a molecule due to interaction between the electromagnetic wave and the molecule is completely established. In the case of VOA, not only does the electric-dipole moment but also the magnetic-dipole moment induced by the electromagnetic wave need to be taken into account. The transition moment induced by the electric and magnetic fields of circularly polarized radiation oscillating in the direction perpendicular to the propagating direction of radiation contains the effect of polarization involving the electric dipole and the magnetic dipole. This term appears as an imaginary part in the mathematical expression of the transition moment induced by circularly polarized radiation.

The absorbance for circularly polarized infrared radiation is proportional to a quantity called the *transition strength B* expressed by the following equation, where superscripts $+$ and $-$ on B, shown as $+/-$, indicate, respectively, L and R in Equation (22.1), and the corresponding symbol of the same sign ($+$ or $-$) is used on the right-hand side at $\pm$ of the equation.

$$B^{+/-} = \frac{8\pi^3}{3h^2}[\boldsymbol{\mu}_{\mathrm{sn}} \cdot \boldsymbol{\mu}_{\mathrm{ns}} + \boldsymbol{m}_{\mathrm{sn}} \cdot \boldsymbol{m}_{\mathrm{ns}} \pm 2\mathrm{Im}(\boldsymbol{\mu}_{\mathrm{sn}} \cdot \boldsymbol{m}_{\mathrm{ns}})] \tag{22.2}$$

where μ_{sn} and m_{sn} refer, respectively, to the induced electric and magnetic dipole-moment vectors relating to the initial (s) and final (n) states. The three terms on the right-hand side of the above equation are re-expressed as

$$\mu_{sn} \cdot \mu_{ns} = D_{ns}, \quad m_{sn} \cdot m_{ns} = M_{ns}, \quad Im(\mu_{sn} \cdot m_{ns}) = R_{ns} \qquad (22.3)$$

They are called, respectively, *electric-dipole transition strength* (D_{ns}), *magnetic-dipole transition strength* (M_{ns}), and *rotational strength* (R_{ns}). Then, the total transition strength is given as ($D_{ns} + M_{ns} \pm 2R_{ns}$). This indicates that the effects for left- and right-circularly polarized infrared radiation differ because of the term $\pm 2R_{ns}$. Among the above three terms, M_{ns} is much smaller than the other two terms and can be neglected. If the molar absorption coefficients for left- and right-circularly polarized infrared radiation are denoted, respectively, by ε_L and ε_R, the difference between them $\Delta\varepsilon$ and their average ε are given, respectively, by $\Delta\varepsilon = \varepsilon_L - \varepsilon_R$ and $\varepsilon = (\varepsilon_L + \varepsilon_R)/2$. The ratio between $\Delta\varepsilon$ and ε, which is called the *anisotropy factor* and expressed by g, is given as

$$g = \frac{\Delta\varepsilon}{\varepsilon} = \frac{4R}{D} = \frac{4Im(\mu \cdot m)}{\mu^2} \qquad (22.4)$$

The factor g is a measure of the magnitude of circular dichroism (CD); CD is more easily measured the larger the value of g. The magnitude of g is roughly proportional to the ratio of the molecular size to the wavelength of radiation used for the measurement. This means that the measurement of CD becomes more difficult at longer wavelengths. In Table 22.1, characteristics of CD spectroscopies associated with electronic transitions (ECD) and vibrational transitions (VCD) are compared.

The intensity of an infrared absorption band is determined by the square of the electric-dipole transition moment representing the magnitude of linear displacement of the charge due to the molecular vibration, giving rise to the absorption band. On the other hand, VCD is due to interactions of linear displacement of the charge represented by the electric-dipole transition moment and circular movement of the charge associated with the molecular vibration, where the magnetic-dipole transition moment is perpendicular to the plane of charge circulation.

Table 22.1 *Comparison of the characteristics of electronic circular dichroism (ECD) and vibrational circular dichroism (VCD).*

Characteristic	ECD	VCD
Wavelength region	Ultraviolet and visible	Infrared
Related transition	Electronic transition	Vibrational transition
Related electronic state(s)	Ground and excited states	Ground state
Origin of phenomenon	Chromophore	Molecular vibration
Information content	Low	High
Anisotropy factor	$\sim 10^{-3}$	10^{-4} to 10^{-5}
Sensitivity	High	Low
Spectral analysis	Difficult	Less difficult

22.3   Instrumentation for VCD Measurement

In this section, the definition of linearly and circularly polarized radiation and how to produce them are first briefly described. Descriptions relating to this subject are also found in Sections 2.5.4, 11.3, and B.2. Following this, matters relating to the instrument for measuring VCD are explained.

22.3.1   Linearly Polarized Radiation

Infrared radiation is a transverse electromagnetic wave. Its electric and magnetic fields oscillate perpendicularly to the direction of propagation, and the electric and magnetic fields are perpendicular to each other. Infrared radiation emitted from a thermal source has its electric-field vectors in all directions perpendicular to the direction of propagation. In other words, the infrared beam from a thermal source is not polarized. To obtain linearly polarized radiation from unpolarized radiation, a wire-grid polarizer is used in the infrared region. By placing a wire-grid polarizer in an infrared beam, the transmitted radiation is polarized perpendicular to the plane of the wire grid as shown in Figure 22.1. In a wire-grid polarizer, fine metal wires are placed parallel to each other with short spacing between them onto an infrared transparent window material. The grid spacing should be sufficiently smaller than the infrared wavelengths used for the measurement. Because the radiation, the electric field of which oscillates parallel to the wire grid, interacts with free electrons in the wire grid, a part of it is absorbed by exciting free electrons, and the remaining part is reflected. On the other hand, the incident radiation, the electric field of which is perpendicular to the wire-grid direction, does not interact with free electrons in the wire grid. Therefore, only the radiation polarized in this direction can pass through the wire grid.

22.3.2   Circularly Polarized Radiation

The speed of light in a material is determined by the interaction between the incident radiation and the material. The speed of radiation in the material (c) is slower than that in

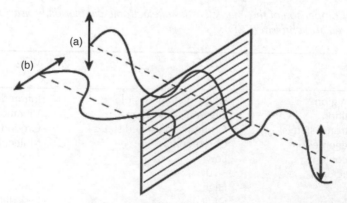

Figure 22.1   *Wire-grid polarizer and linearly polarized radiation. (a) Radiation polarized perpendicularly to the wire-grid direction and (b) radiation polarized parallel to the wire-grid direction.*

vacuum (c_0), and the following relation holds; that is, $c = c_0/n$, where n is the refractive index of the material. It is assumed here that the material has no absorption. When linearly polarized radiation enters a crystal, the speed of the polarized radiation depends on both the direction of polarization and the optical property of the crystal. As shown in Figure 22.2, when radiation polarized at 45° to the x (or y) axis travels in the z direction and enters an isotropic crystal (i.e., $n_x = n_y$), the electric field of the radiation E is divided equally into the components polarized along the x axis (E_x) and the y axis (E_y), and both components pass through the crystal with exactly the same speed. As a result, the radiation leaves the crystal without changing its polarization.

If the crystal is anisotropic (i.e., $n_x \neq n_y$; this property is called *birefringence*), the peaks and troughs of E_x and E_y are shifted relative to each other because of the different speeds within the crystal, and the polarization direction of radiation leaving the crystal is determined by mixing the electric-field components of different phases. If the shift between E_x and E_y in the crystal along the z axis is one quarter of the wavelength of the radiation ($\lambda/4$), the electric field of the emerging radiation rotates along the z axis, as shown in Figure 22.3a. Such a crystal is called a *quarter-wave plate* (*QWP*) or a *quarter-wave retarder*. By using a QWP, linearly polarized radiation can be transformed to circularly polarized radiation.

The unit electric-field vectors in the x, y-plane of circularly polarized radiation traveling along the z axis are described as $u_x = \cos \omega(t - z/c)$ and $u_y = \pm \sin \omega(t - z/c)$, where $+\sin$ and $-\sin$ correspond, respectively, to the left- and right-circularly polarized radiation.

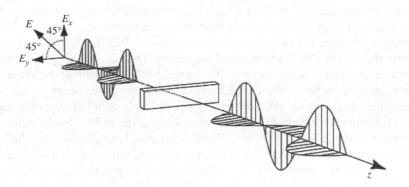

Figure 22.2 *Polarized radiation emerging from an isotropic crystal.*

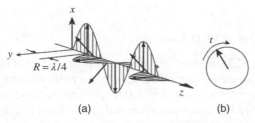

Figure 22.3 *Role of a quarter plate in producing circularly polarized radiation. (a) The phase of x-polarized radiation is delayed by 90° from that of y-polarized radiation (or 1/4 wavelength delayed). (b) Definition of right-circularly polarization, with t indicating time.*

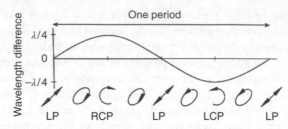

Figure 22.4   *Role of a photoelastic modulator (PEM) and the polarization of output radiation. LP, linear polarization; RCP, right-circularly polarization; LCP, left-circularly polarization.*

The clockwise or right-hand turning movement of the electric-field vector at a particular point on the z axis looking back from the traveling direction is shown in Figure 22.3b for right-circularly polarized radiation.

An infrared transparent QWP is a special crystal (e.g., ZnSe plate) that can give rise to a phase difference of 90° between E_x and E_y if the crystal is stressed. This is realizable by applying pressure along one direction of the crystal. Practically, periodically oscillating birefringence is generated in the crystal by applying an alternating voltage onto a piezo-electric device that is attached to an infrared transparent isotropic crystal, such as CaF_2 or ZnSe. If the phase difference or retardation between E_x and E_y is changed in the range from $+\lambda/4$ to $-\lambda/4$, the radiation coming out from the QWP shows the variation in polarization as shown schematically in Figure 22.4. If the phase difference is shifted from $\pm\lambda/4$, various shapes of elliptically polarized radiation may be produced.

The device described is called a *photoelastic modulator* (PEM), which is one of the most important elements of a VCD spectrometer. As seen in Figure 22.4, during one period of the PEM operation, linearly polarized radiation emerges twice with each of the left- and right-circularly polarized radiation being produced once. This means that linearly polarized radiation is detected at twice the PEM driving frequency, while both left- and right-circularly polarized radiation are detected at the same frequency as the PEM driving frequency.

As explained, the polarization plane of linearly polarized radiation can be tilted when it passes through a crystal that has any anisotropy or a medium in which molecules are oriented. In general, the rotation of the plane of polarization is a function of the thickness of the crystal or medium and the difference between the refractive indices along the two orthogonal axes perpendicular to the direction of incident radiation. CD spectroscopy detects differences in the absorption of a chiral molecule for left- and right-circularly polarized radiation.

22.3.3   Measurement of VCD Spectra

A block diagram of a VCD spectrometer is shown in Figure 22.5. The units to generate circularly polarized radiation (an optical filter, a linear polarizer, and a PEM) are inserted between a Fourier transform-infrared (FT-IR) spectrometer and the detector. Left- and right-circularly polarized radiation are generated with a fixed frequency by the PEM, and the signal synchronized with the PEM is detected.

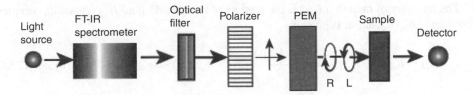

Figure 22.5   *Block diagram of a VCD spectrometer.*

22.3.4   Detection of the VCD Signal

The detection of the VCD signal is performed in essentially the same way as the polarization-modulation measurements described in Chapter 11. Briefly, a high-frequency modulation frequency that is more than one order of magnitude faster than that of the Michelson interference modulation frequency is applied to the sample, and the ac signal of the higher frequency component is selectively detected. This gives the difference spectrum, and the dc signal as an average of the modulated signals gives the normal FT-IR spectrum. In Chapter 11, the advantage of this double-modulation Fourier transform spectroscopy [13, 14] is described with some applications in reflection–absorption spectroscopy using linearly polarized light. In this chapter, application of the double-modulation method to VCD spectroscopy is explained with some overlapping with the descriptions in Chapter 11.

The linearly polarized radiation entering the PEM with its polarization plane at 45° to the x and y axes of the PEM emerges from the PEM with a phase difference of $\delta = \delta_x - \delta_y$. The electric-field vector of the modulated infrared beam is expressed as

$$E_m(\tilde{v}_i) = \left(\frac{E_0(\tilde{v}_i)}{\sqrt{2}} \right) (e_x + e_y e^{i\delta}) \tag{22.5}$$

where $E_0(\tilde{v}_i)$ is the amplitude of the electric field of the incident infrared beam, and e_x and e_y are unit vectors along the x and y axes. Equation (22.5) may be rewritten as

$$E_m(\tilde{v}_i) = \left(\frac{E_0(\tilde{v}_i)}{\sqrt{8}} \right) [(1 - ie^{i\delta})(e_x + ie_y) + (1 + ie^{i\delta})(e_x - ie_y)] \tag{22.6}$$

where $(e_x + ie_y)$ and $(e_x - ie_y)$ represent, respectively, the unit vectors for the right- and left-circularly polarized radiation.

If circularly polarized radiation passes through an optically active medium and absorptions for right- and left-circularly polarized radiation are different, the resultant electric-field vector becomes

$$E_m(\tilde{v}_i) = \left(\frac{E_0(\tilde{v}_i)}{\sqrt{8}} \right) \left[\left(1 - ie^{i\delta}\right) (e_x + ie_y)e^{-\frac{a_R}{2}} + (1 + ie^{i\delta})(e_x - ie_y)e^{-\frac{a_L}{2}} \right] \tag{22.7}$$

where a_R and a_L are associated, respectively, with absorbances for right- and left-circular polarized radiation.

The intensity of radiation $I(\tilde{\nu}_i)$ is the product of the electric field $\boldsymbol{E}_\mathrm{m}(\tilde{\nu}_i)$ and its complex conjugate $\boldsymbol{E}_\mathrm{m}^*(\tilde{\nu}_i)$, and it is given as

$$I(\tilde{\nu}_i) = \boldsymbol{E}_\mathrm{m}(\tilde{\nu}_i) \cdot \boldsymbol{E}_\mathrm{m}^*(\tilde{\nu}_i) = \left(\frac{I_0(\tilde{\nu}_i)}{2}\right)[(e^{-a_\mathrm{R}} + e^{-a_\mathrm{L}}) + (e^{-a_\mathrm{R}} - e^{-a_\mathrm{L}})\sin\delta] \qquad (22.8)$$

The phase difference δ is a function of the PEM frequency, which is expressed as

$$\sin\delta = \sin(\delta_0^i \sin\omega_\mathrm{m}t) = 2\sum_{n=1}^{\infty} J_{2n-1}(\delta_0^i)\sin(2n-1)\omega_\mathrm{m}t \qquad (22.9)$$

where δ_0^i is the maximum phase difference for $\tilde{\nu}_i$ caused by the PEM, ω_m is the angular frequency of the PEM, and $J_{2n-1}(\delta_0^i)$ is the Bessel function with integer n. The detector receives both the dc component I_dc and the ac component I_ac, and their ratio at ω_m is given as

$$\frac{I_\mathrm{ac}(\tilde{\nu}_i)}{I_\mathrm{dc}(\tilde{\nu}_i)} = 2J_1(\delta_0^i)\left(\frac{e^{-a_\mathrm{R}} - e^{-a_\mathrm{L}}}{e^{-a_\mathrm{R}} + e^{-a_\mathrm{L}}}\right)G(\tilde{\nu}_i) \qquad (22.10)$$

where $G(\tilde{\nu}_i)$ is the specific constant for the instrument used for the measurement including the effects from the filter, lock-in amplifier, and the gain of the electronic system. This should be determined before the measurement is made. If both the numerator and denominator of the right-hand side of the above equation are multiplied by $e^{(a_\mathrm{R}+a_\mathrm{L})/2}$, and the relation $(e^x - e^{-x})/(e^x + e^{-x}) = \tanh x \approx x$ for a small x is taken into account, Equation (22.10) is rewritten as

$$\frac{I_\mathrm{ac}}{I_\mathrm{dc}} = 2J_1(\delta_0^i)\left[\frac{(a_\mathrm{L}(\tilde{\nu}_i) - a_\mathrm{R}(\tilde{\nu}_i))}{2}\right]G(\tilde{\nu}_i)$$

$$= 2J_1(\delta_0^i)\left[\frac{2.303(A_\mathrm{L}(\tilde{\nu}_i) - A_\mathrm{R}(\tilde{\nu}_i))}{2}\right]G(\tilde{\nu}_i) \qquad (22.11)$$

In this equation, $A_\mathrm{L}(\tilde{\nu}_i)$ and $A_\mathrm{R}(\tilde{\nu}_i)$ indicate, respectively, absorbances for right- and left-circularly polarized radiation at wavenumber $\tilde{\nu}_i$. To determine $A_\mathrm{L}(\tilde{\nu}_i) - A_\mathrm{R}(\tilde{\nu}_i)$, the value of $2J_1(\delta_0^i)G(\tilde{\nu}_i)$ should be known in advance of the VCD measurement by the method to be described later.

22.3.5   Calibration of the VCD Spectrometer

Close to the position of the sample in Figure 22.5, a birefringent plate and an analyzer (the same as a linear polarizer) are placed (the combination of these two kinds of plates are sometimes called a *stress plate*). The relation of the axes of the polarizer and analyzer can be either parallel or perpendicular (denoted by $A_\parallel$ and $A_\perp$, respectively). Similarly, the relation of the axes of the birefringent plate and the PEM can also be either parallel or perpendicular (denoted by $B_\parallel$ and $B_\perp$, respectively). Consequently, four different combinations of the axes are possible; however, only the case where all the axes are parallel is considered here.

Since the phase difference δ_0^i is a function of the wavenumber of incident radiation, the phase difference arising in the PEM that oscillates with an angular frequency of ω_m is given by

$$\delta_m^i = \delta_0^i(\tilde{v}_i) \sin \omega_m t \tag{22.12}$$

The optimum value of δ_0^i which corresponds to the maximum of $J_1[\delta^i(\tilde{v}_i)]$ is a function of retardation angle $\delta^i(\tilde{v}_i)$, and $J_1(\delta_0^i) \approx 0.57$ when $\tilde{v}_i$ is equal to a quarter wavelength of the PEM. However, because the PEM is designed to satisfy this condition for a fixed wavenumber $\tilde{v}_q$ of the incident radiation, the phase difference deviates from the optimum value for the other wavenumbers. In addition to the δ_m^i, the birefringent plate adds a time-independent phase difference of $\delta_B^i = 2\pi\tilde{v}_i d\Delta n$, where d is the thickness of the birefringent plate and Δn is the difference between refractive indices. As a result, the electric-field vector at the exit of the analyzer is given as

$$\mathbf{E}_m(\tilde{v}_i) = \left(\frac{E_0(\tilde{v}_i)}{\sqrt{2}} \right) (\mathbf{e}_x + \mathbf{e}_y e^{i(\delta_m^i + \delta_B^i)}) \tag{22.13}$$

By a similar mathematical treatment, the intensity ratio of the ac and dc components at the detector is given as

$$f(\tilde{v}_i) = \frac{I_{ac}}{I_{dc}} = \frac{(\mp \text{ or } \pm)2J_1(\delta_0^i) \sin \delta_B^i}{1 \pm J_0(\delta_0^i) \cos \delta_B^i} G(\tilde{v}_i) \tag{22.14}$$

Here, four different combinations of signs in the numerator and denominator correspond to the four sets of axes described earlier. By using these results, the correct VCD intensities can be obtained from Equation (22.11) by appropriate calibration procedures.

22.3.6 Points to Note in VCD Measurements

In ordinary infrared measurements, the requirements for the optical properties of cell windows and other materials (flatness, transparency, etc.) are not as severe as needed for ultraviolet and visible spectroscopy. Sometimes it is even necessary to make the window surface slightly coarse or to make its surfaces slightly deviated from parallel to avoid interference effects. However, such measures should never be taken in VCD measurements in order to ensure that circularly polarized radiation may be generated without fail. In VCD measurements, cell windows must be flat and parallel. A mirror cannot be used for any purpose. Samples should be transparent. VCD will not be observed in samples such as colloids because of scattering.

In order to keep sufficient absorption intensity and to detect VCD corresponding to a difference on the order of 10^{-4} to 10^{-5} of the peak intensity, it is known empirically that the optimum absorbance of a band is around 0.4. The absorbance should never exceed 1.0 and not fall below 0.1. To obtain a high-quality VCD spectrum, it is most important to prepare the sample at an appropriate optimum concentration and to use a cell of appropriate pathlength.

22.4   Theoretical Prediction of VCD Spectra

To calculate VCD spectra theoretically, the following quantities must be evaluated.

$$\mu_{01} = \frac{\partial \mu}{\partial Q}, \quad m_{01} = \frac{\partial m}{\partial P} \tag{22.15}$$

where Q and P represent, respectively, the normal coordinate of a molecular vibration and the momentum associated with it. The method for calculating $\partial m / \partial P$ is found in Ref. [15], and comprehensive discussions of VCD (and ROA) are given in Refs [16, 17].

Recently, quantum-chemical calculations have been shown to give results that agree with the observed infrared spectra and corresponding VCD spectra reasonably well for molecules of a medium size. In Figure 22.6, the observed infrared spectrum of β-pinene is compared with the results of calculations at various theoretical levels. β-Pinene is used as a reference sample in VCD measurements. In Figure 22.7, the observed VCD spectrum of this sample is compared with the calculated results. It is clear that the results of calculations at higher theoretical levels as shown in infrared spectrum (b) in Figure 22.6 and VCDspectrum (b) in Figure 22.7 are in better agreement with the observed shown as (a) of both figures than the results of low-level calculations shown in (d) and (e) of both figures, which do not take into account electron correlation. The results shown in Figures 22.6 and 22.7 indicate that the use of reasonably large basis sets and inclusion of electron correlation as provided by density functional theory (DFT) are necessary to obtain satisfactory results from the calculations. The calculated wavenumbers are generally higher than the observed. This is partly due to the fact that the calculations have been performed using the harmonic approximation; that is, vibrational anharmonicities have not been taken into account.

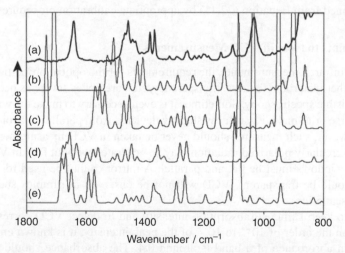

Figure 22.6   *Observed and calculated IR spectra of (S)-β-pinene. (a) Observed spectrum, (b)–(e) calculated spectra. Methods of calculation are: (b) B3LYP/6-31+G**, (c) B3LYP/6-31, (d) HF/6-32+G*, (e) HF/4-31G.*

22.5   Models for Understanding VCD Activity

Practical prediction of VCD spectra is now possible for molecules of a medium size as discussed in the previous section. However, *ab initio* calculations for a macromolecule such as a protein molecule are still impracticable. Even if the calculated results are in agreement with the observed, it is not possible to understand the sign and intensity of the VCD signal through chemical intuition. In such a situation, it is desirable to have an empirical model for understanding the relationship between molecular properties (molecular structure or its partial structure and related vibrational mode) and the observed VCD spectrum.

The coupled-oscillator model [18], which is sometimes used for the need described above, is introduced here.

22.5.1   Coupled-Oscillator Model

In an asymmetric molecule, let two oscillators μ_1 and μ_2 be placed at a distance d between them. The two vectors will be coupled to make two orthogonal motions, that is, $\xi_+ = \mu_1 + \mu_2$ and $\xi_- = \mu_1 - \mu_2$ as shown in Figure 22.8a,b, respectively. Here, normalization for ξ_+ and ξ_- is neglected.

Let us consider the interaction of circularly polarized radiation with ξ_+. The relation of ξ_+ with right-circularly polarized radiation (RCL) and left-circularly polarized radiation (LCL) are illustrated in Figure 22.8c,d, respectively, where μ_2 is separated from μ_1 by $d = \lambda/4$ and rotated by 90° clockwise when viewed along the z direction from μ_1. It is clear from these figures that RCL is in phase with ξ_+ and can be stimulated by it, whereas LCL cannot. As a result, this case gives a negative VCD signal. This model requires at least two equivalent oscillators to exist in a molecule at an appropriate geometrical relation.

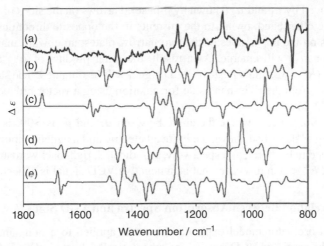

Figure 22.7 *Observed and calculated VCD spectra of (S)-β-pinene. (a) Observed spectrum, (b)–(e) calculated spectra. Methods of calculation are: (b) B3LYP/6-31+G**, (c) B3LYP/6-31, (d) HF/6-32+G*, (e) HF/4-31G.*

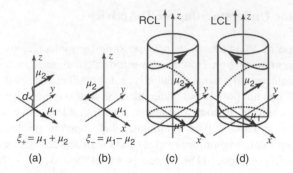

$$\xi_+ = \mu_1 + \mu_2 \qquad \xi_- = \mu_1 - \mu_2$$

(a)   (b)   (c)   (d)

Figure 22.8 *Coupled-oscillator model.*

There is another graphical representation of the coupled-oscillator model. The magnitude and sign of a VCD signal is determined by the rotational strength $R_{ns} = \mathrm{Im}(\mu_{sn} \cdot m_{sn})$. This is rewritten for the coupled-oscillator model as

$$R^{\pm} = \mathrm{Im}(\mu_{\pm} \cdot m_{\pm}) = \mp \frac{\pi}{2} \tilde{v} r_{12} (\mu_1 \times \mu_2) \qquad (22.16)$$

where $\pm$ means whether two transition dipole moments μ_1 and μ_2 are moving in phase (+) or out of phase (−), and r_{12} is the distance vector from μ_1 to μ_2. It is difficult to visualize the term $\mathrm{Im}(\mu_{\pm} \cdot m_{\pm})$ in Equation (22.16), but its image in the coupled-oscillator model is shown in Figure 22.9.

Electric currents are expected to be generated by the transition dipole moments. As shown in Figure 22.9a, if both μ_1 and μ_2 are directed upward but μ_2 is rotated counterclockwise when viewed from μ_1, and the angle between the two vectors is smaller than 90°, the induced dipole-moment vector μ_{ind} (or μ_+) will be directed upward as a sum of the electric currents produced by μ_1 and μ_2 (shown by thin solid arrows on the sides of the two narrow plates). On the other hand, owing to the currents in the opposite directions shown by the broken arrows on the upper sides of the two narrow plates, an upward magnetic moment vector m_{ind} (or m_+) will emerge. As a result, the inner product of μ_{ind} and m_{ind} has a positive value ($R^+ > 0$), and a positive VCD signal will be obtained. The same conclusion is obtained also from the right-hand side formulation using a vector product of μ_1 and μ_2 in Equation (22.16).

In Figure 22.9b, the case where the angle between μ_1 and μ_2 is >90° is shown. By the same reasoning as described above, the induced electric and magnetic dipole-moment vectors are expected to be directed, respectively, as indicated, right backward and left forward. Consequently, $R^+ < 0$ in this case; that is, a negative VCD signal is expected.

22.5.2 Correlation Between Absorption Spectra and VCD Signs

If the coupled-oscillator model explained above is applied to a nonplanar molecule of BAAB type, the predicted VCD spectra are schematically shown in Figure 22.10, together with their corresponding infrared absorption spectra. In this figure, v_s and v_a represent, respectively, the symmetric and antisymmetric AB stretching vibrations. It is common that v_a is at a higher wavenumber than v_s.

To apply Equation (22.16) to the model molecule shown in Figure 22.10, μ_1 and μ_2 are assumed, respectively, to be along the AB bonds shown schematically at the front and

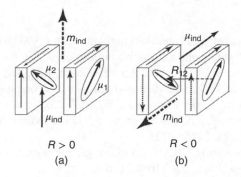

R > 0
(a)

R < 0
(b)

Figure 22.9 *Graphical representation of Im(μ · m) by the coupled-oscillator model.*

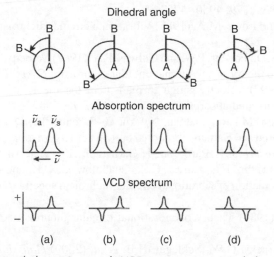

Dihedral angle

Absorption spectrum

VCD spectrum

(a)   (b)   (c)   (d)

Figure 22.10 *Infrared-absorption and VCD spectra expected for a molecule of the BAAB type.*

back. Here, the dihedral angles shown by counterclockwise arrows in (a) and (b) are taken as negative values and those shown by clockwise arrows in (c) and (d) as positive values. When the absolute value of the dihedral angle is smaller than 90°, the absorption intensity of v_s is stronger than that of v_a, as shown in (a) and (d). The reverse is the case where the absolute value of the dihedral angle is larger than 90°, as shown in (b) and (c). The sign of the VCD signal, on the other hand, follows a different rule; when the dihedral angle is negative as in (a) and (b), the VCD of v_s is positive and that of v_a negative, and when the dihedral angle is positive as in (c) and (d), the VCD of v_s is negative and that of v_a positive.

It is said that this model is valid when the stretching modes of two equivalent (or nearly so) AB bonds or the bending modes of two equivalent (or nearly so) AAB groups exist in a molecule, particularly if B is a hydrogen atom. However, this model is not necessarily applicable to other cases.

Efforts to obtain a better model, which will lead to an intuitive understanding of observed VCD results, are still under way [17, 19].

References

1. Barron, L.D., Bogaard, M.P. and Buckingham, A.D. (1973) Raman scattering of circularly polarized light by optically active molecules. *J. Am. Chem. Soc.*, **95**, 603–605.
2. Holzwarth, G., Hsu, E.C., Mosher, H.S., Faulkner, T.R., and Moscowitz, A. (1974) Infrared Circular dichroism of carbon–hydrogen and carbon–deuterium stretching modes. Observations. *J. Am. Chem. Soc.*, **96**, 251–252.
3. Nafie, L.A., Keiderling, T.A. and Stephens, P.J. (1976) Vibrational circular dichroism. *J. Am. Chem. Soc.*, **98**, 2715–2723.
4. Nafie, L.A. and Freedman, T.B. (1983) Vibronic coupling theory of infrared vibrational intensities. *J. Chem. Phys.*, **78**, 7108–7116.
5. Nafie, L.A. (1983) Adiabatic behavior beyond the Born-Oppenheimer approximation. Complete adiabatic wavefunctions and vibrationally induced electronic current density. *J. Chem. Phys.*, **79**, 4950–4957.
6. Stephens, P.J. and Lowe, M.A. (1985) Vibrational circular dichroism. *Annu. Rev. Phys. Chem.*, **36**, 213–241.
7. Buckingham, A.D., Fowler, P.W. and Galwas, P.A. (1987) Velocity-dependent property surfaces and the theory of vibrational circular dichroism. *Chem. Phys.*, **112**, 1–14.
8. Nafie, L.A. (1992) Velocity-gauge formalism in the theory of vibrational circular dichroism and infrared absorption. *J. Chem. Phys.*, **96**, 5687–5702.
9. Nafie, L.A., Cita, M., Ragunathan, N., Yu, G.-S., and Che, D. (1994) Instrumental method of infrared and Raman vibrational optical activity, in *Analytical Applications of Circular Dichroism* (eds N. Purdie and H.G. Brittain), Elsevier, Amsterdam, pp. 53–90.
10. Stephens, P.J., Devlin, F.J., Ashvar, C.S., Chabalowski, C.F., and Frisch, M.J. (1994) Theoretical calculation of vibrational circular dichroism spectra. *Faraday Discuss.*, **99**, 103–119.
11. Stephens, P.J. (1985) Theory of vibrational circular dichroism. *J. Phys. Chem.*, **89**, 748–752.
12. Frisch, M.J., Trucks, G.W., Schlegel, H.B. *et al.* (2009) *Gaussian 09, Revision A.1.* Gaussian, Inc., Wallingford, CT.
13. Nafie, L.A. and Vidrine, D.W. (1982) Double modulation Fourier transform spectroscopy, in *Fourier Transform Infrared Spectroscopy*, Vol. **3** (eds J.R. Ferraro and L.J. Basile), Academic Press, New York, pp. 83–123.
14. Nafie, L.A., Dukor, R.K. and Freedman, T.B. (2002) Vibrational circular dichroism, in *Handbook of Vibrational Spectroscopy*, Vol. **1** (eds J.M. Chalmers and P.R. Griffiths), John Wiley & Sons, Ltd, Chichester, pp. 731–744.
15. Jalkanen, K.J. and Stephens, P.J. (1988) Gauge dependence of vibrational rotational strength: NHDT. *J. Chem. Phys.*, **92**, 1781–1785.
16. Polavarapu, P.L. (1998) *Vibrational Spectra: Principles and Applications with Emphasis on Optical Activity*, Elsevier, Amsterdam.
17. Nafie, L.A. (2011) *Vibrational Optical Activity. Principles and Applications*, John Wiley & Sons, Ltd, Chichester.
18. Holzwarth, G. and Chabay, I. (1972) Optical activity of vibrational transitions: a coupled oscillator model. *J. Chem. Phys.*, **57**, 1632–1635.
19. Freedman, T.B., Lee, E. and Nafie, L.A. (2000) Vibrational transition current density in (*S*)-methyl lactate: visualizing the origin of the methine-stretching vibrational circular dichroism intensity. *J. Phys. Chem. A*, **104**, 3944–3951.

Part III

Appendices

Appendix A

The Speed, Frequency, Wavelength, and Wavenumber of an Electromagnetic Wave

Mitsuo Tasumi

Professor Emeritus, The University of Tokyo, Japan

If the speed, frequency, wavelength, and wavenumber of an electromagnetic wave, including a light wave, in a medium is denoted, respectively, by c, v, λ, and $\tilde{v}$, the following relations exist between them.

$$v\lambda = c \tag{A1}$$

$$\tilde{v} = \frac{1}{\lambda} = \frac{v}{c} \tag{A2}$$

The speed of an electromagnetic wave in a medium c and that in vacuum c_0 are related as

$$c = \frac{c_0}{n} \tag{A3}$$

where n is the refractive index of the medium. Strictly speaking, n is the real part of the complex refractive index (see Section 1.2.4 and Appendix B). The refractive index of vacuum is unity by definition, and that of air is very close to unity. Although most infrared spectroscopic measurements are performed in air, Fourier transform infrared (FT-IR) spectrometers usually provide wavenumbers in vacuum by calibration procedures.

It is clear from Equation (A3) that the speed c of an electromagnetic wave in a medium changes from that in vacuum c_0, but the dependence of the frequency v, wavelength λ, and wavenumber $\tilde{v}$ on the medium is not self-explanatory in Equations (A1) and (A2). It should be remembered that the wavelength and wavenumber depend on the medium but

---

Introduction to Experimental Infrared Spectroscopy: Fundamentals and Practical Methods,
First Edition. Edited by Mitsuo Tasumi and Akira Sakamoto.
© 2015 John Wiley & Sons, Ltd. Published 2015 by John Wiley & Sons, Ltd.

the frequency does not. The minimum amount of the energy of an electromagnetic wave is determined by the energy of a photon $h\nu$ (h is the Planck constant) and this energy does not depend on the medium. Accordingly, the frequency ν also does not depend on the medium.

An example is given to show numerically the relation between the speed, frequency, wavelength, and wavenumber of an infrared radiation. For the sake of simplicity, the speed in vacuum c_0 is taken to be $3 \times 10^8 \, \mathrm{m\,s^{-1}}$ instead of the more precise value of $299\,792\,458 \, \mathrm{m\,s^{-1}}$. By Equation (A1), the electromagnetic wave with $\nu = 3 \times 10^{13} \, \mathrm{Hz\,(s^{-1})}$ corresponds to an infrared radiation with $\lambda = 10 \, \mu\mathrm{m}$. By Equation (A2), $\tilde{\nu}$ corresponding to this wavelength is $10^5 \, \mathrm{m^{-1}} = 1000 \, \mathrm{cm^{-1}}$. In a medium of $n = 2.0$, $c = 1.5 \times 10^8 \, \mathrm{m\,s^{-1}}$, $\lambda = 5 \, \mu\mathrm{m}$, and $\tilde{\nu} = 2000 \, \mathrm{cm^{-1}}$.

As described in Section 1.2.1, the abscissa axis of an infrared spectrum should be designated as "Wavenumber/$\mathrm{cm^{-1}}$." In assigning absorption bands in an observed infrared spectrum to molecular vibrations giving rise to the absorption bands, it has been the usual practice to express molecular vibrational frequencies in units of $\mathrm{cm^{-1}}$. However, it is not logical to use $\mathrm{cm^{-1}}$ as the unit for frequencies. Therefore, the term *vibrational wavenumber*/$\mathrm{cm^{-1}}$ is increasingly used instead of *vibrational frequency* ($\mathrm{cm^{-1}}$). However, this terminology may appear strange since a molecular vibration does not propagate as a wave. Then, it should be understood that the term *vibrational wavenumber*/$\mathrm{cm^{-1}}$ is defined as "vibrational frequency divided by c_0." Under this understanding, "vibrational wavenumber/$\mathrm{cm^{-1}}$" seems to be more acceptable than "vibrational frequency ($\mathrm{cm^{-1}}$)."

Appendix B

Formulae Expressing the Electric Field of an Electromagnetic Wave and Related Subjects

Mitsuo Tasumi
Professor Emeritus, The University of Tokyo, Japan

In this appendix, formulae expressing the electric field of an electromagnetic wave and their physical meaning are discussed. Attention is paid to the relation between the formulae of the electric field of an electromagnetic wave and formulae expressing the complex refractive index. Formulae for circularly polarized radiations are also examined.

B.1 Formulae Expressing a Propagating Electromagnetic Wave

In Section 1.2.4, the electric field E of a light wave (generally an electromagnetic wave) propagating in the x direction is expressed by the following formula without giving any explanation. In this appendix, the z axis is chosen, instead of the x axis, as the direction of the propagation of the electromagnetic wave.

$$E = A \exp 2\pi i \left(\frac{z}{\lambda} - vt \right) \tag{B1}$$

In this equation, t is time, and A, v, and λ are, respectively, the amplitude vector, frequency, and wavelength of the electromagnetic wave. This equation may be found in any standard textbook on optics and electromagnetism. In many books, for the sake of the simplicity of expression, the right-hand side is given as $A \exp i(kz - \omega t)$, where k and ω stand for $2\pi/\lambda$ and $2\pi v$, respectively. As an alternative form of Equation (B1), the following equation is

Introduction to Experimental Infrared Spectroscopy: Fundamentals and Practical Methods,
First Edition. Edited by Mitsuo Tasumi and Akira Sakamoto.

also often found.

$$E = A \exp 2\pi i \left(vt - \frac{z}{\lambda} \right) \tag{B2}$$

where the sign of the exponent is reversed from that of Equation (B1), but, as shown later, Equations (B1) and (B2) are not different from each other in their physical contents. It is worth pointing out, however, that the forms of the right-hand sides of Equations (B1) and (B2) determine the definitions of the complex refractive index $\hat{n}$. If the form of Equation (B1) is chosen, it is natural to define the complex refractive index as $\hat{n} = n + ik$ (as in Equation (1.8)). If the form of Equation (B2) is chosen, the alternative type of definition, $\hat{n} = n - ik$, should be used to derive Equation (1.9). The former definition, $\hat{n} = n + ik$, is recommended by IUPAC [1]. This implicitly means that the electric field should now be expressed in the form of Equation (B1).

The electric field of an electromagnetic wave is usually expressed by an exponential function, mainly because the exponential function is convenient for various calculations, but it is better to express the electric field in the following form by using the cosine and sine functions in order to understand the propagation of the electromagnetic wave.

$$E_y = e_y A \left[\cos 2\pi \left(\frac{z}{\lambda} - vt \right) + i \sin 2\pi \left(\frac{z}{\lambda} - vt \right) \right] \tag{B3}$$

where e_y is the unit vector in the y direction, so this equation expresses the electric field vector in the y direction. The angle part $2\pi(z/\lambda - vt)$ of the cosine (and sine) function is called its *phase*.

Equation (B3) expresses a wave propagating in the z direction. To understand this point, it is best to draw figures corresponding to this equation. The cosine part of Equation (B3) is depicted for $t = 0$ and $t = 1/(8v)$ in Figure B.1a, and the sine part is depicted in a similar way in Figure B.1b. In these figures, a wavelength is divided into eight, so that how the cosine and sine waves advance with time can be clearly shown. With a time progress of $1/(8v)$, both the cosine and sine waves move to the right-hand side by $\lambda/8$. Thus, it is clear that Equation (B3) expresses a wave advancing in the positive direction of the z axis.

Equation (B2) may be rewritten in the following form.

$$E_y = e_y A \left[\cos 2\pi \left(vt - \frac{z}{\lambda} \right) + i \sin 2\pi \left(vt - \frac{z}{\lambda} \right) \right] \tag{B4}$$

The phases of the cosine and sine functions in this equation are reversed from those of Equation (B3). However, the cosine part is equal to that of Equation (B3), as the cosine function is an even function, and the sine part differs from that of Equation (B3) only in sign, as the sine function is an odd function. Thus, both the cosine and sine parts of Equation (B4) express waves advancing in the positive direction of the z axis. In other words, Equation (B4) (or Equation (B2)) has the same physical meaning as Equation (B3) (or Equation (B1)).

The electric field of an electromagnetic wave advancing in the direction opposite to that of the wave expressed in Equation (B1) (or Equation (B3)) may be expressed in the following form.

$$E_y = e_y A \exp 2\pi i \left(\frac{z}{\lambda} + vt \right)$$
$$= e_y A \left[\cos 2\pi \left(\frac{z}{\lambda} + vt \right) + i \sin 2\pi \left(\frac{z}{\lambda} + vt \right) \right] \tag{B5}$$

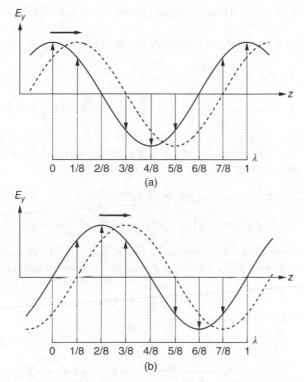

Figure B.1 *Electric field of an electromagnetic wave advancing in the positive direction of the z axis. (a) The shapes of* $\cos 2\pi(z/\lambda - vt)$ *at* $t = 0$ *(solid curve) and* $t = 1/(8v)$ *(dashed curve) and (b) the shapes of* $\sin 2\pi(z/\lambda - vt)$ *at* $t = 0$ *(solid curve) and* $t = 1/(8v)$ *(dashed curve).*

If figures corresponding to the cosine and sine functions in this equation are depicted in the same way as in Figure B.1, the results show that the cosine and sine functions of this equation express waves advancing in the negative direction of the z axis.

B.2   Linearly and Circularly Polarized Radiations

The electric field vector of the electromagnetic wave expressed by Equation (B3) is parallel to the y axis, and the wave propagates in the yz plane. Such an electromagnetic wave is called *linearly polarized* or *plane-polarized radiation*. An electromagnetic wave polarized parallel to the x axis also exists, and its electric vector is denoted by E_x. If a sample for an infrared absorption measurement has any molecular orientation, it generally shows different absorptions for the x-polarized and y-polarized infrared radiations. For example, a stretched (uniaxially oriented) thin polymer film usually shows different absorptions for infrared radiations polarized parallel and perpendicular to the stretching direction. This difference is called *infrared dichroism*.

The electric field of an electromagnetic wave advancing in the z direction, denoted by E, is generally defined as $E = E_x + E_y$, where the x, y, and z axes form a right-handed system. Let us consider a case where E_x and E_y have amplitudes of an equal magnitude and have a phase difference of $\pm\pi/2$. These electromagnetic waves are called *circularly polarized radiations*.

In the case where E_y has a phase difference of $-\pi/2$ against E_x, E_x and E_y are given as

$$E_x = e_x A \cos 2\pi \left(\frac{z}{\lambda} - vt \right) \tag{B6}$$

$$E_y = e_y A \cos 2\pi \left(\frac{z}{\lambda} - vt - \frac{1}{4} \right) = e_y A \sin 2\pi \left(\frac{z}{\lambda} - vt \right) \tag{B7}$$

If $t = -1/(4v) + t_0$, the phase of the cosine wave of E_y in Equation (B7) is equal to $2\pi[(z/\lambda) - vt_0]$. This means that the phase of the cosine wave of E_y is in advance of that of E_x (by $\pi/2$ as postulated). The electric vector E is given as

$$E = E_x + E_y$$

$$= A \left[e_x \cos 2\pi \left(\frac{z}{\lambda} - vt \right) + e_y \sin 2\pi \left(\frac{z}{\lambda} - vt \right) \right] \tag{B8}$$

E_x and E_y in this equation are shown in Figure B.2a for $t = 0$. E synthesized from E_x and E_y is shown in Figure B.2b, where the E vectors (shown by arrows) are projected on

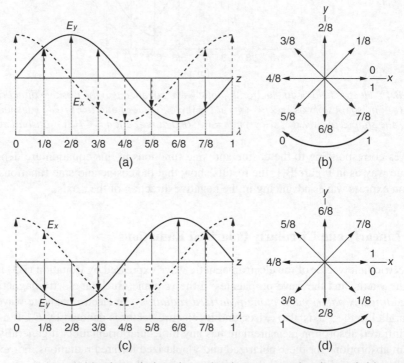

Figure B.2 *Electric fields of circularly polarized radiations. (a) The electric field components of a right-circular radiation at $t = 0$ depicted on the same plane. E_x (dashed curve) is perpendicular to the plane of the page and directed toward its back side, and E_y (solid curve) is on the page plane, (b) the electric field $E = E_x + E_y$ of a right-circular radiation projected on the xy plane, (c) the electric field components of a left-circular radiation at $t = 0$ depicted in the same way as in (a), and (d) the electric field E of a left-circular radiation projected on the xy plane.*

the *xy* plane. If the E vectors advancing along the z axis are looked back by an observer toward the source of radiation, they show a right-handed (clockwise) rotation on going from $z = \lambda$ to $z = 0$. Such an electromagnetic wave is called *right-circularly polarized radiation* (abbreviated as *right-circular radiation*).

A left-circularly polarized (or left-circular) radiation is obtained if E_y has a phase difference of $+\pi/2$ in relation to E_x. In this case, E_y and E are given as

$$E_y = e_y A \cos 2\pi \left(\frac{z}{\lambda} - vt + \frac{1}{4} \right) = -e_y A \sin 2\pi \left(\frac{z}{\lambda} - vt \right) \tag{B9}$$

$$E = E_x + E_y$$

$$= A \left[e_x \cos 2\pi \left(\frac{z}{\lambda} - vt \right) - e_y \sin 2\pi \left(\frac{z}{\lambda} - vt \right) \right] \tag{B10}$$

E_y in Equation (B9), together with E_x, is shown for $t = 0$ in Figure B.2c. The E vectors projected on the *xy* plane are shown in Figure B.2d, where they show a left-handed (counterclockwise) rotation on going from $z = \lambda$ to $z = 0$, as expected for a left-circular radiation.

The curve formed by connecting the tips of the E vectors along the z axis gives a helix. The sense of rotation is reversed in helices corresponding to right- and left-circular radiations. The difference between infrared absorptions for right- and left-circular infrared radiations is called *infrared circular dichroism* or *vibrational circular dichroism* (*VCD*). Measurements of VCD are described in Chapter 22.

Reference

1. International Union of Pure and Applied Chemistry and Physical and Biophysical Chemistry Division (2007) *Quantities, Units and Symbols in Physical Chemistry*, 3rd edn, RSC Publishing, London.

Appendix C

Coherence of the Thermal Radiation

Mitsuo Tasumi
Professor Emeritus, The University of Tokyo, Japan

The explanation of the mechanism of the Michelson interferometer given in Section 4.3.1 does not specify the source of infrared radiation but it is clear that the discussion implicitly assumes the "coherence" of the infrared radiation emitted from the source; that is, the discussion is based on the assumption that the interference of the two beams divided by the beamsplitter occurs at any optical path difference (OPD) even if it is very long. It is generally believed that thermal radiation is "incoherent," whereas radiation from a laser is "coherent." An ordinary source, which emits "incoherent" infrared radiation at a high temperature, is used with any Fourier transform infrared (FT-IR) spectrometer utilizing a Michelson interferometer. Then, a question may arise as to why interference is expected to occur between the two beams of "incoherent" radiation within the Michelson interferometer. The simplest answer to this question would be to state that the thermal radiation has a considerable degree of coherence, which ensures the interference between the two beams in the Michelson interferometer. Experimentally, there is no question at all about the fact that the two beams interfere with each other at an OPD of 1 m or shorter lengths, which are those usually employed for routine infrared measurements. FT-IR spectrometers with Michelson interferometers which can realize much longer OPDs are commercially available.

It is beyond the scope of this book to discuss in depth the degree of coherence of the thermal radiation. Readers interested in the theory of coherence are advised to consult Refs [1–3].

Introduction to Experimental Infrared Spectroscopy: Fundamentals and Practical Methods,
First Edition. Edited by Mitsuo Tasumi and Akira Sakamoto.
© 2015 John Wiley & Sons, Ltd. Published 2015 by John Wiley & Sons, Ltd.

References

1. Wolf, E. (2007) *Introduction to the Theory of Coherence and Polarization of Light*, Cambridge University Press, Cambridge.
2. Born, M. and Wolf, E. (2006) *Principles of Optics*, 7th (expanded) edn, Cambridge University Press, Cambridge.
3. Hecht, E. (2002) *Optics*, 4th edn, Addison-Wesley, San Francisco, CA.

Appendix D

Mathematical Methods in FT-IR Spectrometry

Mitsuo Tasumi
Professor Emeritus, The University of Tokyo, Japan

It is desirable to have some knowledge of the mathematical methods associated with Fourier transforms to understand completely the descriptions in this book, particularly the descriptions in Chapters 4 and 6. A basic description of the mathematical equations used in these chapters is given in this appendix. Extensive information on Fourier transforms and related subjects is found in Refs [1, 2].

D.1 The Fourier Transform

D.1.1 Definition of Fourier Transforms

In defining the Fourier transform of a function representing a physical property, two variables having reciprocal dimensions are involved. These are called *conjugate variables*. For example, if a variable has the dimension of length (L), the other has the dimension of inverse length (L^{-1}). If s and t stand for the two variables, the most general way of defining the Fourier transform and its inverse is to use complex exponentials such as

$$F(t) = \int_{-\infty}^{\infty} f(s)e^{-2\pi i s t}\mathrm{d}s \tag{D1a}$$

$$f(s) = \int_{-\infty}^{\infty} F(t)e^{2\pi i s t}\mathrm{d}t \tag{D1b}$$

Introduction to Experimental Infrared Spectroscopy: Fundamentals and Practical Methods,
First Edition. Edited by Mitsuo Tasumi and Akira Sakamoto.
© 2015 John Wiley & Sons, Ltd. Published 2015 by John Wiley & Sons, Ltd.

In these equations, $F(t)$ is the Fourier transform of $f(s)$, and $f(s)$ is the inverse Fourier transform of $F(t)$. These two functions are called the *Fourier transform pair*. It is to be noted that a function of one of the two variables is transformed into another function of the other variable. From a physical viewpoint, the Fourier transform is a mathematical operation to extract a function representing a physical property buried in the other function representing another physical property.

Strictly speaking, the right-hand side of Equation (D1b) should be multiplied by $1/2\pi$. In another way of defining the Fourier transform, the right-hand sides of both Equations (D1a) and (D1b) are multiplied by $1/\sqrt{2\pi}$. However, these constants are omitted not only here but also throughout this book because the numerical values of $F(t)$ and $f(s)$ are not important for understanding the issues discussed in this book.

Let us assume that $f(s)$ is an even function and note the relation that $\exp(\pm 2\pi i s t) = \cos 2\pi s t \pm i \sin 2\pi s t$. Then, $\exp(\pm 2\pi i s t)$ in Equations (D1a) and (D1b) can be replaced with $\cos 2\pi s t$, because the part of the integrand containing $\sin(2\pi s t)$ is an odd function with respect to either s or t, and the integration of this part from $-\infty$ to ∞ tends to zero. As a result, the following two equations are obtained, and the Fourier transform in this expression is called the *Fourier cosine transform*. In this expression, no distinction exists between the Fourier transform and its inverse transform.

$$F(t) = \int_{-\infty}^{\infty} f(s) \cos 2\pi s t \, ds \qquad (D2a)$$

$$f(s) = \int_{-\infty}^{\infty} F(t) \cos 2\pi s t \, dt \qquad (D2b)$$

It is clear that Equations (4.11) and (4.12) have the same form as Equations (D2a) and (D2b). In Equations (4.11) and (4.12), the two variables x and $\tilde{v}$ have reciprocal dimensions and are usually given in units of cm and cm^{-1}, respectively.

A variety of notations are in use to indicate the operation of a Fourier transform. In this book, two kinds of notations are used. One is to put a bar above a function as in $\overline{f(s)}$, which is equivalent to the right-hand side of Equation (D1a). The other is to put a symbol $\mathfrak{F}$ before a function as in $\mathfrak{F}\,[f(s)]$, which is again equivalent to the right-hand side of Equation (D1a). A symbol $\mathfrak{F}^{-1}$ is used to indicate the operation of an inverse transform. Thus, $\mathfrak{F}^{-1}[F(t)]$ is equivalent to the right-hand side of Equation (D1b). An advantage of using $\mathfrak{F}$ and $\mathfrak{F}^{-1}$ is that this notation makes it possible to specify the Fourier transform and its inverse transform, if necessary. $\mathfrak{F}$ and $\mathfrak{F}^{-1}$ may be regarded as operators, and the following relations hold: $\mathfrak{F}^{-1}\mathfrak{F} = 1$ and $\mathfrak{F}\,\mathfrak{F}^{-1} = 1$.

D.1.2   Useful Properties Relating to the Fourier Transform

There are many useful properties relating to the Fourier transform, but only the following properties are introduced here, because they are related to the discussion in Section D.3.3.

Let us examine what happens to the Fourier transform $\mathfrak{F}\,[f(s)] = F(t)$, if s is shifted by an arbitrary magnitude s_0 to become $s - s_0$. In the following derivation, $s - s_0$ and $t - t_0$ are substituted, respectively, with s' and t'.

$$\mathfrak{F}\,[f(s - s_0)] = \int_{-\infty}^{\infty} f(s - s_0)\, e^{-2\pi i s t} ds = \int_{-\infty}^{\infty} f(s')\, e^{-2\pi i (s' + s_0)t} ds'$$

$$= e^{-2\pi i s_0 t} \int_{-\infty}^{\infty} f(s') e^{-2\pi i s' t} ds' = e^{-2\pi i s_0 t} F(t) \qquad (D3a)$$

Next, the Fourier transform of $e^{2\pi i s t_0} f(s)$ is considered.

$$\mathfrak{F}\left[e^{2\pi i s t_0} f(s)\right] = \int_{-\infty}^{\infty} f(s) e^{2\pi i s t_0} e^{-2\pi i s t} ds$$

$$= \int_{-\infty}^{\infty} f(s) e^{-2\pi i s(t-t_0)} ds = \int_{-\infty}^{\infty} f(s) e^{-2\pi i s t'} ds$$

$$= F(t') = F(t - t_0) \tag{D3b}$$

This equation is used in Section D.3.2. It means that the inverse Fourier transform of $F(t - t_0)$ is $e^{2\pi i s t_0} f(s)$.

D.2 Convolution and the Convolution Theorem

Convolution and the convolution theorem are related to various descriptions in Chapters 4 and 6.

Convolution of two functions $f(s)$ and $g(s)$, usually expressed as $f(s) * g(s)$, is an integral defined as

$$f(s) * g(s) = \int_{-\infty}^{\infty} f(p) \cdot g(s - p) dp \tag{D4}$$

The commutative rule holds for convolution. That is

$$f(s) * g(s) = g(s) * f(s) \tag{D5}$$

The convolution theorem is important for understanding basic procedures used in Fourier transform infrared (FT-IR) spectrometry. It is concerned with the Fourier transforms of a product of two functions $f(s) \cdot g(s)$ and convolved two functions $f(s) * g(s)$. It is given as the following two equations.

$$\overline{f(s) \cdot g(s)} = \overline{f(s)} * \overline{g(s)} \tag{D6a}$$

$$\overline{f(s) * g(s)} = \overline{f(s)} \cdot \overline{g(s)} \tag{D6b}$$

These are of identical format with Equations (4.13a) and (4.13b). Equation (D6a) is used in deriving a few key relations in Chapter 4, and Equation (D6b) is used in Chapter 6 for explaining the method of Fourier self-deconvolution.

To prove Equation (D6a), it is easier to derive the inverse Fourier transform of the right-hand side of this equation than to derive the Fourier transform of the left-hand side of this equation. By definition, the following equations hold.

$$\mathfrak{F}[f(s)] = \int_{-\infty}^{\infty} f(s) e^{-2\pi i s t} ds = F(t) \tag{D7a}$$

$$\mathfrak{F}[g(s)] = \int_{-\infty}^{\infty} g(s) e^{-2\pi i s t} ds = G(t) \tag{D7b}$$

$$\mathfrak{F}^{-1}[F(t)] = \int_{-\infty}^{\infty} F(t) e^{2\pi i s t} dt = f(s) \tag{D8a}$$

$$\mathfrak{F}^{-1}[G(t)] = \int_{-\infty}^{\infty} G(t) e^{2\pi i s t} dt = g(s) \tag{D8b}$$

Then, the right-hand side of Equation (D6a) is equal to $F(t) * G(t)$, which is by definition

$$F(t) * G(t) = \int_{-\infty}^{\infty} F(q)G(t-q)dq \tag{D9}$$

The inverse Fourier transform of $F(t) * G(t)$ is calculated in the following way.

$$\mathfrak{F}^{-1}[F(t) * G(t)] = \int_{-\infty}^{\infty} \left[\int_{-\infty}^{\infty} F(q)\,G(t-q)dq \right] e^{2\pi ist} dt \tag{D10}$$

By changing variable as $t - q = y$ and $dt = dy$, the right-hand side of Equation (D10) is rewritten as

$$\int_{-\infty}^{\infty} F(q) \left[\int_{-\infty}^{\infty} G(y)\,e^{2\pi is(q+y)}dy \right] dq$$

$$= \int_{-\infty}^{\infty} F(q)e^{2\pi iqs} \left[\int_{-\infty}^{\infty} G(y)\,e^{2\pi isy}dy \right] dq$$

$$= \left[\int_{-\infty}^{\infty} F(q)\,e^{2\pi iqs}dq \right] \left[\int_{-\infty}^{\infty} G(y)\,e^{2\pi isy}dy \right] \tag{D11}$$

The integrals in the first and second square brackets in Equation (D11) have the forms identical with the integrals in Equations (D8a) and (D8b), respectively; if q and y are changed to t, the first and second integrals become equal to the integrals in Equations (D8a) and (D8b), respectively. This means that the product of integrals in Equation (D11) are equal to $f(s) \cdot g(s)$. Finally, the following equations, both of which are identical with Equation (D6a), are obtained.

$$\mathfrak{F}^{-1}[F(t) * G(t)] = f(s) \cdot g(s) \tag{D12a}$$

$$\mathfrak{F}[f(s) \cdot g(s)] = F(t) * G(t) \tag{D12b}$$

Equation (D6b) can be derived in a way similar to the procedure described above.

D.3   Functions Used in FT-IR Spectrometry and their Fourier Transforms

D.3.1   The Rectangular Function and its Fourier Transform

The rectangular function and its Fourier transform are referred to in Section 4.4.1.1. The rectangular function $\Pi_a(x)$ on the x axis is defined in the following way.

$$\Pi_a(x) = 0 \left(-\infty < x < -\frac{a}{2}\right), \ 1\left(-\frac{a}{2} \leq x \leq \frac{a}{2}\right), \ \text{and} \ 0\left(\frac{a}{2} < x < \infty\right) \tag{D13}$$

The shape of this function with $a = 2D$ is shown in Figure 4.8. This function has other names such as the boxcar, gate, rect, rectangle, top-hat, and window functions. It is noted that the following relation holds.

$$\left(\frac{1}{a}\right) \int_{-\infty}^{\infty} \Pi_a(x)dx = 1 \tag{D14}$$

The Fourier transform of this function is obtained in the following way by taking $\tilde{v}$ as the conjugate variable of x.

$$\overline{\Pi_a(x)} = \Phi_a(\tilde{v}) = \int_{-\infty}^{\infty} \Pi_a(x)e^{-2\pi i \tilde{v} x}dx = \int_{-a/2}^{a/2} e^{-2\pi i \tilde{v} x}dx$$

$$= \frac{1}{2\pi i \tilde{v}}(e^{\pi i \tilde{v} a} - e^{-\pi i \tilde{v} a}) = a\left(\frac{\sin \pi \tilde{v} a}{\pi \tilde{v} a}\right) \qquad (D15)$$

The function having a general form of $(1/x)\sin(x)$ is called the *sinc function* and expressed by sinc(x). Then,

$$\Phi_a(\tilde{v}) = (a)\,\text{sinc}(\pi \tilde{v} a) \qquad (D16)$$

$\Pi_{2D}(x)$ used in Chapter 4 is obtained by substituting $2D$ for a in $\Pi_a(x)$. Its Fourier transform $\Phi_{2D}(\tilde{v})$ is given as $\Phi_{2D}(\tilde{v}) = (2D)\text{sinc}(2\pi \tilde{v} D)$. The shape of this function is shown in Figure 4.9.

D.3.2   The Dirac Delta Function and Its Properties

The Dirac delta function and its properties are related to various descriptions in Chapters 4 and 6. They are also essential in the description to be given in Section D.3.3.

The Dirac delta function or simply the delta function $\delta(x)$ is related to the rectangular function $\Pi_a(x)$ as

$$\delta(x) = \lim_{a \to 0} \left(\frac{1}{a}\right) \Pi_a(x) \qquad (D17)$$

From this equation and Equation (D14), the following properties of the delta function are derived. It is common to define the delta function by these properties.

$$\delta(x) = 0 \quad \text{for} \quad x \neq 0 \quad \text{and} \quad \delta(0) = \infty \qquad (D18)$$

$$\int_{-\infty}^{\infty} \delta(x)dx = 1 \qquad (D19)$$

The shape of $\delta(x)$ is shown in Figure D.1.

The delta function has many important properties, but only a few of them are mentioned here.

$$\delta(x - a) = 0 \quad \text{for} \quad x \neq a \text{ and } \infty \text{ for } x = a \qquad (D20a)$$

$$\int_{-\infty}^{\infty} \delta(x - a)dx = 1 \qquad (D20b)$$

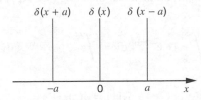

Figure D.1 *Schematic illustration of delta functions. Each vertical line represents an infinite length with an infinitely narrow width.*

These two equations mean that the delta function is 0 unless the variable in parentheses is 0. It should be noted that the following equations hold.

$$\delta(x - a) = \delta(-x + a) = \delta(a - x) \tag{D21}$$

The shapes of $\delta(x - a)$ and $\delta(x + a)$ $(a > 0)$ are shown in Figure D.1.

The delta function has a property of sifting a value from a function in the following way.

$$\int_{-\infty}^{\infty} f(x)\delta(x - a)\mathrm{d}x = f(a) \tag{D22}$$

As the integrand of the left-hand side of Equation (D22) is nonzero only for $x = a$, the result of integration becomes $f(a)$ because of Equation (D20b).

Equation (D22) can be used to derive the Fourier transform of $\delta(x - a)$ by taking $\tilde{v}$ as the conjugate variable of x.

$$\overline{\delta(x - a)} = \int_{-\infty}^{\infty} \delta(x - a)\mathrm{e}^{-2\pi\mathrm{i}\tilde{v}x}\mathrm{d}x = \mathrm{e}^{-2\pi\mathrm{i}\tilde{v}a} \tag{D23a}$$

Then,

$$\overline{\delta(x)} = 1 \tag{D23b}$$

It is noted that Equations (D23a) and (D23b) are consistent with Equation (D3a). It follows from Equation (D23b) that

$$\mathfrak{F}^{-1}[1] = \int_{-\infty}^{\infty} 1 \cdot \mathrm{e}^{2\pi\mathrm{i}\tilde{v}x}\mathrm{d}\tilde{v} = \delta(x) \tag{D23c}$$

which may be rewritten as

$$\delta(x) = \int_{-\infty}^{\infty} \mathrm{e}^{2\pi\mathrm{i}\tilde{v}x}\mathrm{d}\tilde{v} \tag{D23d}$$

This equation is a special case of the following equation, which is obtained as the inverse transform of Equation (D23a).

$$\delta(x - a) = \mathfrak{F}^{-1}[\overline{\delta(x - a)}] = \int_{-\infty}^{\infty} \mathrm{e}^{-2\pi\mathrm{i}\tilde{v}a}\mathrm{e}^{2\pi\mathrm{i}\tilde{v}x}\mathrm{d}\tilde{v}$$

$$= \int_{-\infty}^{\infty} \mathrm{e}^{2\pi\mathrm{i}\tilde{v}(x-a)}\mathrm{d}\tilde{v} \tag{D23e}$$

Equation (D23d) has an important physical meaning (see Appendix E).

Next, $\mathfrak{F}[1]$ is considered.

$$\mathfrak{F}[1] = \int_{-\infty}^{\infty} 1 \cdot \mathrm{e}^{-2\pi\mathrm{i}\tilde{v}x}\mathrm{d}\tilde{v} = \int_{-\infty}^{\infty} 1 \cdot \mathrm{e}^{2\pi\mathrm{i}\tilde{v}(-x)}\mathrm{d}\tilde{v} = \delta(-x) \tag{D23f}$$

As $\delta(-x) = \delta(x)$,

$$\mathfrak{F}[1] = \delta(x) \tag{D23g}$$

Thus, both the Fourier transform and the inverse Fourier transform of 1 are the same delta function.

By combining Equation (D3b) with Equation (D23g), the following relation, which will become important in Section D.3.3, is obtained.

$$\mathfrak{F}[e^{2\pi i \tilde{v}_0 x} \cdot 1] = \mathfrak{F}[e^{2\pi i \tilde{v}_0 x}] = \delta(\tilde{v} - \tilde{v}_0) \tag{D24}$$

The following relation is called the *shift theorem* for convolution with the delta function.

$$f(x) * \delta(x - a) = f(x - a) \tag{D25}$$

This equation is derived by using the commutative rule for convolution in Equation (C4) and the sifting property in Equation (D22).

$$f(x) * \delta(x - a) = \delta(x - a) * f(x) = \int_{-\infty}^{\infty} \delta(p - a) f(x - p) \mathrm{d}p$$

$$= \int_{-\infty}^{\infty} f(x - p)\delta(p - a)\mathrm{d}p = f(x - a)$$

It is noted that the convolution of $f(x)$ with a delta function $\delta(x - a)$ translates $f(x)$ by a distance of a in the x direction.

D.3.3   The Dirac Delta Comb and its Fourier Transform

The Dirac delta comb and its Fourier transform are referred to in Section 4.4.2. An infinite train of the Dirac delta functions at intervals of a on the x axis is called the *Dirac delta comb* and denoted by $\unicode{x1E37}\unicode{x1E37}_a(x)$. This function is defined as

$$\unicode{x1E37}\unicode{x1E37}_a(x) = \sum_{n=-\infty}^{\infty} \delta(x - na) \tag{D26}$$

The shapes of $\unicode{x1E37}\unicode{x1E37}_a(x)$ and its Fourier transform $(1/a)\unicode{x1E37}\unicode{x1E37}_{1/a}(\tilde{v})$ are illustrated in Figure 4.13.
It is known that $\unicode{x1E37}\unicode{x1E37}_a(x)$ can be expressed as the following Fourier series.

$$\unicode{x1E37}\unicode{x1E37}_a(x) = \left(\frac{1}{a}\right) \sum_{n=-\infty}^{\infty} e^{2\pi i n\left(\frac{1}{a}\right)x} \tag{D27}$$

The Fourier transform of $\unicode{x1E37}\unicode{x1E37}_a(x)$ in the above equation is derived by using the relation in Equation (D24). In the following derivation, $\tilde{v}_0$ in Equation (D24) is substituted with $n(1/a)$.

$$\overline{\unicode{x1E37}\unicode{x1E37}_a(x)} = \mathfrak{F}\left[\left(\frac{1}{a}\right) \sum_{n=-\infty}^{\infty} e^{2\pi i n\left(\frac{1}{a}\right)x}\right] = \left(\frac{1}{a}\right) \sum_{n=-\infty}^{\infty} \mathfrak{F}\left[e^{2\pi i n\left(\frac{1}{a}\right)x}\right]$$

$$= \left(\frac{1}{a}\right) \sum_{n=-\infty}^{\infty} \delta\left(\tilde{v} - n\left(\frac{1}{a}\right)\right) = \left(\frac{1}{a}\right)\unicode{x1E37}\unicode{x1E37}_{1/a}(\tilde{v}) \tag{D28}$$

D.3.4   The Exponential Decay and the Lorentz Profile

The exponential decay and the Lorentz profile are referred to in the descriptions of the method of Fourier self-deconvolution in Chapter 6. The function $f(x)$ expressing an exponential decay is defined as

$$f(x) = e^{-2\pi\sigma x} \quad (\sigma > 0) \text{ for } x \geq 0, \quad 0 \text{ for } x < 0 \tag{D29}$$

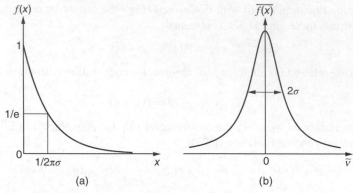

Figure D.2 *(a) An exponential decay function and (b) the Lorentz profile corresponding to the Fourier transform of the function in (a).*

The shape of this function is shown in Figure D.2a. The Fourier transform of this function is derived in the following way by taking $\tilde{v}$ as the conjugate variable of x.

$$\overline{f(x)} = \int_{-\infty}^{\infty} f(x)e^{-2\pi i \tilde{v} x} dx = \int_{0}^{\infty} e^{-2\pi \sigma x} e^{-2\pi i \tilde{v} x} dx$$

$$= \int_{0}^{\infty} e^{-2\pi(\sigma + i\tilde{v})x} dx = \left(-\frac{1}{2\pi}\right) \left[\frac{e^{-2\pi(\sigma + i\tilde{v})x}}{\sigma + i\tilde{v}}\right]_{0}^{\infty}$$

$$= \left(\frac{1}{2\pi}\right)\left(\frac{1}{\sigma + i\tilde{v}}\right) \tag{D30}$$

The real part (and the imaginary part as well) of Equation (D30) is important. If the constant $1/2\pi$ is disregarded, the real part is given as

$$\mathrm{Re}[\overline{f(x)}] = \frac{\sigma}{\sigma^2 + \tilde{v}^2} \tag{D31}$$

The right-hand side of this equation represents a curve centering about $\tilde{v} = 0$, which is shown in Figure D.2b. This curve is called the *Lorentz profile* which is known to fit approximately the shapes of infrared absorption bands. The integrated area under this curve is π, and its full width at half maximum (FWHM) is 2σ.

D.3.5 The Fourier Transform of the Cosine Function

The Fourier transform of the cosine function is mentioned in the description of the method of Fourier self-deconvolution in Chapter 6.

The Fourier transform of $\cos 2\pi \tilde{v}_0 x$ is derived by using the relation in Equation (D23e).

$$\overline{\cos 2\pi \tilde{v}_0 x} = \int_{-\infty}^{\infty} (\cos 2\pi \tilde{v}_0 x)e^{-2\pi i \tilde{v} x} dx = \int_{-\infty}^{\infty} \left(\frac{e^{2\pi i \tilde{v}_0 x} + e^{-2\pi i \tilde{v}_0 x}}{2}\right) e^{-2\pi i \tilde{v} x} dx$$

$$= \left(\frac{1}{2}\right)\left(\int_{-\infty}^{\infty} e^{2\pi i (\tilde{v}_0 - \tilde{v})x} + \int_{-\infty}^{\infty} e^{2\pi i (-\tilde{v}_0 - \tilde{v})x}\right) dx$$

$$= \left(\frac{1}{2}\right)[\delta(\tilde{v}_0 - \tilde{v}) + \delta(-\tilde{v}_0 - \tilde{v})]$$

$$= \left(\frac{1}{2}\right)[\delta(\tilde{v} - \tilde{v}_0) + \delta(\tilde{v} + \tilde{v}_0)] \tag{D32}$$

Thus, the Fourier transform of $\cos 2\pi\tilde{v}_0 x$ consists of two delta functions symmetrically located about $\tilde{v} = 0$ on the $\tilde{v}$ axis, but it is a single delta function at $\tilde{v} = \tilde{v}_0$ in the positive region of $\tilde{v}$.

D.3.6   The Fourier Transform of the Triangular Function

The triangular function $A(x)$ shown in Figure D.3a is defined as

$$A(x) = 1 - \frac{x}{a} \quad (0 \le x \le a), \quad 0 \ (x < 0, \ a < x) \tag{D33}$$

This function is used in the description of the method of Fourier self-deconvolution in Chapter 6, where a is replaced by X. The Fourier transform of this function can be calculated in the same way as was done for the exponential decay in Section D.3.4, and the result obtained is a complex function. The real part of the Fourier transform is given as

$$\mathrm{Re}[\overline{A(x)}] = \left(\frac{a}{2}\right)\frac{\sin^2(\pi\tilde{v}a)}{(\pi\tilde{v}a)^2} = \left(\frac{a}{2}\right)\mathrm{sinc}^2\pi\tilde{v}a \tag{D34}$$

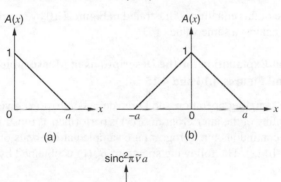

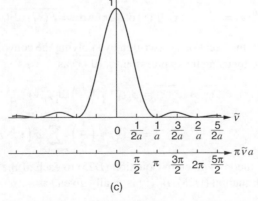

Figure D.3   *(a) A triangular function, (b) a symmetric triangular function, and (c) the Fourier transform of the symmetric triangular function in (b).*

In Figure D.3b, another triangle function is shown. This function consists of the triangle function in Equation (D33) and one symmetric to it with respect to $x = 0$. This symmetric triangular function corresponds to the apodizing function shown in Figure 4.11A(c). Its Fourier transform has no imaginary part and is simply given as $(a)\mathrm{sinc}^2\pi\tilde{v}a$. The shape of $\mathrm{sinc}^2\pi\tilde{v}a$ is shown in Figure D.3c.

D.4 Additional Explanations for a Few Descriptions in Chapters 4 and 6

D.4.1 Additional Explanation for the Description of Wavenumber Resolution and Figure 4.10

A monochromatic ray at wavenumber $\tilde{v}_1$ with intensity I_m may be expressed as $I_m\delta(\tilde{v} - \tilde{v}_1)$. If $B_e(\tilde{v})$ consists of only two monochromatic rays at $\tilde{v}_1$ and $\tilde{v}_2$, each of which has an equal intensity I_m, $B'_e(\tilde{v})$ in the region of $\tilde{v} > 0$ may be expressed by using Equations (4.16) and (D20b) as

$$B'_e(\tilde{v}) \quad (\tilde{v} > 0) = [I_m\delta(\tilde{v} - \tilde{v}_1) + I_m\delta(\tilde{v} - \tilde{v}_2)] * \Phi_{2D}(\tilde{v}) \tag{D35}$$

This equation may be rewritten in the following form by using the shift theorem for convolution with the delta function in Equation (C25).

$$B'_e(\tilde{v}) \quad (\tilde{v} > 0) = I_m\Phi_{2D}(\tilde{v} - \tilde{v}_1) + I_m\Phi_{2D}(\tilde{v} - \tilde{v}_2) \tag{D36}$$

The right-hand side of this equation is illustrated in Figure 4.10 by changing the separation between $\tilde{v}_1$ and $\tilde{v}_2$ against a same value of D.

D.4.2 Additional Explanation for the Descriptions of Measurable Wavenumber Region and Figures 4.14 and 4.15

For expressing the sampling operation, the Dirac delta comb $\text{Ш}_a(x)$ given in Section D.3.3 is useful. The sampling of the interferogram $F(x)$ is performed as illustrated in Figure 4.14, where points on a continuous interferogram are sampled at intervals of a. This process is expressed as $F(x) \cdot \text{Ш}_a(x)$. The following spectrum $B''_e(\tilde{v})$ is obtained by the Fourier transform of $F(x) \cdot \text{Ш}_a(x)$.

$$B''_e(\tilde{v}) = \int_{-\infty}^{\infty} F(x)\text{Ш}_a(x)\cos 2\pi\tilde{v}x \, \mathrm{d}x = \overline{F(x) \cdot \text{Ш}_a(x)} \tag{D37}$$

The right-hand side of this equation is rewritten by applying the convolution theorem given in Equation (D6a) and by using the expression for $\text{Ш}_a(x)$ as

$$\overline{F(x) \cdot \text{Ш}_a(x)} = \overline{F(x)} * \overline{\text{Ш}_a(x)} = B_e(\tilde{v}) * \left(\frac{1}{a}\right)\text{Ш}_{1/a}(\tilde{v})$$

$$= B_e(\tilde{v}) * \left(\frac{1}{a}\right) \sum_{n=-\infty}^{\infty} \delta\left(\tilde{v} - n\left(\frac{1}{a}\right)\right) \tag{D38}$$

By applying the shift theorem given in Equation (D25) to each of the delta functions with different n values in Equation (D28), $B''_e(\tilde{v})$ is finally given as

$$B''_e(\tilde{v}) = \left(\frac{1}{a}\right) \sum_{n=-\infty}^{\infty} B_e\left(\tilde{v} - n\left(\frac{1}{a}\right)\right) \tag{D39}$$

This equation indicates that the spectrum B_e repeatedly appears at intervals of $(1/a)$ as illustrated in Figure 4.15.

D.4.3 Additional Explanation for the Descriptions of Fourier Self-Deconvolution and Figure 6.6

In the descriptions of Fourier self-deconvolution and Figure 6.6, the convolution theorem in Equation (D6b) and the shift theorem for convolution with the delta function in Equation (D25) are repeatedly used. In Equation (6.2), the Lorentz profile centering at $\tilde{\nu} = 0$ is shifted to the same profile centering at $\tilde{\nu} = \tilde{\nu}_0$ by the shift property of the delta function.

The relation between the exponential decay and the Lorentz profile and that between the cosine function and the delta function as the Fourier transform pairs are described, respectively, in Sections D.3.4 and D.3.5. The Fourier transform of the triangular function shown in Figure 6.6f is a sinc function squared as described in Section D.3.6.

References

1. James, J.F. (2011) *A Student's Guide to Fourier Transforms*, 3rd edn, Cambridge University Press, Cambridge.
2. Bracewell, R.N. (2000) *The Fourier Transform and Its Applications*, 3rd edn, McGraw-Hill, Boston, MA.

Appendix E

Electromagnetic Pulse on the Time Axis and its Spectrum

Mitsuo Tasumi
Professor Emeritus, The University of Tokyo, Japan

Let us denote time by t and its conjugate variable, frequency, by ν. The delta function on the time axis $\delta(t)$ means a pulse of an electromagnetic wave with an infinitely short width. As indicated in Section D.3.2, $\delta(t)$ may be expressed in the following form:

$$\delta(t) = \int_{-\infty}^{\infty} e^{2\pi i \nu t} \, d\nu \qquad (E1)$$

This equation indicates that an infinitely short pulse on the time axis consists of electromagnetic waves of all frequencies ranging from $\nu = -\infty$ to ∞ (in reality from 0 to ∞). This statement is synonymous with the relation $\overline{\delta(t)} = 1$ (see Equation (D23b)), which means that the Fourier transform of $\delta(t)$ is unity everywhere on the ν axis; that is, its spectrum on the frequency axis has a flat shape. In other words, an electromagnetic pulse with an infinitely short width in the time domain has a spectrum with an infinitely large width in the frequency domain.

In practice, an electromagnetic pulse with an infinitely short width does not exist, but ultrashort laser pulses are now used for various spectroscopic measurements. Terahertz spectrometry described in Chapter 19 is based on femtosecond laser pulses. In Chapter 20, time-resolved infrared spectroscopic methods using picosecond to femtosecond laser pulses are described. Such ultrashort laser pulses have large spectral widths in the frequency domain. Let us discuss the relation between the pulse width in the time domain and its spectral width expressed in either frequency or wavenumber.

The theoretical answer to the above relation is given by the uncertainty principle in quantum mechanics. If the uncertainties of time and energy are denoted, respectively, by Δt and

Introduction to Experimental Infrared Spectroscopy: Fundamentals and Practical Methods,
First Edition. Edited by Mitsuo Tasumi and Akira Sakamoto.
© 2015 John Wiley & Sons, Ltd. Published 2015 by John Wiley & Sons, Ltd.

ΔE (Δ in this case means the standard deviation), the uncertainty relation with respect to time and energy is given as follows:

$$\Delta t \; \Delta E \geq \left(\frac{1}{2}\right) \frac{h}{2\pi} \tag{E2}$$

As the energy of an electromagnetic wave E is related with its frequency ν and its wavenumber $\tilde{\nu}$ by Equation (1.1), the following relation holds; that is, $\Delta E = h\Delta\nu = hc_0\Delta\tilde{\nu}$. By substituting this relation with $c_0 \approx 3 \times 10^{10}\,\mathrm{cm\,s^{-1}}$ into Relation (E2), the following relation is obtained for the pulse width Δt (in s) and the spectral width $\Delta\tilde{\nu}$ (in $\mathrm{cm^{-1}}$).

$$\Delta t \, \Delta \tilde{\nu} \geq 2.65 \times 10^{-12}\,\mathrm{cm^{-1}\,s} \tag{E3}$$

This means that, if the width of a laser pulse is 10 fs (10^{-14} seconds), it has a spectral width broader than $265\,\mathrm{cm^{-1}}$.

In time-resolved spectroscopic measurements using short laser pulses, it is necessary to know the accurate pulse shape. The Fourier transform of the pulse shape on the time axis gives its spectrum on the frequency axis. In practice, the electric field strength of a pulse on the time axis $E_p(t)$ is often approximated by a Gaussian profile, and the electric field strength on the frequency axis $E_p(\nu)$, the Fourier transform of $E_p(t)$, is also expressed by a Gaussian function as follows:

$$E_p(t) = E_0 e^{-at^2} \tag{E4}$$

$$E_p(\nu) = \sqrt{\frac{\pi}{a}} E_0 e^{-\frac{\pi^2 \nu^2}{a}} \tag{E5}$$

E_0 and a are constants, and a is given in units of $\mathrm{s^{-2}}$. The intensity on the time axis $I_p(t)$ and that on the frequency axis $I_p(\nu)$ are proportional, respectively, to the square of $E_p(t)$ and $E_p(\nu)$. They are expressed in the following way if the proportionality constants are disregarded:

$$I_p(t) = E_0^2 e^{-2at^2} \tag{E6}$$

$$I_p(\nu) = \frac{\pi}{a} E_0^2 e^{-\frac{2\pi^2 \nu^2}{a}} \tag{E7}$$

Both $I_p(t)$ and $I_p(\nu)$ in the above equations are expressed by the Gaussian functions. If their full widths at half maxima are denoted by Δt^G (in s) and $\Delta \nu^G$ (in Hz or $\mathrm{s^{-1}}$), respectively, $\Delta t^G = 2\sqrt{\ln 2/2a}$ and $\Delta \nu^G = \sqrt{2a \ln 2}/\pi$. Then,

$$\Delta t^G \Delta \nu^G = \frac{2\ln 2}{\pi} \approx 0.441 \tag{E8}$$

Equation (E8) represents an important relation called the *transform limit*. This relation indicates that there is a limit on decreasing simultaneously both Δt^G and $\Delta \nu^G$, which are inversely proportional to each other. Therefore, Equation (E8) has the same content as Relation (E3). The value of $\Delta \nu^G$ calculated by using Equation (E8) for the case of $\Delta t^G = 10\,\mathrm{fs}$ is $4.41 \times 10^{13}\,\mathrm{Hz}$ ($\mathrm{s^{-1}}$), which corresponds to $1470\,\mathrm{cm^{-1}}$. This value, while much larger, is not inconsistent with the earlier conclusion that $\Delta\bar{\nu}$ is broader than $265\,\text{cm}^{-1}$. The difference between the two values is due to the different definitions of Δ between Relation (E3) and Equation (E8), so the difference between the two values is not a serious matter. What is important in the present discussion is to understand that the same conclusion is derived from the two considerations (the uncertainty principle and the Fourier transform of a pulse shape); that is, an ultrashort electromagnetic pulse has a broad spectral width. This point should be borne in mind when ultrashort laser pulses are used in spectroscopic measurements.

Appendix F

Basic Concept of Two-Dimensional Correlation Spectroscopy

Isao Noda

*Department of Materials Science and Engineering, University of Delaware,
USA*

Computation of two-dimensional (2D) correlation spectra from a set of systematically varying spectral data obtained under a perturbation applied to the sample is by itself a relatively straightforward procedure [1, 2]. Useful insight into the dynamics of spectral intensity changes provided by a set of simple sign rules applied to correlation peaks also makes this technique attractive [1, 3]. However, because of the ease and convenience of routine use, the simple question of how or why 2D correlation actually works is often overlooked [4]. Thus, it is helpful to go over the derivation of fundamental equations and related basic mathematics to fully appreciate the logic behind the formal approach to correlation analysis.

F.1   Formal Development of Correlation Spectra

F.1.1   Dynamic Spectrum

The interaction pattern of an electromagnetic probe, like infrared photons, and the pertinent constituents of the sample under study, for example, electric dipole transition moments associated with various molecular vibrations, is recorded as a form of spectrum $x(v)$. The spectrum can often be modified in a systematic manner if the sample is placed under the influence of an appropriate external perturbation, such as stress, electric field, chemical reaction, or temperature change. The spectral intensity $x(v, u)$ thus becomes a function of two separate variables: spectral index variable "v" of the probe and additional variable "u"

Introduction to Experimental Infrared Spectroscopy: Fundamentals and Practical Methods,
First Edition. Edited by Mitsuo Tasumi and Akira Sakamoto.
© 2015 John Wiley & Sons, Ltd. Published 2015 by John Wiley & Sons, Ltd.

reflecting the effect of the applied perturbation. For convenience, here we will refer to the variable v as wavenumber $\tilde{v}$ and the variable u as time t. We apply the well-established mathematical tool of time series analysis to the time-dependent spectrum $x(\tilde{v}, t)$ to formally derive two-dimensional correlation spectra [3]. However, it should be understood that the two-dimensional correlation technique is generally applicable to many analytical methods other than vibrational spectroscopy and t may be a physical quantity other than chronological time, such as temperature, pressure, or voltage.

The spectral intensity $x(\tilde{v}, t)$ is measured only for given portions of variables, which are accessible for the specific experiment. The observable wavenumber range is determined by the specific spectrometer used for the measurement. More importantly, the other variable t is also bound by the "observation interval" of the measurement between t_{min} and t_{max}. Given the above, it is necessary to define the "dynamic spectrum" $y(\tilde{v}, t)$ as

$$y(\tilde{v}, t) = \begin{cases} x(\tilde{v}, t) - \overline{x}(\tilde{v}) & \text{for } t_{min} \leq t \leq t_{max} \\ 0 & \text{otherwise} \end{cases} \tag{F1}$$

In this equation, the reference spectrum $\overline{x}(\tilde{v})$ is selected such that the dynamic spectrum may be regarded as the deviation induced by the applied perturbation from the presumed reference state, such as a known stationary state, initial equilibrium state prior to the application of perturbation, and so forth. Without the specific *a priori* knowledge of the reference state, it is customary to select the spectrum averaged over the observation interval given by

$$\overline{x}(\tilde{v}) = \frac{1}{t_{max} - t_{min}} \int_{t_{min}}^{t_{max}} x(\tilde{v}, t)\mathrm{d}t \tag{F2}$$

as the reference spectrum. It is noted that the dynamic spectrum is defined explicitly over the entire range of t from $-\infty$ to $+\infty$ even though the actual measurement is carried out only within the observation interval.

F.1.2   Complex Cross Correlation Function

The two-dimensional correlation spectrum is obtained by comparing the variation patterns of dynamic spectra observed at two wavenumbers, $\tilde{v}_1$ and $\tilde{v}_2$. This comparison is carried out mathematically by calculating the complex cross correlation function between $y(\tilde{v}_1, t)$ and $y(\tilde{v}_2, t)$ [3]. The complex correlation function is formally given by

$$\Phi(\tilde{v}_1, \tilde{v}_2) + \mathrm{i}\, \Psi(\tilde{v}_1, \tilde{v}_2) = \frac{2}{t_{max} - t_{min}} \int_0^\infty Y(\tilde{v}_1, s) \cdot Y^*(\tilde{v}_2, s)\mathrm{d}s \tag{F3}$$

The real part $\Phi(\tilde{v}_1, \tilde{v}_2)$ of the complex cross correlation function is referred to as the *synchronous* spectrum, while the imaginary part $\Psi(\tilde{v}_1, \tilde{v}_2)$ is called the *asynchronous* spectrum. The practical significance of the correlation spectra related to applications in physical science and the step-by-step derivation of the above equation central to the two-dimensional correlation analysis are provided in this appendix.

The term $\tilde{Y}(\tilde{v}_1, s)$ represents the Fourier transform of the dynamic spectrum $y(\tilde{v}_1, t)$ with respect to t observed at wavenumber $\tilde{v}_1$:

$$Y(\tilde{v}_1, s) = \int_{-\infty}^{\infty} y(\tilde{v}_1, t)\, \mathrm{e}^{-2\pi \mathrm{i} s t}\mathrm{d}t = \mathrm{Re}\{Y(\tilde{v}_1, s)\} + \mathrm{i} \cdot \mathrm{Im}\{Y(\tilde{v}_1, s)\} \tag{F4}$$

and $Y^*(\tilde{v}_2, s)$ corresponds to the conjugate of the Fourier transform of the dynamic spectrum $y(\tilde{v}_2, t)$ observed at $\tilde{v}_2$:

$$Y^*(\tilde{v}_2, s) = \int_{-\infty}^{\infty} y(\tilde{v}_2, t) e^{+2\pi i s t} dt = \text{Re}\{Y(\tilde{v}_2, s)\} - i \cdot \text{Im}\{Y(\tilde{v}_2, s)\} \qquad (F5)$$

The notations "Re" and "Im" represent, respectively, the real and imaginary components of the Fourier transform. Detailed discussion on the Fourier transform is found in Appendix D. It should be pointed out that the reciprocal variable s is adopted here for the Fourier frequency instead of the angular frequency $\omega = 2\pi s$ used in the previous literature [1–3].

F.1.3   Cross Correlation Function

The "cross correlation function" $C(\tau)$ between two signals, such as the dynamic spectrum intensities measured at two different wavenumbers $\tilde{v}_1$ and $\tilde{v}_2$, is given by the bound integral of the product of two signals within the observation interval, with one signal being shifted by a finite amount on the time axis.

$$C(\tau) = \frac{1}{t_{max} - t_{min}} \int_{t_{min}}^{t_{max}} y(\tilde{v}_1, t) \cdot y(\tilde{v}_2, t + \tau) dt \qquad (F6)$$

The time shift τ is referred to as the *correlation time*. The cross correlation function is a measure of the close matching or similarity of functional forms or patterns between the two signals. As the time axis of the signal measured at $\tilde{v}_2$ is shifted forward by τ, the development of a significant value for $C(\tau)$ indicates that the matched portion of the signal measured at $\tilde{v}_2$ is lagging behind that measured at $\tilde{v}_1$ by τ.

The cross correlation function given in Equation (F6) may be further modified to a form better suited for the two-dimensional correlation analysis with the help of the Wiener–Khintchine theorem [5]. This theorem conveniently relates the cross correlation function with the corresponding Fourier transforms. In the first step, the expression for the dynamic spectrum $y(\tilde{v}_2, t + \tau)$ in Equation (F6) is rewritten in terms of the inverse of Fourier transform of $Y(\tilde{v}_2, s)$.

$$C(\tau) = \frac{1}{t_{max} - t_{min}} \int_{t_{min}}^{t_{max}} y(\tilde{v}_1, t) \left[\int_{-\infty}^{\infty} Y(\tilde{v}_2, s) e^{+2\pi i s (t + \tau)} ds \right] dt \qquad (F7)$$

After the appropriate rearrangement of the order of integrations, the cross correlation function becomes

$$C(\tau) = \frac{1}{t_{max} - t_{min}} \int_{-\infty}^{\infty} \left[\int_{t_{min}}^{t_{max}} y(\tilde{v}_1, t) e^{+2\pi i s t} dt \right] Y(\tilde{v}_2, s) e^{+2\pi i s \tau} ds \qquad (F8)$$

By taking advantage of the fact that the dynamic spectrum $y(\tilde{v}_1, t)$ outside of the observation interval is set to zero, the integration limit within the bracket of Equation (F8) can be extended over the range from $-\infty$ to ∞. The content of the bracket then becomes the conjugate Fourier transform of $y(\tilde{v}_1, t)$, which gives the final expression for the cross correlation function.

$$C(\tau) = \frac{1}{t_{max} - t_{min}} \int_{-\infty}^{\infty} Y^*(\tilde{v}_1, s) \cdot Y(\tilde{v}_2, s) e^{+2\pi i s \tau} ds \qquad (F9)$$

We now introduce the quantities called *cospectrum* $\phi_s(\tilde{\nu}_1, \tilde{\nu}_2)$ and *quad-spectrum* $\psi_s(\tilde{\nu}_1, \tilde{\nu}_2)$ of cross correlation, which are given by

$$\phi_s(\tilde{\nu}_1, \tilde{\nu}_2) = \frac{2 \operatorname{Re}\{ Y^*(\tilde{\nu}_1, s) \cdot Y(\tilde{\nu}_2, s)\}}{t_{max} - t_{min}}$$

$$= \frac{2[\operatorname{Re}\{Y(\tilde{\nu}_1, s)\} \cdot \operatorname{Re}\{Y(\tilde{\nu}_2, s)\} + \operatorname{Im}\{Y(\tilde{\nu}_1, s)\} \cdot \operatorname{Im}\{Y(\tilde{\nu}_2, s)\}]}{t_{max} - t_{min}} \quad \text{(F10)}$$

$$\psi_s(\tilde{\nu}_1, \tilde{\nu}_2) = \frac{-2 \operatorname{Im}\{ Y^*(\tilde{\nu}_1, s) \cdot Y(\tilde{\nu}_2, s)\}}{t_{max} - t_{min}} = \frac{2 \operatorname{Im}\{ Y(\tilde{\nu}_1, s) \cdot Y^*(\tilde{\nu}_2, s)\}}{t_{max} - t_{min}}$$

$$= \frac{2[\operatorname{Im}\{Y(\tilde{\nu}_1, s)\} \cdot \operatorname{Re}\{Y(\tilde{\nu}_2, s)\} - \operatorname{Re}\{Y(\tilde{\nu}_1, s)\} \cdot \operatorname{Im}\{Y(\tilde{\nu}_2, s)\}]}{t_{max} - t_{min}} \quad \text{(F11)}$$

The cospectrum and quad-spectrum are well-known quantities in the field of time-series analysis [6, 7]. They represent, respectively, the in-phase correlation and $\pi/2$ out-of-phase (i.e., quadrature) correlation of individual Fourier components of signals at a specific Fourier frequency s.

The cross correlation function expressed in terms of the cospectrum and quad-spectrum is given by

$$C(\tau) = \int_0^\infty \{\phi_s(\tilde{\nu}_1, \tilde{\nu}_2) \cos 2\pi s\tau + \psi_s(\tilde{\nu}_1, \tilde{\nu}_2) \sin 2\pi s\tau\} ds$$

$$= \int_0^\infty \{\phi_s(\tilde{\nu}_1, \tilde{\nu}_2) - i \, \psi_s(\tilde{\nu}_1, \tilde{\nu}_2)\} e^{+2\pi i s\tau} ds \quad \text{(F12)}$$

The integration limit of the above equation is reset to the range from 0 to $+\infty$ by taking advantage of the fact that the entire integrand is an even function of s. The generalized expressions of the synchronous and asynchronous correlation function for signals with arbitrary waveforms are now obtained by setting the correction time as either $\tau = 0$ or $\tau = 1/(4s)$.

$$\Phi(\tilde{\nu}_1, \tilde{\nu}_2) = C(0) = \int_0^\infty \phi_s(\tilde{\nu}_1, \tilde{\nu}_2) ds \quad \text{(F13)}$$

$$\Psi(\tilde{\nu}_1, \tilde{\nu}_2) = C\left(\frac{1}{4s}\right) = \int_0^\infty \psi_s(\tilde{\nu}_1, \tilde{\nu}_2) ds \quad \text{(F14)}$$

Combining Equations (F13) and (F14) into one in terms of a complex number leads to the formal definition of the complex correlation function found in Equation (F3).

$$\Phi(\tilde{\nu}_1, \tilde{\nu}_2) + i\Psi(\tilde{\nu}_1, \tilde{\nu}_2) = \int_0^\infty \{\phi_s(\tilde{\nu}_1, \tilde{\nu}_2) + i \, \psi_s(\tilde{\nu}_1, \tilde{\nu}_2)\} ds$$

$$= \frac{2}{t_{max} - t_{min}} \int_0^\infty Y(\tilde{\nu}_1, s) \cdot Y^*(\tilde{\nu}_2, s) ds \quad \text{(F15)}$$

It should be pointed out that the correlation time $\tau = 1/(4s)$ used for the determination of the asynchronous spectrum in Equation (F14) is not a single fixed quantity on a time axis but a function of the Fourier frequency s. The asynchronous correlation function is a special case of the cross correlation function, where the time axis of every Fourier component of the dynamic spectrum $y(\tilde{v}_2, t)$ with a positive frequency is individually shifted forward by one quarter of period, that is, $1/(4s)$. Such a shift results in the forward shift of the phase of each Fourier component by $\pi/2$. Because the quad-spectrum is an odd function of s, the negative Fourier frequency component is shifted backward by the similar amount. The contribution of the negative Fourier frequency components does not appear explicitly here, because the integration is confined to the positive Fourier frequency region.

F.1.4 Alternative Expressions for Correlation Spectra

The formal definition of synchronous spectrum and asynchronous correlation spectrum given in Equation (F3) is mathematically concise and rigorous. However, the requirement for obtaining the Fourier transforms of signals with respect to the variable t at every point of wavenumber $\tilde{v}$ for a given dynamic spectrum makes the computation of correlation spectra rather cumbersome, even with the aid of the fast Fourier transform (FFT) algorithm. Fortunately, there is a simple way to circumvent the use of the Fourier transforms to efficiently compute the desired correlation spectra [2].

It is straightforward to obtain the expression for the synchronous spectrum without the need for the Fourier transform directly from the cross correlation function given in Equation (F6). By setting the correlation time as $\tau = 0$, we have

$$\Phi(\tilde{v}_1, \tilde{v}_2) = C(0) = \frac{1}{t_{\max} - t_{\min}} \int_{t_{\min}}^{t_{\max}} y(\tilde{v}_1, t) \cdot y(\tilde{v}_2, t) \mathrm{d}t \tag{F16}$$

In contrast, the derivation of asynchronous spectrum requires several more steps, as the correlation time $\tau = 1/(4s)$ used in this case depends on the Fourier frequency s. This process is carried out by once again using the Wiener–Khintchine theorem. The proper application of this theorem requires the expansion of the integration range used in Equation (F3) from the half range between 0 and $+\infty$ to the full range between $-\infty$ and $+\infty$. As the quad-spectrum is an odd function, sign change is required for the portion of the integrand in the range $s < 0$ to obtain a meaningful integration result. We then have

$$\Psi(\tilde{v}_1, \tilde{v}_2) = \frac{-\mathrm{i}}{t_{\max} - t_{\min}} \int_{-\infty}^{\infty} \mathrm{sgn}(s) \cdot Y(\tilde{v}_1, s) \cdot Y^*(\tilde{v}_2, s) \, \mathrm{d}s \tag{F17}$$

The selective sign change of the integrand is accomplished by multiplying the integrand with the "signum function" $\mathrm{sgn}(s)$ defined by

$$\mathrm{sgn}(s) = \begin{cases} 1 & s > 0 \\ 0 & \text{if } s = 0 \\ -1 & s < 0 \end{cases} \tag{F18}$$

We now introduce the Hilbert transform $h(\tilde{v}_2, t)$ of the dynamic spectrum $y(\tilde{v}_2, t)$, which is given by

$$h(\tilde{v}_2, t) = \frac{1}{\pi} pv \int_{-\infty}^{\infty} \frac{y(\tilde{v}_2, t')}{t' - t} \mathrm{d}t' \tag{F19}$$

The notation pv indicates that the Cauchy principal value is taken, such that the singularity point at $t' = t$ is excluded from the integration. Although the formal definition of the Hilbert transform requires the integration over the complete range from $-\infty$ to $+\infty$, as the dynamic spectrum $y(\tilde{\nu}_2, t')$ is zero outside of the observation interval, the actual integration needs to be carried out only from t_{min} to t_{max}.

The close inspection of the Hilbert transform defined in Equation (F19) reveals that it is simply a convolution integral of the reciprocal function and the dynamic spectrum. The Fourier transform of the reciprocal function $1/(t'-t)$ is the imaginary signum function $i \, \text{sgn}(s)$, and the Fourier transform of the dynamic spectrum is $Y(\tilde{\nu}_2, s)$. The convolution theorem (see Appendix D) dictates that the Fourier transform $H(\tilde{\nu}_2, s)$ of the Hilbert transform of $y(\tilde{\nu}_2, t)$ is obtained by the product of the two Fourier transforms.

$$H(\tilde{\nu}_2, s) = i \, \text{sgn}(s) \cdot Y(\tilde{\nu}_2, s) \tag{F20}$$

Substitution of Equation (F20) into Equation (F17) yields

$$\Psi(\tilde{\nu}_1, \tilde{\nu}_2) = \frac{1}{t_{max} - t_{min}} \int_{-\infty}^{\infty} Y(\tilde{\nu}_1, t) \cdot H^*(\tilde{\nu}_2, t) \mathrm{d}s \tag{F21}$$

where $H^*(\tilde{\nu}_2, t)$ is the complex conjugate of $H(\tilde{\nu}_2, t)$, that is, the sign of the imaginary component is changed. The above equation can be treated with the usual procedure based on the Wiener–Khintchine theorem to eliminate the need for the Fourier transforms. Rewriting the term $H^*(\tilde{\nu}_2, s)$ appearing in Equation (F21) with the expression for the conjugate of the Fourier transform of $h(\tilde{\nu}_2, t)$ yields

$$\Psi(\tilde{\nu}_1, \tilde{\nu}_2) = \frac{1}{t_{max} - t_{min}} \int_{-\infty}^{\infty} Y(\tilde{\nu}_1, s) \cdot \left[\int_{-\infty}^{\infty} h\left(\tilde{\nu}_2, t\right) \mathrm{e}^{+2\pi i s t} \mathrm{d}t \right] \mathrm{d}s \tag{F22}$$

By rearranging the order of integrations, we now have

$$\Psi(\tilde{\nu}_1, \tilde{\nu}_2) = \frac{1}{t_{max} - t_{min}} \int_{-\infty}^{\infty} \left[\int_{-\infty}^{\infty} Y\left(\tilde{\nu}_1, s\right) \mathrm{e}^{+2\pi i s t} \mathrm{d}s \right] h(\tilde{\nu}_2, t) \mathrm{d}t \tag{F23}$$

As the term in the brackets is the inverse Fourier transform of $Y(\tilde{\nu}_1, s)$, it becomes $y(\tilde{\nu}_1, s)$. Furthermore, because the value of the dynamic spectrum is set to zero outside of the observation interval in Equation (F1), the integration with respect to t in Equation (F23) can be limited only to the observation interval. Thus, we obtain the final simplified expression for the asynchronous spectrum as

$$\Psi(\tilde{\nu}_1, \tilde{\nu}_2) = \frac{1}{t_{max} - t_{min}} \int_{-\infty}^{\infty} y(\tilde{\nu}_1, t) \cdot h(\tilde{\nu}_2, t) \mathrm{d}t \tag{F24}$$

This expression does not contain any Fourier transform terms either. The practical advantage of Equation (F24) over the form containing the Fourier transforms arises from the fact that it is not necessary to evaluate the Hilbert transform outside of the observation interval to obtain the desired result.

F.2   Practical Significance of Correlation Spectra

F.2.1   Correlation of Cosine Functions

A useful insight into the nature of the cross correlation function is revealed by examining a special case when the signals compared by the correlation analysis may be expressed in the form of a simple periodic function [8]. Let us assume the dynamic variation of spectral intensity is given by a cosine function of time t with a fixed single positive frequency S, exhibiting the wavenumber-dependent amplitude $\hat{y}(\tilde{v})$ and finite phase angle $\beta(\tilde{v})$.

$$y(\tilde{v}, t) = \begin{cases} \hat{y}(\tilde{v})\cos\{2\pi St + \beta(\tilde{v})\} & \text{for } t_{\min} \leq t \leq t_{\max} \\ 0 & \text{otherwise} \end{cases} \qquad \text{(F25)}$$

Alternatively, this cosine function with a phase angle may be expressed as a combination of cosine and sine functions with the same frequency.

$$y(\tilde{v}, t) = \begin{cases} \hat{y}^c(\tilde{v})\cos 2\pi St - \hat{y}^s(\tilde{v})\sin 2\pi St & \text{for } t_{\min} \leq t \leq t_{\max} \\ 0 & \text{otherwise} \end{cases} \qquad \text{(F26)}$$

The amplitudes of the cosine and sine components, respectively, are given by the trigonometric identities, $\hat{y}^c(\tilde{v}) = \hat{y}(\tilde{v})\cos\beta(\tilde{v})$ and $\hat{y}^s(\tilde{v}) = \hat{y}(\tilde{v})\sin\beta(\tilde{v})$. For the purpose of demonstration, the observation interval is set to be sufficiently long compared to the periodicity of the cosine function, that is, $t_{\max} - t_{\min} \gg 1/S$. We also focus our attention only to the range of the cross correlation function with a relatively small value of correlation time, $\tau \ll t_{\max} - t_{\min}$. The cross correlation function vanishes to zero as τ approaches the value of $t_{\max} - t_{\min}$.

The substitution of Equation (F25) into Equation (F6) yields the expression for the cross correlation function between two cosine functions:

$$C(\tau) = \frac{1}{t_{\max} - t_{\min}} \int_{t_{\min}}^{t_{\max}} \hat{y}(\tilde{v}_1)\cos\{2\pi St + \beta(\tilde{v}_1)\} \cdot \hat{y}(\tilde{v}_2)\cos\{2\pi S(t+\tau) + \beta(\tilde{v}_2)\} dt$$

$$\cong \frac{\hat{y}(\tilde{v}_1) \cdot \hat{y}(\tilde{v}_2)}{2}[\cos\{\beta(\tilde{v}_1) - \beta(\tilde{v}_2)\}\cos 2\pi S\tau + \sin\{\beta(\tilde{v}_1) - \beta(\tilde{v}_2)\}\sin 2\pi S\tau] \quad \text{(F27)}$$

The synchronous and asynchronous correlation spectrum pair are obtained from the correlation function by setting the correlation time as $\tau = 0$ and $\tau = 1/(4S)$.

$$\Phi(\tilde{v}_1, \tilde{v}_2) = C(0) = \frac{1}{2}\hat{y}(\tilde{v}_1) \cdot \hat{y}(\tilde{v}_2)\cos\{\beta(\tilde{v}_1) - \beta(\tilde{v}_2)\}$$

$$= \frac{1}{2}\{\hat{y}^c(\tilde{v}_1) \cdot \hat{y}^c(\tilde{v}_2) + \hat{y}^s(\tilde{v}_1) \cdot \hat{y}^s(\tilde{v}_2)\} \qquad \text{(F28)}$$

$$\Psi(\tilde{v}_1, \tilde{v}_2) = C\left(\frac{1}{4S}\right) = \frac{1}{2}\hat{y}(\tilde{v}_1) \cdot \hat{y}(\tilde{v}_2)\sin\{\beta(\tilde{v}_1) - \beta(\tilde{v}_2)\}$$

$$= \frac{1}{2}\{\hat{y}^s(\tilde{v}_1) \cdot \hat{y}^c(\tilde{v}_2) - \hat{y}^c(\tilde{v}_1) \cdot \hat{y}^s(\tilde{v}_2)\} \qquad \text{(F29)}$$

The synchronous and asynchronous spectrum, especially those expressed in terms of the amplitudes of cosine and sine function, clearly reveal the close resemblance of the functional forms to the ones given for the cospectrum and quad-spectrum in Equations (F10) and (F11). The amplitudes of cosine and sine component, respectively, of the dynamic spectrum with a single frequency S reflect the real and imaginary parts of the Fourier transform of the dynamic spectrum at the Fourier frequency of $s = S$. Alternatively, the more general synchronous spectrum and asynchronous spectrum in Equation (F15) derived for the dynamic spectrum with arbitrary waveforms may be viewed as the collective sum totals of individual correlation spectra obtained for the corresponding Fourier components.

F.2.2   Amplitude and Sign of Correlation Spectra

Examination of the result obtained for the cross correlation of simple cosine functions provides valuable insight into how features in correlation spectra become useful to the analysis of spectral signals varying under a perturbation [8]. Equations (F28) and (F29) reveal that correlation spectra are proportional to the product of the amplitudes of dynamically changing spectral intensities measured at two different wavenumbers, $\tilde{v}_1$ and $\tilde{v}_2$. In short, the magnitude of the correlation intensity directly reflects the extent of spectral intensity changes induced by the applied perturbation. Furthermore, the signs of correlation spectra provide the relative directions of changes and temporal phase relations between signals detected at different wavenumbers.

The term $\cos\{\beta(\tilde{v}_1) - \beta(\tilde{v}_2)\}$ in Equation (F28) becomes the maximum when the phase angles of two signals are matched, such that signals are varying totally in phase with each other, and minimum if the phase difference is exactly $\pm\pi$. The sign of the synchronous spectrum becomes positive if the phase difference is confined to the range between $-\pi/2$ and $\pi/2$, which corresponds to the situation when two cosine signals are mostly changing in the same direction. If the phase difference falls outside of this range, the sign of the synchronous spectrum becomes negative. The two signals are then changing in opposite directions, that is, one is increasing while the other is decreasing.

The asynchronous spectrum assumes a nonzero value only if there is some difference between the two phase angles, that is, one signal leads ahead of the other. The sequential order of intensity changes simply reflects this phase angle difference between the two cosine functions [8]. The sign of asynchronous spectrum becomes positive if the difference in $\beta(\tilde{v}_1)$ and $\beta(\tilde{v}_2)$ lies between 0 and π, while it becomes negative if the phase difference falls outside of this range. The dynamic spectrum measured at $\tilde{v}_1$ leads ahead of that measured at $\tilde{v}_2$, as long as the phase angle difference is confined either between 0 and $\pi/2$ or between π and $3\pi/2$ (i.e., between $-\pi/2$ and $-\pi$). The order is reversed if the phase difference falls outside of these ranges. In short, if the signs of the synchronous spectrum and asynchronous spectrum at the spectral coordinate $(\tilde{v}_1, \tilde{v}_2)$ are the same, the dynamic signal at $\tilde{v}_1$ leads ahead of that at $\tilde{v}_2$. If the signs of correlation spectra are different, the reversed order is found.

The above insight into the significance of phase relationship and intensity signs of correlation spectra obtained for a simple cosine or sine function may be readily extended to the analysis of dynamic spectrum signals even having arbitrary waveforms, as long as they are signals induced by the same perturbation [3]. Correlation spectra for a dynamic spectrum having waveforms other than a simple cosine or sine function can be directly obtained from Equations (F13) and (F14), by using the Fourier transform of the dynamic spectrum with respect to t. This general procedure of calculating the correlation spectra in Equation (F3) actually comprises three distinct mathematical steps combined into a seamless single operation [4]:

1. The Fourier transformation of a dynamic spectrum with respect to t can be viewed as a mathematical operation to decompose the signal of a particular form into the sum of a number of sine and cosine functions, each with a distinct frequency s.
2. A correlation analysis is applied to the individual sine and cosine functions in the manner similar to the derivation for Equations (F28) and (F29), which in turn generates a number of synchronous and asynchronous spectra, that is, cospectra ϕ_s and quad-spectra ψ_s, for individual Fourier components with different frequencies s.

3. The collective summation by integration of all cospectra for the entire positive range of *s* yields the synchronous spectrum in Equation (F13), and the same for quad-spectra to obtain the asynchronous spectrum in Equation (F14).

The interpretation rules associated with signs of correlation spectra, based on the phase angle difference for a simple cosine function with a single frequency, is now extended to the situation for multiple Fourier frequency components. In this case, the dominant Fourier components of the signals determine the overall trends of the relative directions and sequences of signals. For typical dynamic spectral signals, generated under the influence of a common perturbation applied to the sample, the dominant Fourier components measured at different wavenumbers tend to share a similar frequency range. This causal constraint of the perturbation-induced dynamic spectral signals makes it possible to extend the complex two-dimensional correlation analysis to general waveforms other than a simple cosine function [4].

F.3 Numerical Computation of Correlation Spectra

F.3.1 Discretely Sampled Data

The explicit analytical expressions given by Equations (F16) and (F24), obtained for the synchronous and asynchronous spectrum, are well suited for the efficient machine computation of correlation intensities from discretely sampled and digitized spectral data. If a discretely sampled dynamic spectrum $y(\tilde{v}_j, t_i)$ with the total of n points of wavenumber $\tilde{v}_j$ is obtained for m times at each point of time t_i, with a constant time increment, that is, $t_{i+1} - t_{i-1} = \Delta t$, the integrations in Equations (F16) and (F24) can be replaced with summations.

$$\Phi(\tilde{v}_p, \tilde{v}_q) = \frac{1}{m-1} \sum_{i=1}^{m} y(\tilde{v}_p, t_i) \cdot y(\tilde{v}_q, t_i) \qquad (F30)$$

$$\Psi(\tilde{v}_p, \tilde{v}_q) = \frac{1}{m-1} \sum_{i=1}^{m} y(\tilde{v}_p, t_i) \cdot h(\tilde{v}_q, t_i) \qquad (F31)$$

The discrete form of the Hilbert transform for the portion within the observation interval is obtained as

$$h(\tilde{v}_q, t_i) = \sum_{j=1}^{m} N_{ij} y(\tilde{v}_q, t_j) \qquad (F32)$$

with N_{ij} being the *i*th row and *j*th column element of the "Hilbert–Noda transformation matrix" N [2] given by

$$N_{ij} = \begin{cases} 0 & \text{if } i = j \\ \dfrac{1}{\pi(j-i)} & \text{otherwise} \end{cases} \qquad (F33)$$

F.3.2 Matrix Notation

It is often convenient to manipulate the discretely sampled array of data points in terms of the matrix algebra notation [9]. A set of spectra recorded as a function of wavenumber and time can be stored in an $m \times n$ data matrix Y. The element of the matrix in the ith row and jth column correspond to the dynamic spectrum value at wavenumber $\tilde{v}_j$ and time t_i, that is, $y_{ij} = y(\tilde{v}_j, t_i)$.

The synchronous spectrum and asynchronous correlation spectrum are now obtained as $n \times n$ correlation matrices, Φ and Ψ. Their matrix elements correspond, respectively, to the value of correlation spectra $\Phi_{pq} = \Phi(\tilde{v}_p, \tilde{v}_q)$ and $\Psi_{pq} = \Psi(\tilde{v}_p, \tilde{v}_q)$. Equations (F30) and (F31) are now presented concisely by the matrix notation as

$$\Phi = \frac{1}{m-1} Y^T Y \tag{F34}$$

$$\Psi = \frac{1}{m-1} Y^T N Y \tag{F35}$$

with Y^T denoting the transpose matrix of Y. The matrix representation of correlation spectra is especially useful when used in conjunction with the various tools of multivariate analyses, such as matrix projection operation [9], for enhancing the information quality of correlation spectra.

F.3.3 Unevenly Sampled Data

The sampling of dynamic spectrum does not always have to occur with a constant increment of t. For unevenly sampled discrete data [10], some modifications are then required for Equations (F30), (F31), and (F33). The modified form of correlation spectra are given by

$$\Phi(\tilde{v}_p, \tilde{v}_q) = \frac{1}{2(t_m - t_1)} \sum_{i=1}^{m} y(\tilde{v}_p, t_i) \cdot y(\tilde{v}_q, t_i) \cdot (t_{i+1} - t_{i-1}) \tag{F36}$$

$$\Psi(\tilde{v}_p, \tilde{v}_q) = \frac{1}{2(t_m - t_1)} \sum_{i=1}^{m} y(\tilde{v}_p, t_i) \cdot \sum_{j=1}^{m} N_{ij} y(\tilde{v}_q, t_j) \cdot (t_{i+1} - t_{i-1}) \tag{F37}$$

$$N_{ij} = \begin{cases} 0 & \text{if } i = j \\ \frac{t_{i+1} - t_{i-1}}{2\pi(t_j - t_i)} & \text{otherwise} \end{cases} \tag{F38}$$

with two additional time points, $t_0 = 2t_1 - t_2$ and $t_{m+1} = 2t_m - t_{m-1}$, which actually lie outside of the observation interval.

References

1. Noda, I. and Ozaki, Y. (2004) *Two-Dimensional Correlation Spectroscopy. Applications in Vibrational and Optical Spectroscopy*, John Wiley & Sons, Ltd, Chichester.
2. Noda, I. (2000) Determination of two-dimensional spectra using the Hilbert transform. *Appl. Spectrosc.*, **54**, 994–999.
3. Noda, I. (1993) Generalized two-dimensional correlation method applicable to infrared, Raman, and other types of spectroscopy. *Appl. Spectrosc.*, **47**, 1329–1336.
4. Noda, I. (2012) Close-up view on the inner workings of two-dimensional correlation spectroscopy. *Vib. Spectrosc.*, **60**, 146–153.

5. Noda, I. (1990) Two-dimensional infrared (2D IR) spectroscopy: theory and applications. *Appl. Spectrosc.*, **44**, 550–561.
6. Chatfield, C. (2009) *The Analysis of Time Series–An Introduction*, 6th edn, Chapman & Hall/CRC Press, Boca Raton, FL.
7. Bloomfield, P. (2000) *Fourier Analysis of Time Series: An Introduction*, 2nd edn, John Wiley & Sons, Inc., New York.
8. Brillinger, D.R. (2001) *Time Series: Data Analysis and Theory*, New edn, Society for Industrial and Applied Mathematics, Philadelphia, PA.
9. Noda, I. (2010) Projection two-dimensional correlation analysis. *J. Mol. Struct.*, **974**, 116–126.
10. Noda, I. (2003) Two-dimensional correlation analysis at unevenly spaced spectral data. *Appl. Spectrosc.*, **57**, 1049–1051.

Index

Figures are indicated by *italic page numbers*; tables by **bold-face numbers**.

Acronyms: ATR = attenuated total reflection; CLS = classical least-squares; ER= external reflection; FT-IR = Fourier transform infrared; IRE = internal-reflection element; LB = Langmuir–Blodgett [film]; NIR = near-infrared; PA = photoacoustic; THz = terahertz; VCD = vibrational circular dichroism.

Introduction to Experimental Infrared Spectroscopy: Fundamentals and Practical Methods,
First Edition. Edited by Mitsuo Tasumi and Akira Sakamoto.
© 2015 John Wiley & Sons, Ltd. Published 2015 by John Wiley & Sons, Ltd.

femtosecond time-resolved infrared
  spectrometer 301, *302*, 303
ferroelectric materials
  monitoring of production 282
  see also bismuth titanate
film-forming apparatus 21
filter function 119
focal-plane array (FPA) 243
folding 56, 74
forensic studies 205, 207, 223, 249–51
Fourier cosine transform 48, 77, 348
Fourier self-deconvolution (FSD) 88–90,
  357
  example of use 90, *91*
Fourier transform 48, 77, 347–8
  definition 347–8
  notations to indicate operation of 348
  properties relating to 348–9
Fourier transform infrared (FT-IR)
  microspectrometric imaging 235,
  241–51
  see also infrared microspectroscopic
    imaging
Fourier transform infrared (FT-IR)
  spectrometer
  beamsplitter *68*, 69, 270
  characteristics 29–31
  computer system 75–6
  data processing 76–81, 83–95
  detector 37, 69–72, 271
  emission measurements 213
  factors that disturb FT-IR
    measurements 73–5
  flow chart for signal processing *62, 76*
  hardware 59–76
  interferometer component 41, 44–7,
    63–9
  optical system 60–75
  PA measurements 202–4
  radiation source *68*, 69, 270
  sample compartment 72–3
  time-resolved measurements using
    291–6
  wavenumber measured by 29–30
Fourier transform infrared (FT-IR)
  spectrometry 12

characteristics 48–57
compared with dispersive spectrometry
  12, 48, 55
Connes advantage 57, 153, 291
in far-IR region 270–1
functions used in 50, *53*, 350–6
mathematical methods 347–57
measurable spectral region 55–7, 356
multiplex/Fellgett's advantage 48, 291
optical throughput/Jacqinot's
  advantage 55, 69, 72, 246, 291
principles 44–8
wavenumber determination 57
wavenumber resolution 30–1, 36–7,
  49–55
Fourier transform pair 348
FT-IR *see* Fourier transform infrared
full width at half maximum (FWHM) 30,
  51
fundamental frequencies/tones 9, 10

gallium arsenide (GaAs) wafer, cadmium
  stearate film on 129, *130*, 135
gas cell(s) 17, *18*
gaseous samples 17–18
germanium, as IRE material 182, **183**,
  185, 187
gold-coated mirror [in time-resolved IR
  measurement system] 303
gradient-shaving method, depth profiling
  by 229, 235–7
grazing-incidence reflection *see*
  external-reflection spectrometry;
  reflection–absorption spectroscopy
group theoretical analysis 9–10, 25
group vibration(s) 10
  characteristic absorption band(s) 10, **11**

Hansen's approximation 134–5, 137
harmonic vibrations 9
helium–neon (He–Ne) laser 57, 64, 67,
  277
hemp, NIR absorption spectrum 264, *265*
high-density polyethylene (HDPE)
  ATR spectra 186, *187*
  correlation spectra 314, *315*